Excitatory Amino Acids – Clinical Results with Antagonists

Excitatory Amino Acids – Clinical Results with Antagonists

Edited by

P.L. HERRLING

Head of Corporate Research,
Sandoz Pharma, Basle, Switzerland

ACADEMIC PRESS
Harcourt Brace & Company, Publishers
San Diego London Boston
New York Sydney Tokyo Toronto

This book is printed on acid-free paper.

Academic Press, Inc.
525 B Street, Suite 1900, San Diego, California 92101-4495, USA
http://www.apnet.com

Academic Press Limited
24–28 Oval Road, London NW1 7DX, UK
http://www.hbuk.co.uk/ap/

ISBN 0-12-546820-2

A catalogue record for this book is available from the British Library

Typeset by Phoenix Photosetting, Chatham, Kent
Printed and bound by CPI Antony Rowe Ltd, Eastbourne
Transferred to digital printing 2006

97 98 99 00 01 02 EB 9 8 7 6 5 4 3 2 1

Contents

Contributors

A. Arthur, Preclinical Safety, Pharmaceuticals Division, Ciba K-125.11.08, Basle, CH-4002, Switzerland.

P. Avenet, Synthélabo Recherche, 31 avenue Paul Vaillant-Couturier, 92220 Bagneux, France.

J. Benavides, Synthélabo Recherche, 31 avenue Paul Vaillant-Couturier, 92220 Bagneux, France.

F. Besnard, Synthélabo Recherche, 10 rue des Carrières, BP248, 92405 Rueil Malmaison, France.

B. Biton, Synthélabo Recherche, 31 avenue Paul Vaillant-Couturier, 92220 Bagneux, France.

C. Carter, Synthélabo Recherche, 10 rue des Carrières, BP248, 92405 Rueil Malmaison, France.

A. Cudennec, Synthélabo Recherche, 31 avenue Paul Vaillant-Couturier, 92220 Bagneux, France.

D. Duverger, Synthélabo Recherche, 31 avenue Paul Vaillant-Couturier, 92220 Bagneux, France.

M. Emre, Clinic of Research and Development of the Central Nervous System, Sandoz Pharma Ltd., Basle, CH-4002, Switzerland.

H. Faleck, Research and Development, Pharmaceuticals Division, Ciba, Summit, NJ 07901, USA.

J. Frost, Synthélabo Recherche, 31 avenue Paul Vaillant-Couturier, 92220 Bagneux, France.

E. R. Gamzu, Cambridge Neuroscience, Inc., 1 Kendall Square, Building 700, Cambridge, MA 02139, USA.

C. Giroux, Synthélabo Recherche, 31 avenue Paul Vaillant-Couturier, 92220 Bagneux, France.

D. Graham, Synthélabo Recherche, 31 avenue Paul Vaillant-Couturier, 92220 Bagneux, France.

P. L. Herrling, Sandoz Pharma Ltd., Basle, CH-4002, Switzerland.

T. Honoré, Department of the Central Nervous System, Sandoz Pharma Ltd., Basle, CH-4002, Switzerland.

J. B. Hutchison, Department of Medical Affairs, Astra Charnwood, Bakewell Road, Loughborough, Leicestershire LE11 0HR, UK.

G. Karlsson, Research and Development, Pharmaceuticals Division, Ciba K-125.11.08, Basle, CH-4002, Switzerland.

A. Kotake, Research and Development, Pharmaceuticals Division, Ciba, Summit, NJ 07901, USA.

A. G. Knapp, Cambridge Neuroscience, Inc., 1 Kendall Square, Building 700, Cambridge, MA 02139, USA.

J. D. Kristensen, Department of Anaesthesiology and Intensive Care, University Hospital, DK-5000 Odense, Denmark.

S. Z. Langer, Synthélabo Recherche, 31 avenue Paul Vaillant-Couturier, 92220 Bagneux, France.

L. Lantwicki, Research and Development, Pharmaceuticals Division, Ciba, Summit, NJ 07901, USA.

L. LaRue, Research and Development, Pharmaceuticals Division, Ciba, Summit, NJ 07901, USA.

D. Lodge, Lilly Research Centre Ltd., Eli Lilly and Company, Erl Wood Manor, Windlesham, Surrey, UK.

F. Louvel, Rhône-Poulenc Rorer Japan, Research and Development Division, 13-1, Kachidoki 1-chome, Chuo-Ku, Tokyo 104, Japan.

M-L. Maccecchini, Symphony Pharmaceuticals, Inc., 3624 Market Street, Philadelphia, PA 19104, USA.

S. Markabi, Research and Development, Pharmaceuticals Division, Ciba, 92506 Rueil Malmaison Cedex, France.

L. I. Mathews, Cambridge Neuroscience, Inc., 1 Kendall Square, Building 700, Cambridge, MA 02139, USA.

D. Murphy, Research and Development, Pharmaceuticals Division, Ciba, Summit, NJ 07901, USA.

L. Nordholm, Novo Nordisk, Krogshoejvej 29, DK-2880 Bagsvaerd, Denmark.

J. P. Nowicki, Synthélabo Recherche, 31 avenue Paul Vaillant-Couturier, 92220 Bagneux, France.

A. Oblin, Synthélabo Recherche, 31 avenue Paul Vaillant-Couturier, 92220 Bagneux, France.

G. C. Palmer, Department of Biology, Astra Arcus USA, PO Box 20890, Rochester, NY 14602, USA.

G. Perrault, Synthélabo Recherche, 31 avenue Paul Vaillant-Couturier, 92220 Bagneux, France.

S. Pigasse, Synthélabo Recherche, 31 avenue Paul Vaillant-Couturier, 92220 Bagneux, France.

M. Powell, Research and Development, Pharmaceuticals Division, Ciba, Summit, NJ 07901, USA.

P. Rosen, Synthélabo Recherche, 31 avenue Paul Vaillant-Couturier, 92220 Bagneux, France.

D. Sanger, Synthélabo Recherche, 31 avenue Paul Vaillant-Couturier, 92220 Bagneux, France.

D. Sauer, Research and Development, Pharmaceuticals Division, Ciba K-125.11.08, Basle, CH-4002, Switzerland.

B. Scatton, Synthélabo Recherche, 31 avenue Paul Vaillant-Couturier, 92220 Bagneux, France.

M. Schmutz, Research and Development, Pharmaceuticals Division, Ciba K-125.11.08, Basle, Ch-4002, Switzerland.

H. Schoemaker, Synthélabo Recherche, 31 avenue Paul Vaillant-Couturier, 92220 Bagneux, France.

D. D. Schoepp, Lilly Research Laboratories, Eli Lilly and Company, Indianapolis, IN 46285, USA.

M. Sheardown, Novo Nordisk, Novo Nordisk Park, DK-2760 Måløv, Denmark.

J. P. Thénot, Synthélabo Recherche, 31 avenue Paul Vaillant-Couturier, 92220 Bagneux, France.

J. C. Watkins, University of Bristol, Department of Pharmacology, School of Medical Sciences, University Walk, Bristol BS8 1TD, UK.

Preface

The excitatory effect of glutamate and aspartate on central nervous system neurons described in the early 1960s (Curtis and Watkins, 1960; Krnjevic and Phillis, 1963) initiated a massive research effort aimed at proving these compounds as important neurotransmitters.

During the 1980s, glutamate gained the status of the major excitatory neurotransmitter in the brain. Receptors were classified as *N*-methyl-D-aspartate (NMDA) and non-NMDA receptors, based on electrophysiological studies (Hicks *et al.*, 1978; Davies *et al.*, 1979; Watkins and Evans, 1981; McLennan and Liu, 1982).

NMDA receptors are stimulated most potently by NMDA itself. Currents activated by NMDA receptors are carried by Na^+, K^+, and Ca^{2+} (Dingledine, 1983; MacDermott *et al.*, 1986).

The non-NMDA receptors were later divided into three groups, α-amino-3-hydroxy-5-methyl-4-isoxazolepropionic acid (AMPA) receptors, kainate receptors, and metabotropic glutamate receptors, based on the preferred agonists, AMPA, kainate, and glutamate (for a review, see Watkins *et al.*, 1990).

AMPA and kainate receptors are ion channel coupled mainly Na^+ channels, whereas metabotropic glutamate receptors mediate responses via G-protein-coupled second messengers.

In the 1990s, glutamate receptor subtypes were cloned, and each group was shown to consist of a number of different subunits, which are different in both structure and function (Hollmann *et al.*, 1989; Keinänen *et al.*, 1990; Houamed *et al.*, 1991; Masu *et al.*, 1991; Monyer *et al.*, 1992; Tanabe *et al.*, 1992).

Dysfunction of glutamate transmission is the likely cause of a number of different diseases, including neurodegeneration followed by cerebral ischemia, Alzheimer's disease, Huntington's chorea, and Amyotrophic lateral sclerosis, as well as epilepsy, spasticity, emesis, Parkinson's disease, chronic pain, and schizophrenia (see chapters in this volume). Excitatory amino acid receptor agonists and antagonists are therefore of major interest as potential drugs for central nervous system disorders.

The first glutamate antagonists described were α-amino adipate (Hicks *et al.*, 1978), D-amino-2-phosphono-valeric acid (2-APV) (Davies *et al.*, 1981; Evans *et al.*, 1982; Perkins *et al.*, 1982; Childs *et al.*, 1988) and 4-(3-phosphonopropyl)piperazine-2-carboxylic acid (CPP) (Davies *et al.*, 1986; Lehman *et al.*, 1987), which were shown to be selective and potent competitive NMDA antagonists.

Later the noncompetitive NMDA antagonists ketamine, phencyclidine, and MK-801 were found (Anis *et al.*, 1983; Wong *et al.*, 1986). These antagonists blocked the NMDA-gated channels (Thomson *et al.*, 1985; Fagg, 1987). Modulatory sites such as the glycine (Johnson and Ascer,

1987) and the polyamine (Ransom and Stec, 1988) sites were also described as targets of noncompetitive NMDA antagonism.

Until the late 1980s only a very limited number of non-NMDA antagonists had been described, and compounds as glutamic acid diethylester (GDEE), γ-D-glutamylaminomethylsulphonate (GAMS), and kynurenate were weak and/or nonselective.

The first potent and selective competitive AMPA antagonists were the quinoxalinediones 6-cyano-7-nitroquinoxaline-2,3-dione (CNQX) and 2,3-dihydroxy-6-nitro-7-sulfamoyl-benzo(1)-quinoxaline (NBQX). NBQX has been especially useful to distinguish AMPA responses from other excitatory amino acid responses (for a review, see Watkins *et al.*, 1990).

The existence of noncompetitive antagonism has been demonstrated using GYKI52466 (Donevan and Rogawski, 1993).

Presently, no selective antagonists for kainate and metabotropic glutamate receptors are described.

No clear correlation between the different clones of AMPA, kainate, NMDA, and metabotropic glutamate receptors and the pharmacological activities of compounds has been found until now.

The historical development of the field is reflected in the status of the compounds in clinical development. The most clinically advanced compounds are those with a mixed action such as ifenprodil/eliprodil (see Chapter 6), riluzole (see Chapter 9), and remacemide (see Chapter 10). The compounds have, to some extent, obtained proof of efficacy.

The competitive NMDA antagonists selfotel (see Chapter 1) and D-(–)-(*E*)-4-(3-phosphonoprop-2-enyl)piperazine-2-carboxylic acid (D-CPPene) (see Chapter 2) are in the clinical phases, with efficacy studies ongoing.

The glycine antagonist 7-aminocyclopropane-carboxylic acid (ACPC) (see Chapter 5) and the AMPA antagonist LY293558 (see Chapter 7) are in the late preclinical stages.

Development of NBQX (see Chapter 8) has been terminated due to unfavorable pharmacokinetic properties, but the compound is still used as the standard AMPA antagonist in animal experimental work.

The present volume describes future hopes as well as disappointments in one of the most exciting fields of neuroscience during the ‘decade of the brain’.

Tage Honoré

Acknowledgments

This book was edited partly during the Novartis pre-merger activities. It could never have been completed within a reasonable time without the invaluable help of Marjan Tavangar.

1 Selfotel (CGS19755)

MARKUS SCHMUTZ[1], A. ARTHUR[1], H. FALECK[2], G. KARLSSON[1], A. KOTAKE[2], L. LANTWICKI[2], L. LARUE[2], S. MARKABI[3], D. MURPHY[2], M. POWELL[2] AND D. SAUER[1]

[1]*Research and Development, Pharmaceuticals Division, Ciba, Ch-4002 Basle, Switzerland*
[2]*Research and Development, Pharmaceuticals Division, Ciba, Summit, NJ 07901, USA*
[3]*Research and Development, Pharmaceuticals Division, Ciba, 92506 Rueil Malmaison Cedex, France*

1 SUMMARY

Selfotel (CGS19755) is a potent selective and competitive *N*-methyl-D-aspartate (NMDA) antagonist.

Preclinically, selfotel reduced ischemia-induced infarct size and neuronal cell death, antagonized the effects of excitotoxic lesions in the brain, and attenuated neuronal damage following traumatic brain injury at intraperitoneal or intravenous doses ranging from 3 to 40 mg kg^{-1}. In addition to these neuroprotectant properties, selfotel also exhibited anticonvulsant and anxiolytic activity. Behavioral central nervous system (CNS) effects include ataxia and increased locomotor activity. The compound was not mutagenic, clastogenic, or teratogenic in rats or rabbits. Similar to other competitive and noncompetitive NMDA antagonists, selfotel produced Olney-type vacuoles in a dose-related manner in rat brain. At present, the clinical significance of these findings in rats is unknown.

In humans, the plasma pharmacokinetics of selfotel were linear in the dose range evaluated, and elimination half-life, clearance and volume of distribution at steady state were independent of dose. There was no appreciable protein binding. Preliminary human data indicate that selfotel rapidly crosses the blood–brain barrier and remains in the cerebral spinal fluid (CSF) for an extended period of time. Doses of up to 3 mg kg^{-1} i.v. have been evaluated in conscious healthy male subjects, with non-psychotomimetic CNS adverse experiences being the dose-limiting factors. Administration of single doses of up to 2 mg kg^{-1} of selfotel did not impact the management of neurosurgical patients. However, a number of these patients also experienced CNS effects. In patients who were conscious following an acute ischemic stroke, doses up to and including 1.5 mg kg^{-1} i.v. were found safe and were tolerated. Dose-limiting adverse experiences in conscious patients included transient agitation,

EXCITATORY AMINO ACIDS – CLINICAL RESULTS WITH ANTAGONISTS
ISBN 0-12-546820-2

hallucinations, and confusion at higher doses. In unconscious patients treated with selfotel following severe closed traumatic brain injury, four bolus doses of up to 5 mg kg^{-1}, administered over a 72 h period, appeared safe and well tolerated.

Four international well-controlled clinical trials including approximately 3600 patients, two trials each in acute ischemic stroke and severe traumatic brain injury, are currently underway to evaluate the definitive safety and efficacy of selfotel in improving the functional outcome of patients with these disorders.

2 OVERVIEW OF THE PHARMACOLOGY

2.1 Physical and chemical properties

Selfotel (CGS19755; systematic name (±)-*cis*-4-(phosphonomethyl)-2-piperidinecarboxylic acid) is an analog of AP5 (2-amino-5-phosphonopentanoic acid). It is a racemic mixture of CGS20281 ((+) isomer), which has been shown to be the biologically active enantiomer, and CGS20282 ((–) isomer). It is not feasible to separate or synthesize large quantities of the biologically active enantiomer. Therefore, the racemate has been selected for development.

Chemical structure:

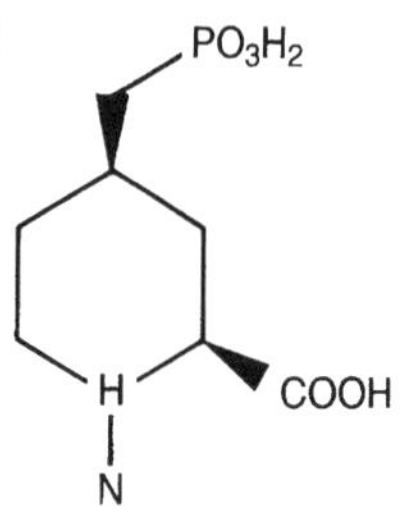

Molecular formula:	$C_7H_{14}NO_5P$
Molecular weight:	223.16
Water solubility (~25 °C)	~18 mg ml^{-1}
Melting range:	Melts with decomposition at ~296 °C

Selfotel is relatively stable in the solid state and in solution under normal conditions.

2.2 Preclinical pharmacology

2.2.1 *Biochemical and electrophysiological characterization*

The selectivity of selfotel for NMDA receptors was evaluated using *in vivo* radioligand-binding assays (Lehmann *et al.*, 1988b; Murphy *et al.*, 1988). Selfotel exhibited high affinity at the NMDA receptor recognition site ([^{3}H]-3-(2-carboxypiperazin-4-yl) propyl-1-phosphoric acid and [^{3}H]selfotel binding: K_d = 9 nM). In contrast to MK-801 and phencyclidine (PCP), which bind to the NMDA receptor-linked ion channel in a noncompetitive manner, selfotel binds competitively to the NMDA receptor. Selfotel neither significantly interacted with non-NMDA excitatory amino acid receptors (e.g. kainate and AMPA) nor with about 20 other binding sites investigated. The *in vitro* receptor binding properties of selfotel and its enantiomers with excitatory amino acid receptors are shown in Table 1 (M. Sills, unpublished).

Table 1 *In vitro* receptor binding of selfotel and its isomers

Assay	Selfotel	CGS20281	CGS20282
[^{3}H]CGP39653 (NMDA receptor antagonist ligand; Sills *et al.* (1991), IC_{50}	137 nM	57 nM	Inactive
[^{3}H]Kainate	Inactive	Inactive	Inactive
[^{3}H]AMPA	Inactive	Inactive	Inactive

Unlike PCP and related drugs, selfotel at a total dose of 40 mg kg^{-1} i.v. did not affect the electrical activity of ventral tegmental A_{10} dopamine neurons in the rat (French *et al.*, 1993) and at 10 mg kg^{-1} i.p. did not affect dopamine metabolism in brain. These results suggest that selfotel may lack the adverse effects associated directly with dopaminergic transmission.

2.2.2 *Neuroprotective properties*

During cerebral ischemia, excessive glutamate release initiates a cascade of events leading to neuronal cell death. Direct injection of the excitatory amino acids NMDA and quinolinic acid into animal brain is used as a mechanism-based *in vivo* model to evaluate potential neuroprotective drug efficacy. Intraperitoneal administration of selfotel over the dose range of 10–50 mg kg^{-1} significantly reduced NMDA- and quinolinic acid-induced neurodegeneration in rats (Schoepp *et al.*, 1989; Sauer *et al.*, 1992).

In cortical cell cultures derived from fetal mice, neuronal injury induced by exogenously added excitotoxins, oxygen or glucose deprivation, or mechanical trauma was attenuated by selfotel treatment (Choi *et al.*, 1989). Additional experiments suggest that selfotel retains its neuroprotective efficacy even under acidic conditions, an important finding as acidosis is typical of hypoxia–ischemia in animals and man (Kaku *et al.*, 1993). In cortical cell cultures of rats, selfotel was neuroprotective under conditions where an altered redox state enhanced NMDA neurotoxicity, suggesting that it may maintain its neuroprotective effectiveness during prolonged ischemic episodes (Aizenman and Hartnett, 1992). Additionally, selfotel prevented the neurotoxicity produced by acute (5 min) as well as prolonged (18–24 h) exposure to NMDA (Aizenman and Hartnett, 1992).

2.2.3 *Anti-ischemic properties*

Many studies in mice, rats, rabbits, and cats have shown that selfotel attenuates the pathological or functional deficits resulting from experimental cerebral ischemia. A summary of published data is provided in Tables 2–4. In

Table 2 Studies with selfotel in rat models of stroke

Model	End-point	Dose (mg kg^{-1})/time	Result	Comments
Permanent MCA occlusion, rats (Simon and Shiraishi, 1990)	Infarct size Cerebral metabolism	10/5 min pre 10/5 min post 10/1 h post	64% protection 50% protection No significant effect	Single intravenous bolus
Permanent MCA occlusion, rats (Sauer *et al.*, 1993)	Infarct volume (magnetic resonance imaging)	40/5 min post	37% protection	Single intravenous bolus
Permanent MCA + bilateral common carotid artery occlusion, rats (Takizawa *et al.*, 1991)	Infarct volume Brain pH Cerebral blood flow	10/immediate post	Significant protection	Intravenous bolus + infusion (5 mg kg^{-1} h^{-1} for 4 h)
Permanent right common carotid and distal MCA occlusion, rats (Simmonds *et al.*, 1993)	Infarct volume	10/5 min post	50% protection	Single intravenous bolus

Table 3 Studies with selfotel in mice and rabbit models of stroke

Model	End-point	Dose (mg kg^{-1})/time	Result	Comments
Permanent MCA occlusion, mice (Gotti *et al.*, 1990)	Cortical ω3 site density	1 3 10	No significant effect 31% protection 46% protection	Multiple intraperitoneal injections
Temporary occlusion, rabbits (Steinberg *et al.*, 1994)	Cerebral edema Infarct volume	40/10 min post	76% protection (overall cortical ischemic neuronal damage)	Single intravenous bolus
Transient (1 h) spinal cord ischemia, rabbits (Madden *et al.*, 1993)	Paraplegia	10/5 min post 30/5 min post 30/30 min post 30/60 min post	No significant effect Significant protection No significant effect No significant effect	Single intravenous bolus

Table 4 Studies with selfotel in animal models of traumatic brain injury

Model	End-point	Dose (mg kg^{-1})/time	Result	Comments
Severe lateral fluid percussion, rats (Panter and Faden, 1992)	Microdialysis Levels of glutamate	30/15 min pre	42% inhibition of presynaptic release	Single intravenous bolus
Fluid percussion + hypoxia, rats (Sanada *et al.*, 1990)	Heat shock protein immunoreactivity	10/10 min post	Significant protection	Intravenous bolus + 2 intraperitoneal doses at 12 + 24 h; combined therapy with lazaroid or lazaroid + nimodipine; individual therapy not evaluated
Lateral fluid percussion, rats (McIntosh *et al.*, 1992)	Cerebral edema Motor performance	10/15 min post	No significant effect	Single intravenous bolus

general, the neuroprotective doses of selfotel in animal models range from 10 to 40 mg kg^{-1} i.p. or i.v. for stroke (e.g. the permanent middle cerebral artery (MCA) occlusion model) and from 3 to 30 mg kg i.v. for traumatic brain injury (e.g. the lateral fluid percussion model).

2.2.4 *Other properties*

When administered intravenously or intraperitoneally, selfotel is a potent anticonvulsant in various animal models of epilepsy (Lehmann *et al.*, 1988a,b; Bennett et al., 1989;

Morimoto *et al.*, 1991). However, the drug was considerably less potent when administered orally. In experiments in mice it was demonstrated that the anticonvulsant activity is attributable to the (+) isomer CGS20281 (M. Schmutz, unpublished), which was very potent (ED_{50} 0.8 mg kg^{-1} i.v. versus ED_{50} of 1.2 mg kg^{-1} i.v. for selfotel) while the (–) isomer CGS20282 was ineffective up to 20 mg kg^{-1}. Anticonvulsant properties of selfotel may be advantageous in the prevention of seizures resulting from embolic stroke or head trauma. In addition to these properties, selfotel exhibited anxiolytic effects in conflict models in rats and pigeons (Bennett *et al.*, 1989; Koek and Colpaert, 1991).

2.2.5 *Preclinical safety overview**

Acute, subacute and subchronic intravenous or subcutaneous studies of up to 13 weeks' duration in mice, rats, dogs, and monkeys revealed slight to marked CNS clinical signs with no specific target organ changes. Selfotel was not mutagenic or clastogenic, or teratogenic in the rat or rabbit.

Similar to other competitive and noncompetitive NMDA antagonists such as dizocilpine or phencyclidine (Olney *et al.*, 1989; Allen and Iversen, 1990; Fix *et al.*, 1993), selfotel has been shown to produce Olney-type vacuoles in rat cingulate and retrosplenial cortical neurons. In rats, a single intravenous administration of selfotel resulted in neuronal vacuolation in a dose-related manner. A dose of 2 mg kg^{-1} appeared to be the no-effect dose, whereas doses of 10 and 100 mg kg^{-1} produced neuronal vacuolization. Following a single dose of 100 mg kg^{-1} in the rat, slightly increased immunohistological staining for glial fibrillary acidic protein (GFAP) was observed in the same brain areas up to 30 days post-dosing. At present, the significance of these rat findings for humans is unknown.

In conclusion, administration of selfotel to mice, rats, dogs, and monkeys produced effects that were considered to be related to the pharmacology of the compound. Similar to other NMDA antagonists, selfotel produced neuronal vacuolation in the posterior cingulate and/or retrosplenial cortices of the rat.

3 RATIONALE FOR CLINICAL TESTING

The preclinical studies described in Section 2.2 have demonstrated that the competitive NMDA antagonist selfotel is a neuroprotectant in various animal models of ischemia. They form the basis for the assumption that treatment with selfotel is a promising new pharmacological approach for disease states resulting from ischemic or hypoxic insults.

The neuroprotective doses of selfotel in animal models range from 3 to 40 mg kg^{-1} i.p. or i.v. These doses are higher than the dose of 1.5 mg kg^{-1} which was found to be safe and well-tolerated in conscious stroke patients and which was selected for further testing in clinical trials in stroke. However, based on animal and preliminary human CSF/brain concentration measurements, this dose should result in selfotel brain concentrations high enough to achieve neuroprotective levels in man. Preliminary human data also indicate that selfotel enters the brain rapidly and remains measurable in the CSF for up to 18 h after a dose of 2 mg kg^{-1} i.v. (Steinberg *et al.*, 1994). A single 1.5 mg kg^{-1} i.v. bolus administration of selfotel was therefore considered sufficient for neuroprotection in stroke patients. In unconscious traumatic brain-injured patients, total doses of up to 20 mg kg^{-1} i.v. administered over 2–4 days were safe and well tolerated. In these patients, primary brain insult is exacerbated by secondary glutamate-induced neuronal damage. Based on the finding that in severe head trauma patients, glutamate levels may be elevated to neurotoxic concentrations for several days (Baker *et al.*, 1993; Choi *et al.*, 1994), dosing for a period of up to 4 days postinjury was selected for this patient population.

4 HUMAN RESULTS*

To date, the clinical program for selfotel has included healthy volunteers, patients requiring neurosurgical procedures for arteriovenous malformation, gliosis, aneurysm or tumors, patients with acute ischemic stroke, and patients with severe traumatic brain injury. Approximately 503 subjects and patients have been dosed with selfotel, and 332 received a placebo.

4.1 Pharmacokinetics

Pharmacokinetic parameters were evaluated in healthy volunteers, neurosurgical patients, and patients with ischemic stroke or traumatic brain injury. In healthy male volunteers and patients with ischemic stroke who received single intravenous selfotel doses of 0.5–1.0 and 1–2 mg kg^{-1}, respectively, the pharmacokinetics of the drug were linear in the dose range evaluated, based on the area under the plasma concentration time curve (AUC). Mean terminal elimination half-life values ranged from 1.8 to 3 h, and were independent of dose, as were the mean selfotel clearance values. However, the CSF elimination kinetics appear different since preliminary data from neurosurgical patients showed that selfotel CSF levels are measurable for as long as 18 h after a dose of 2 mg kg^{-1} i.v. (Steinberg *et al.*, 1994). Selfotel was not bound to plasma proteins at concentrations from 0.5 to 10 μg ml^{-1}. Analysis of plasma, urine, and feces for radioactivity and unchanged drug indicated that selfotel was not metabolized, and was excreted exclusively in the urine, the majority within the first 24 h.

* This section is based on unpublished data from toxicity tests conducted by or for Ciba, where all reports are on file; see also Markabi (1994).

* Section 4 is based in part on unpublished data from clinical trials conducted by or for Ciba, where all reports are on file.

4.2 Healthy volunteers

Twenty-three male volunteers with ages ranging from 19 to 35 years received selfotel at doses of 1–160 mg (unit dose), 2 mg kg^{-1}, and 3 mg kg^{-1}. No clinically significant unwanted effects were seen at single intravenous bolus doses below 160 mg (2 mg kg^{-1}). At doses of 160 mg and above, all volunteers reported at least one adverse experience. Almost all of them were CNS related, lasted in general for less than 24 h, and were completely reversible. They included drowsiness, light-headedness, dizziness, and alterations of smell and taste. No psychotomimetic effects were observed. Selfotel had no significant impact on physical, psychiatric, or ophthalmic examinations, laboratory tests, electrocardiograms (ECGs), pulmonary function tests or vital signs. Local venous tolerance to selfotel injection was excellent. The maximum tolerated single dose was judged to be 160 mg or 2 mg kg^{-1} (see also Markabi, 1994).

4.3 Patients undergoing craniotomy

Thirty-two patients undergoing craniotomy for vascular malformation resection, aneurism clipping, and tumour resection received selfotel in a single intravenous dose (0.5, 1.0, 1.5, or 2.0 mg kg^{-1}) prior to or after induction of anesthesia (Steinberg *et al.*, 1994). A single CSF sample was collected from each patient at times ranging from 2 to 18 h post-dose. After the selfotel dose of 2 mg kg^{-1} (10 patients), CSF drug levels at 1.5–6 h post-dose ranged from 0.20 to 4.76 μM. Selfotel remained measurable in the CSF for up to 18 h post-dose (drug levels at 13–18 h: 0.15–1.17 μM; seven patients). As expected in a neurosurgical population, all patients had nervous system-related adverse experiences. Most of the patients experienced headache, and approximately one-third experienced agitation, dizziness, and/or hallucinations. The symptoms were easily controlled with intravenous haloperidol.

4.4 Patients with acute ischemic stroke

To date, more than 450 patients with acute ischemic stroke have participated in international clinical trials. Two international clinical trials including approximately 1800 patients were recently initiated to evaluate the safety and efficacy of selfotel in patients with acute ischemic stroke.

In a randomized, double-blind, placebo-controlled pilot trial with 141 patients, the tolerability and preliminary efficacy of selfotel were investigated. Thirty-two patients (24 selfotel, 8 placebo) participated to determine the maximum tolerated dose. No apparent dose-related laboratory abnormalities were seen. Overall, a single dose of selfotel of 1.5 mg kg^{-1} i.v. was determined to be the maximum clinically tolerated single dose. Higher doses were associated with an increase in incidence and severity of CNS side-effects. For example, at least one adverse experience (agitation, paranoia, hallucinations, confusion, or delirium) occurred in all six patients treated with either one or two doses of 2 mg kg^{-1}.

Similar but milder adverse experiences occurred in three of six patients receiving a single dose of 1.5 mg kg^{-1} and in one of six patients receiving two doses of 1.0 mg kg^{-1} of selfotel. The duration of these adverse experiences averaged 24 h. Mortality in the selfotel group and the placebo group was equal (Grotta *et al.*, 1995).

Based on the above, 109 patients were treated with a placebo (55 patients) or a single 1.5 mg kg^{-1} selfotel intravenous bolus dose (54 patients) within 6 h following an ischemic stroke (Clark and Coull, 1994). Of the adverse experiences, agitation, confusion, and hallucinations appeared more frequently with selfotel. As judged by investigator assessment, no selfotel-related deaths occurred. As the size of the patient population was not based on statistical considerations, no firm conclusions regarding drug efficacy can be drawn. The data were, however, explored with *post hoc* analysis for administrative reasons and subsequent study design. As expected, due to the limited sample size, no statistically significant differences between selfotel and placebo treatment were seen in the NIH Stroke Scale or total Barthel Index scores. However, in the subgroup of 51 patients with mild to moderate ischemic stroke (by the Scandinavian Stroke Scale prognostic score), the proportion of patients achieving independence, i.e. a Barthel Index total score of at least 70, at day 90 or at the terminal visit was significantly greater for the selfotel-treated group (92 and 88%, respectively) compared to the placebo-treated group (67 and 65%, respectively; no adjustments for multiple comparisons). No significant difference between the two treatment groups was seen in the 58 patients having severe strokes. A total number of 19 deaths (17%) were registered, seven of them under treatment with selfotel (13%) and 12 (22%) under placebo (Markabi, 1994).

4.5 Patients with traumatic brain injury

About 300 patients with traumatic brain injury participated in clinical trials designed to evaluate the safety and tolerability of total selfotel doses ranging from 2 to 20 mg kg^{-1} with a treatment duration of 2–4 days. In a recent double-blind placebo controlled phase II trial, 110 patients with severe traumatic brain injury were enrolled to evaluate the safety and tolerability of 3 or 5 mg kg^{-1} of selfotel given daily for three or four consecutive days, respectively, versus a placebo. The sample size was not based on statistical considerations, as safety and tolerability were the end-points.

A preliminary analysis of selected demographic safety and tolerability data has been conducted. All patients who had a minimum of 5 days of data entered into the database after receiving the last dose of trial drug were included in this analysis. The predominate outcome of all serious adverse experiences (SAEs) reported in this trial across all treatment groups was death. Overall, 12% of the patients randomized into the 3 mg kg^{-1} dose group (four selfotel-treated and two placebo-treated patients) and 28% of the patients randomized to the 5 mg kg^{-1} dose group (10 selfotel-treated and seven placebo-treated patients) died. None of these deaths were considered to be related to the trial drug. The similarity in mortality between patients treated with 5 mg kg^{-1} of selfotel and the placebo suggests that selfotel did not impact negatively on mortality when administered to patients with severe traumatic brain injury.

The five most frequent adverse experiences were fever (selfotel 73%, placebo 64%) intracranial hypertension (selfotel 47%, placebo 36%), pneumonia (selfotel 33%, placebo 22%), agitation (selfotel 29%, placebo 26%), and hypotension (selfotel 26%, placebo 18%). Most of the adverse experiences were not considered to be related to the trial drug.

The most frequent abnormal ECG findings were nonspecific ST and T wave abnormalities and sinus and ventricular tachycardia. The combined frequency of these ECG changes in the selfotel-treated groups did not differ from the placebo-treated group. In addition, summary statistics performed on clinical laboratory data (hematology and blood chemistries) showed no clinically significant differences between the selfotel-treated and the placebo-treated patients for the 3 and 5 mg kg^{-1} treatment groups for any of the laboratory parameters.

A primary concern in the management of traumatic brain injury is the ability to maintain adequate mean arterial blood pressure (MABP) and low intracranial pressure (ICP). Both parameters affect cerebral perfusion pressure (CPP), which provides an indication of adequate blood perfusion of the brain. MABP, ICP, and CPP were recorded just prior to dosing and every 2 h for 24 h after each dose of trial drug. When compared to the placebo, selfotel (3 and 5 mg kg^{-1} dose groups) treatment did not lower mean MABP and CPP or increase ICP. Indeed, the 5 mg kg^{-1} selfotel-treated group had a trend toward higher mean MABP and CPP values and lower mean ICP than the corresponding placebo-treated group.

The heterogeneity of the patient population (age, gender, and severity of injury on computed tomography scan diagnosis of brain injury) and the small patient numbers evaluated to date at each dose and treatment group does not permit definitive conclusions to be made about the efficacy of selfotel in this population. However, the dose regimen of 5 mg $kg^{-1} \times 4$ doses separated by 24 h intervals appears to be safe and well tolerated. Physiological parameters are not adversely affected by selfotel. Indeed, at this dose, there is a trend toward improvement in CPP, MABP and ICP (see also Steward et al, 1993). Two well-controlled international clinical trials with approximately 1800 patients were recently initiated to determine the safety and efficacy of selfotel in patients with severe traumatic brain injury.

5 CONCLUSIONS

Selfotel is a specific and potent competitive NMDA antagonist with neuroprotective, anticonvulsant, and anxiolytic properties in various animal models that offers a promising new approach for the treatment of human ischemic insults. Similar to other NMDA antagonists, selfotel produced neuronal vacuolation in the posterior cingulate and/or retrosplenial cortices of the rat. At present, the significance of these rat findings for humans is unknown, but it is felt that with the severity of the disease states envisaged, the potential risk for patients treated with selfotel is outweighed by the potential benefit of the drug.

Preliminary data from clinical trials show that the drug enters rapidly into the brain and remains measurable in the CSF for up to 18 h following administration. The tolerability profile includes controllable, transient dose-dependent CNS adverse experiences. The maximum tolerable single dose in conscious patients was estimated to be 1.5 mg kg^{-1} i.v., whereas in comatose patients with traumatic brain injury considerably higher doses can be administered. Despite the evidence of some favorable efficacy trends, no definitive comments about clinical outcome can be made at present. Four pivotal clinical trials including approximately 3600 patients, two each in acute ischemic stroke and severe traumatic brain injury, were initiated to evaluate the definitive safety and efficacy of selfotel in these patient populations.

2 D-CPPene (SDZ EAA-494)—A Competitive NMDA Antagonist: Pharmacology and Results in Humans

PAUL L. HERRLING[1], MURAT EMRE[1] AND J. C. WATKINS[2]

[1]*Sandoz Pharma Ltd, CH-4002 Basle, Switzerland*
[2]*Department of Pharmacology, School of Medical Sciences, University Walk, Bristol BS8 1TD, UK*

1 SUMMARY

D-(–)-(*E*)-4-(3-Phosphonoprop-2-enyl)piperazine-2-carboxylic acid (D-CPPene) is one of the most extensively studied potent and pure competitive *N*-methyl-D-aspartate (NMDA) antagonists with systemic activity. It has been developed from the original discovery of neuroactive excitatory amino acids (EAAs) through rational chemical derivatization and structure–activity relationships. Over the course of more than 10 years (1986 to the present) it has been exhaustively characterized in animal studies, giving a very good picture of many physiological effects of the mammalian EAA NMDA receptor as it is plausible that most biological effects seen are exclusively mediated by

EXCITATORY AMINO ACIDS – CLINICAL RESULTS WITH ANTAGONISTS
ISBN 0-12-546820-2

this receptor. More recently, human characterization has also begun to complete the picture. The major findings in intact organisms include: muscle relaxation, inhibition of proconvulsive phenomena, modulation of learning mechanisms, neuroprotective effects, inhibition of some form of centrally mediated pain and interaction with monoaminergic circuits probably leading to the psychotomimetic effects seen in humans at high doses. No organ-related toxicological effects have been seen up to high doses, so that the clinical potential of D-CPPene can be widely explored. The main clinical indication pursued at the time of writing is the reduction of damage following trauma to the central nervous system (CNS).

The scientific story of D-CPPene very clearly illustrates the inestimable value of a pure systemically active pharmacological tool: (i) scientifically it is essential in order to determine the physiology of a particular receptor in all species including humans; (ii) medically it can potentially lead to better treatment in severe conditions such as traumatic injury of the CNS or some forms of chronic pain in the present case.

2 STRUCTURE–ACTIVITY RELATIONSHIPS

Following the first demonstration that L-glutamate directly excites central neurons, structure–activity studies showed that this activity is mediated by a receptor mechanism (Curtis and Watkins, 1960, 1965). Subsequent studies were directed toward the identification of antagonists for this effect with the dual purpose of (a) establishing that glutamate receptors were involved in synaptic excitation—implying that glutamate or a similar EAA functioned as a synaptic transmitter—and (b) differentiation of subtypes of EAA receptors should such a multiplicity exist. In the mid-1970s, three types of EAA antagonists, chemically very different from one another, were recognized: (i) homologs of D-glutamate, typified by D-α-aminoadipate; (ii) magnesium ions; and (iii) the compound 3-amino-1-hydroxypyrrolidone-2 (HA-966) (Watkins and Evans, 1981). All of these antagonists had one feature in common: they antagonized the actions of a range of EAA agonists in a similar, differential, manner. Actions of the glutamate analog NMDA were the most sensitive, the actions of some EAAs were relatively unaffected (particularly those of kainate and quisqualate), while those of other amino acids (including L-glutamate and L-aspartate) showed intermediate sensitivity. This gave rise to the concept of NMDA and non-NMDA EAA receptors (Watkins, 1978, 1980).

Later work confirmed this classification and further differentiated non-NMDA receptors into kainate and quisqualate (now renamed α-amino-3-hydroxy-5-methyl-4-isoxazolepropionic acid (AMPA) subtypes, all classified within the family of ionotropic EAA receptors which function as ligand-gated ion channels (Watkins and Evans, 1981; Monaghan *et al.*, 1989). Ionotropic EAA receptors are thought to be involved in the short-term electrochemical phenomena underlying fast synaptic excitation. Another major family of EAA receptors, metabotropic glutamate receptors (mGluRs), have more recently been identified (Nakanishi, 1992; Knöpfel *et al.*, 1995; Pin and Duvoisin, 1995). mGluRs are coupled to G-proteins and produce longer-term metabolic changes in neurons, thought to be important in synaptic plasticity (Watkins and Collingridge, 1994).

Just as the discovery of the excitatory action of L-glutamate led to detailed structure–activity studies and ultimately resulted in the development of agonists with higher potency and receptor subtype selectivity, so too did the initial discovery of NMDA receptor antagonists lead to the development of a range of more potent antagonists, not only those specific for NMDA receptors but also those selective for other ionotropic and metabotropic EAA receptors (Watkins and Collingridge, 1994). Of the three types of NMDA receptor antagonist initially recognized, only the D-α-aminoadipate type clearly acted at the agonist recognition site by a competitive mechanism (Evans *et al.*, 1979). The actions of Mg^{2+} (channel permeability regulation) and HA-966 (competitive at the so-called glycine recognition site of the NMDA receptor) were established later (Ascher and Johnson, 1994; Lodge *et al.*, 1994). Structure–activity studies on D-α-aminoadipate (see (**I**) in Fig. 1) led to the identification initially of D-α-aminosuberate (**II**) as an NMDA receptor antagonist of similar selectivity and potency (Evans *et al.*, 1979) and later to the considerably more potent ω-phosphono analog of these compounds, D-2-amino-5-phosphonopentanoate (D-AP5, (**III**)) and D-2-amino-7-phosphonoheptanoate (D-AP7, (**IV**)) (Evans *et al.*, 1982; Perkins *et al.*, 1982).

These studies established the principle that compounds bearing an ω-phosphono group and a chain length similar to that in D-AP5 or D-AP7 were likely to be NMDA receptor antagonists. Clearly, it was important to test a further range of such compounds. The principle was borne out with a series of γ-D-glutamyl and β-D-aspartyl peptides containing the ω-phosphono moiety (Jones *et al.*, 1984). We next turned our attention to cyclic compounds. Since we had earlier shown that *trans*-2,3-piperidinedicarboxylic acid (*trans*-2,3-PDA) and *trans*-2,4-piperidinedicarboxylic acid (*trans*-2,4-PDA, (**V**) were potent and selective NMDA receptor agonists (Davies *et al.*, 1982), it seemed logical to incorporate a heterocyclic nucleus into the structure of proposed antagonists. The use of the piperidine nucleus, however, would result in the formation of both *cis* and *trans* geometric isomers, involving the substituents at piperidine ring positions C2 and C4, as well as optical enantiomers involving C2. Such isomers, as with the short-chain agonists 2,3-and 2,4-PDA, could well have different actions (Evans *et al.*, 1979, 1982; Davies *et al.*, 1982) and might be difficult to prepare in pure stereoisomeric form. We therefore selected the piperazine nucleus for our first series of such compounds—which would be devoid of geometric isomerism as regards the

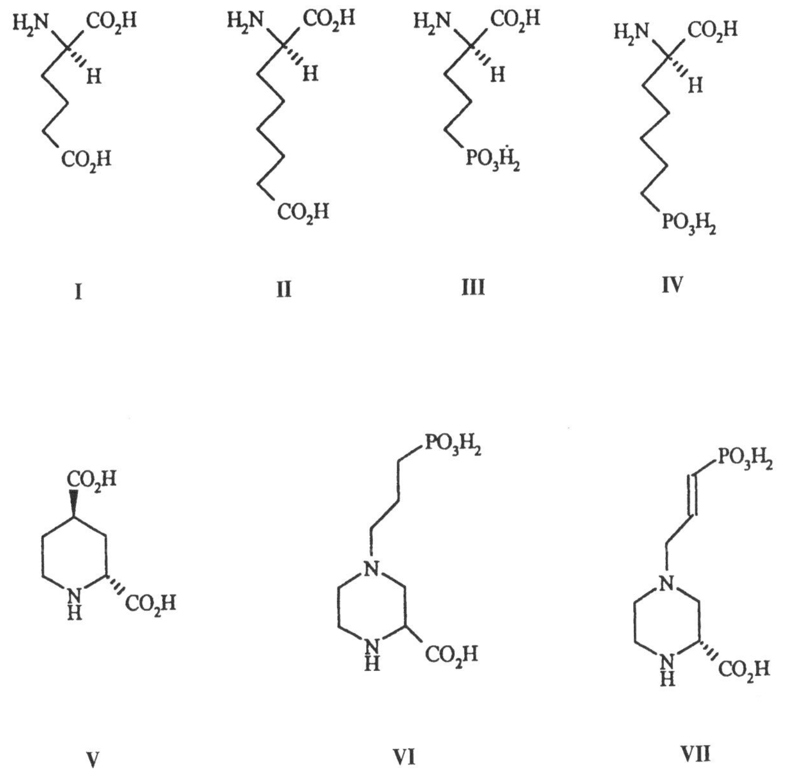

FIG. 1 Structures of NMDA receptor ligands. The acid form is shown in all cases though the active form is assumed to be the predominant ionic species at physiological pH.

two ring substituents—and synthesized a series of homologous 4-substituted piperazine-2-carboxylic acids by attachment of a phosphonoalkyl or phosphonoalkylene chain to the nitrogen atom in the 4-position of the piperazine ring. We also prepared a range of the corresponding carboxy and sulfo analogs for comparison. Other investigators adopted the piperidine nucleus as the basis of a similar series of compounds (Hutchison *et al.*, 1989).

Our first pharmacological tests were conducted on the isolated spinal cord of the neonatal rat, where we determined K_D values for the antagonism of NMDA-induced depolarizations of motoneurons. We also determined the ability of the substances to displace [^{3}H]D-AP5 from rat brain membranes (Olverman *et al.*, 1984, 1988a,b). Later, with the recognition of 4-(3-phosphonopropyl)piperazine-2 carboxylic acid (CPP, (**VI**)) as a potent NMDA receptor antagonist (Davies *et al.*, 1986), we were able to compare the substances as inhibitors of the binding of [^{3}H]CPP to brain membranes (Olverman *et al.*, 1986). Where all three assays were performed on a single compound, results were similar. We therefore screened most of our potential antagonists as inhibitors of the binding of [^{3}H]D-AP5 and, later, [^{3}H]CPP. Some of the results are summarized in Table 1.

It was immediately apparent that the structure–activity relations shown by NMDA receptor antagonists based on the piperazine nucleus were similar to those of open-chain antagonists (Evans *et al.*, 1979, 1982; Jones *et al.*, 1984). When the alkyl group attached to N4 of the piperazine ring was varied in chain length from 1 to 4 carbon atoms, and the ω-acidic group between carboxyl, sulfo, or phosphono, the highest activity was seen with a chain length of one or three carbon atoms, and with phosphono as the ω-acidic group. Moreover, the sulfo group, as in open-chain compounds (Olverman *et al.*, 1988b), was by the far the least effective. Thus, the most potent NMDA receptor antagonists were directly comparable to AP5 and AP7. However, in the piperazine series, the AP7 analog, CPP, was more active than the AP5 analog, (±)-4-phosphonomethylpiperazine-2-carboxylic acid, the reverse of that seen in open-chain compounds. With extension of the ω-carboxylic acid series to include acyl N4 substituents and substances containing unsaturation in the alkyl chain, further conclusions could be drawn: (a) a carbonyl group linked to N4 was less effective than methylene; (b) α,β-unsaturation (olefinic or acetylenic) relative to the terminal acidic group was more effective than a saturated alkyl chain; and (c) a *trans* (*E*) double bond was more effective than a *cis* (*Z*) double bond in an alkylene chain. In accord with all these conclusions, our most potent compound was CPPene (**VII**) and, as in open-chain compounds, the (*R*) enantiomer of both this compound and CPP was the more active isomer in each case (Aebischer *et al.*, 1989).

Table 1 Activity of NMDA receptor antagonists

Compound	Acronym	Form	K_D (μM)[a,b]	K_i (μM)[c]	
				[³H]D-AP5	[³H]CPP
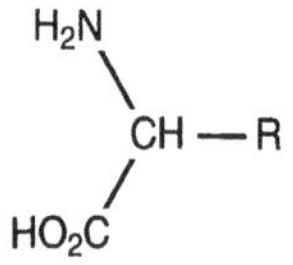					
R =					
$(CH_2)_3CO_2H$	DαAA	(*R*)	42[d]	13[d]	
	LαAA	(*S*)		89[d]	
$(CH_2)_5CO_2H$	DαAS	(*R*)	16[d]	25[d]	
	LαAS	(*S*)		80[d]	
$(CH_2)_3PO_3H_2$	AP5	(*RS*)	1.4[d]	1.2[f]	1.6[g]
	D-AP5	(*R*)	0.7[d]	0.62[d]	0.5[g]
	L-AP5	(*S*)	22[d]	40[d]	
$(CH_2)_5PO_3H_2$	AP7	(*RS*)	3.1[d]	3.1[f]	4.2[g]
	D-AP7	(*R*)		1.7[d]	
	L-AP7	(*S*)		28[d]	
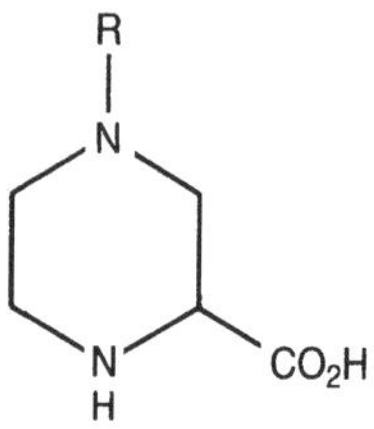					
R =					
CH_2CO_2H		(*RS*)	(27)[h]		28[d]
$(CH_2)_2CO_2H$		(*RS*)	(7.5)[h]		7.9[d]
$(CH_2)_3CO_2H$		(*RS*)	6.5[h]		1.9[d]
$CH_2PO_3H_2$		(*RS*)	(0.8)[h]		0.32[d]
$(CH_2)_2PO_3H_2$		(*RS*)	(39)[h]		30[d]
$(CH_2)_3PO_3H_2$	CPP	(*RS*)	(0.25)[h]	0.48[d]	0.28[d]
	D-CPP	(*R*)	(0.16)[h]		0.14[d]
	L-CPP	(*S*)	(3.0)[h]		2.3[d]
$(CH_2)_2SO_3H$		(*RS*)	(38)[h]		63[d]
$(CH_2)_3SO_3H$		(*RS*)	(24)[h]		42[d]
$CO—(CH_2)_2—CO_2H$		(*RS*)			11[d,e]
$CO—CH=CH—CO_2H$		(*RS*), (*Z*)			18[d,e]
		(*RS*), (*E*)			6.3[d,e]
$CO—C≡C—CO_2H$		(*RS*)			4.5[d,e]
$CH_2—CH=CH—CO_2H$		(*RS*), (*E*)			0.56[d,e]
$CH_2—CH=CH—PO_3H_2$	CPPene	(*RS*), (*E*)	(0.18)[h]		
	D-CPPene	(*R*), (*E*)	(0.09)[h]		0.044[d]
	L-CPPene	(*S*), (*E*)	(1.6)[h]		0.60[d]
$(CH_2)_3—PO(OH)(OC_2H_5)$		(*RS*)			8.4[d]
$CH_2—CH=CH—PO(OH)(OC_2H_5)$		(*RS*)			8.9[d]

[a] Apparent dissociation constant for antagonist–NMDA receptor interaction as determined in electrophysiological experiments in neonatal rat motoneurons (see Evans *et al.*, 1979, 1982).

[b] Values in parentheses are approximate, calculated from electron paramagnetic resonance (EPMR) values relative to D-AP5 and/or CPP.

[c] Inhibition of the binding of [³H]D-AP5 or [³H]CPP to rat brain membranes (Olverman *et al.*, 1988a,b; Olverman *et al.*, 1986).

Values taken from the following references: [d] Jane *et al.* (1994); [e] H.J. Olverman, D.C. Sunter, and J.C. Watkins, unpublished data; [f] Davies *et al.* (1986); [g] Olverman *et al.* (1986); [h] P.C-K. Pook and J.C. Watkins, unpublished data.

One further feature of interest was that, although greatly reducing potency, monoesterification of the phosphono moiety in CPP or CPPene resulted in significant activity still being retained. This accorded with previous findings with AP5 phosphonomonoethyl ester (A. W. Jones, P. C-K. Pook, and J. C. Watkins, unpublished observations), suggesting that although the dibasic phosphono group was preferred, a single dissociable hydrogen in the phosphono moiety was sufficient for activity. It was considered possible that this finding might ultimately prove useful for increasing the ability of such compounds to cross the blood–brain barrier. However, no esterifying alcohol capable of bestowing this property on CPP or CPPene has yet been identified.

3 GENERAL PHARMACOLOGY

Based on the structure–activity relationships described above, it was decided to select D-CPPene for further pharmacological profiling in view of its possible development as a therapeutic agent. In the present chapter we describe only some of the major features thought to be relevant for the therapeutic applications.

3.1 Specificity in binding assays

In order to assess the specificity of D-CPPene a number of binding and functional assays for various transmitter receptors were performed (Aebischer *et al.*, 1989; Lowe *et al.*, 1994).

3.1.1 Non-EAA receptors

Conclusion D-CPPene has no significant affinity for any of the non-EAA binding sites listed in Table 2. From Table 3, D-CPPene interacts in the nanomolar range only with the NMDA receptor.

Table 2 Affinity of D-CPPene to transmitter receptor sites in binding assays (means ± SD(n>2))

^{3}H-labelled ligand	Binding site	pK_i
Prazosine	NE-α_1	<4.0
Clonidine	NE-α_2	<4.0
Serotonin	5HT-1	<5.0
Ketanserin	5HT-2	<4.0
cis-Methyldioxolan	ACh M_1/M_2	<4.0
Pirenzepine	ACh M_1	<4.0
ADTN	DA-1	<4.0
SCH23390	DA-1	<4.0
SDZ201-501	DA-2	<4.0
Spiperone	DA-2	<4.0
Naloxone	Opiate	<4.0
GABA	GABA	<4.0
Flunitrazepam	Benzodiazepine	<4.0
Ditolylguanidine	σ	<5.0

Reproduced, with permission, from Lowe *et al.* (1994).
ACh, acetylcholine; GABA, γ-aminobutyr; 5HT, serotonin; M, muscarinic; NE, norepinephrine.

Table 3 Interaction of D-CPPene with EAA receptors (Lowe *et al.*, 1994)

Assay	Receptor	Effect
Cortical wedge	AMPA	None at 50 µm
Cortical wedge	KA	None at 50 µm
Binding assay	SIGS	IC_{50}>100 µm
Binding assay [^{3}H]CPP	NMDA	K_i = 40 nM
Binding assay, [^{3}H]CGP39653	NMDA	K_i = 30 nM
Hippocampal slice, PI turnover	mGluR	None at 100 µm

KA, kainate; SIGS, strychnine-insensitive glycine site.

3.1.2 Detailed studies at the NMDA receptor

In view of the selectivity of D-CPPene for the NMDA receptor described above, more in-depth studies were performed at the receptor level.

In a kinetic study in voltage-clamped cultured mouse hippocampal neurons (Benveniste and Mayer, 1991), a K_i of 230 nM was determined for D-CPPene against NMDA. As the experiments done in competition with NMDA antagonists described above yielded K_i values of 30 and 40 nM, the lower affinity in the agonist experiment indicates that the NMDA receptor has an agonist and an antagonist conformation for which D-CPPene has different affinities.

The NMDA receptor consists of heterogeneous assemblies of subunits, called NMDAR1 and NMDAR2A-D (Hollmann and Heinemann, 1994). Laurie and Seeburg (1994) determined that D-CPPene displaces glutamate from different combinations of recombinant NMDA receptor subunits with different affinities (Table 4).

Conclusions D-CPPene has a higher affinity for the NMDA receptor in its antagonist conformation and a lower affinity for the agonist conformation. It has the highest affinity to the NMDAR1–NMDAR2A subunit combination, which also displays the agonist/antagonists affinity difference.

Table 4 Interaction of D-CPPene with different NMDA receptor subunits displacing an agonist, glutamate, or an antagonist, [^{3}H]CGP39653 (Laurie and Seeburg, 1994)

Combination of NMDA receptor subunits	D-CPPene affinities, K_i (nM)
NMDAR1–NMDAR2A	510 (agonist)
NMDAR1–NMDAR2A	12 (antagonist)
NMDAR1–NMDAR2B	1200 (agonist)
NMDAR1–NMDAR2C	21000 (agonist)
NMDAR1–NMDAR2D	1900 (agonist)

3.2 Functional assays *in vitro*

D-CPPene was studied in several nonmammalian and mammalian nervous tissue assays to evaluate its electrophysiological effects on neurons.

3.2.1 *Frog hemisected spinal cord*

The experimental methods are described in Herrling (1985). Briefly, the hemisected spinal cords from *Rana temporaria* were placed in a chamber allowing grease gap direct current recordings from the ventral root and stimulation of the dorsal root to evoke dorsal root–ventral root potentials (DR–VR potentials). Experiments were conducted in magnesium-free Ringer solution. The early part of the DR–VR potential is known to be mediated by non-NMDA EAA receptors while the later components are NMDA receptor mediated (Francis *et al.*, 1980). D-CPPene predominantly affected the late components of the DR–VR potential, suggesting a predominant action on NMDA receptors (Lowe *et al.*, 1994). A competitive interaction with the NMDA receptor was indicated by the parallel shift to the right of dose–response curves constructed in the same preparation with multiple doses of bath-applied NMDA resulting in a calculated pA_2 value of 6.8 ± 0.14 (mean $\pm$ SD, $n>3$; Aebischer *et al.*, 1989). D-CPPene did not inhibit depolarizations induced by bath-applied kainate or AMPA at concentrations totally abolishing NMDA responses.

3.2.2 *Rat cortical wedge*

To quantitatively assess the effects of D-CPPene in mammalian tissues the rat cortical wedge preparation was used as described by Harrison and Simmonds (1985) and modified by Lowe *et al.* (1990). As in the frog hemisected spinal cord, a magnesium-free buffer was used, and grease gap recordings of the potential difference between gray and white matter were made.

In this preparation, D-CPPene also caused a parallel shift to the right of dose–response curves constructed by applying variable doses of NMDA (example in Fig. 2). From such experiments a pA_2 value of 6.77 was calculated

The data from 39 separate experiments were pooled to produce a Schild plot (Schild, 1947; Fig. 3) and yielded a pA_2 value of 6.7 for D-CPPene and a slope of approximately 1 over a dose range of 30-fold concentrations. This and the parallel displacement shown in Fig. 2 indicates competitive antagonism of D-CPPene at the NMDA receptor.

Conclusion In these functional *in vitro* assays in amphibian and mammalian neural tissue D-CPPene was a potent competitive and selective antagonist of NMDA receptor-mediated electrophysiological excitations. It also inhibited synaptic components in the amphibian preparation thought to be NMDA receptor mediated.

3.2.3 *Glutamate release in synaptosomes*

D-CPPene does not affect glutamate release in this model for direct release modulation, whereas its L enantiomer has some inhibiting activity (Urwyler and Puente, 1989).

Conclusion The pharmacological effects of D-CPPene are not due to a direct effect of glutamate release in synaptic terminals.

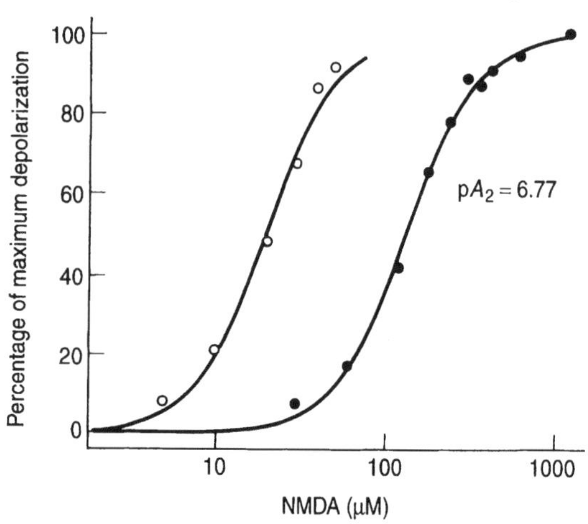

FIG. 2 Parallel displacement to the right of a dose–response curve constructed by variable doses of bath-applied NMDA in the rat cortical wedge preparation.

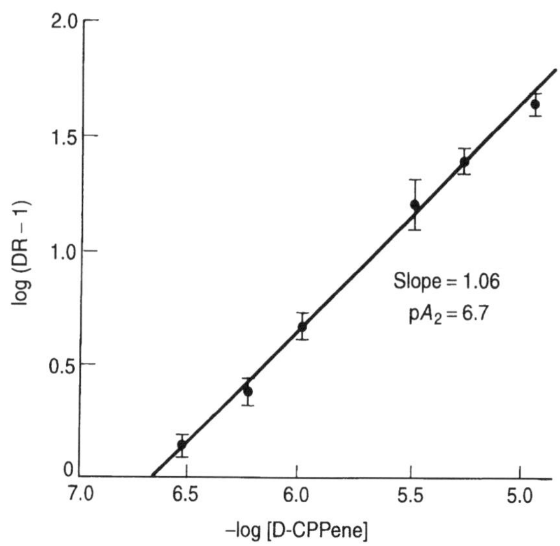

FIG. 3 Schild plot of antagonist experiments in the rat cortical wedge. Dose–response curves were constructed by bath-applying varying doses of NMDA in the presence of varying doses of equally bath-applied D-CPPene. This figure summarizes 39 experiments, and the data are given as means and standard deviations. (Modified, with permission, from Lowe *et al.* (1994).)

3.3 Characterization of D-CPPene in whole animals

3.3.1 *Effects of* D-*CPPene on caudate neurons recorded* in vivo

The responses of the NMDA receptor are known to be very sensitive to the ionic composition of the extracellular environment, e.g. at low Mg^{2+} concentrations of 100 μM or less there is a maximal stimulation of this receptor by agonists whereas higher concentrations inhibit the effects of agonists, and this effect is voltage-dependent (Johnson and Ascher, 1990). It can therefore not be automatically assumed that the results obtained in *in vitro* preparations entirely predict the situation *in vivo*. To clarify some of these aspects a series of experiments were conducted where the electrical activity of cat caudate neurons were recorded *in vivo* where cells are presumed to be in their natural extracellular environment with the exception of the anesthetic agent used (halothane in the present case). The agonists and antagonists to be studied were applied by microiontophoresis, often during simultaneous electrical stimulation of afferent pathways to the caudate (e.g. the corticostriatal pathway). The methods are described in detail in Herrling *et al.* (1983). The results obtained in three cats and six cells confirmed the specificity of D-CPPene for the NMDA receptor (Fig. 4): micro-iontophoretically applied D-CPPene reversibly inhibited NMDA-induced depolarizations and firing while excitatory effects induced by the non-NMDA receptor agonist quisqualate were little affected.

Conclusion The pharmacological specificity of D-CPPene for the NMDA receptor can be found at the cellular level in neurons in their natural environment and in extracellular fluid without manipulation of ionic concentrations.

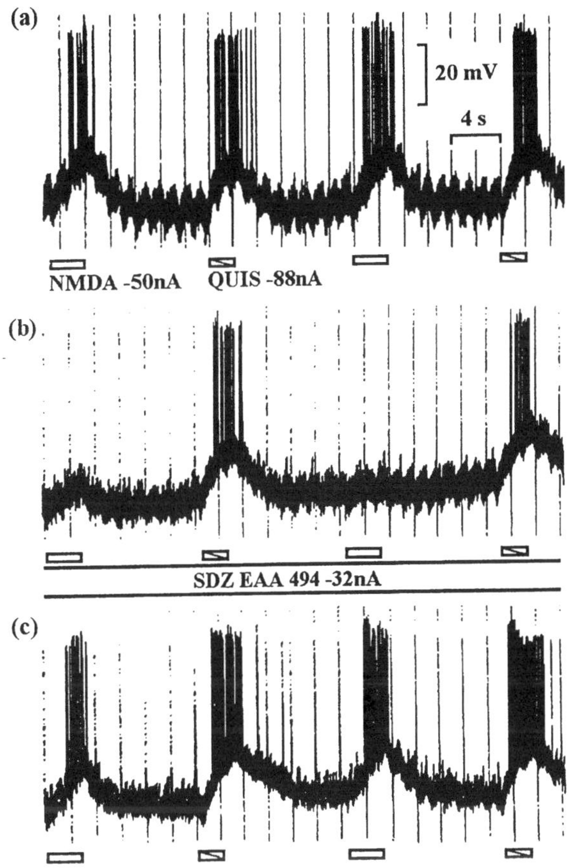

FIG. 4 The interaction of EAA receptor agonists and D-CPPene (SDZ EAA 494) on the membrane and action potentials of caudate neurons during intracellular recordings in halothane-anesthetised cats. (a) Control: the effects of NMDA and quisqualate (QUIS) before the application of D-CPPene. The currents were adjusted to evoke similar depolarizations and firing with both agonists. (b) D-CPPene applied throughout a subsequent recording from the same cell completely abolished NMDA-induced excitations with little effect on QUIS-induced excitations. D-CPPene had no visible effects on cortically evoked synaptic responses (not visible in these time-compressed traces). (c) Recovery after termination of iontophoretic application of D-CPPene: NMDA responses are fully recovered.

NMDA and QUIS, a non-NMDA EAA receptor agonist, were applied by microiontophoresis at the times indicated by the empty and crossed bars, respectively, and at the currents indicated next to the abbreviations; the calibration for voltage fluctuations of the neuronal membrane are indicated by the vertical bar of 20 mV, and time calibration by the horizontal bar of 4 s. The regularly spaced vertical thin lines indicate electrical cortical stimulation. Current controls (not shown) were performed to determine if current effects were contributing to the observed effects of iontophoresis; this was not the case in the present experiments. (Modified, with permission, from Lowe *et al.* (1994).)

3.3.2 *Effects of* D-*CPPene on models for convulsions*

Rodents D-CPPene antagonizes electroshock-induced convulsions, as has been repeatedly shown both in mice and in rats (Lowe *et al.*, 1994; Zarnowski *et al.*, 1994), and has also been shown to inhibit acoustically induced seizures in DBA/2 mice (Chapman *et al.*, 1990) or convulsions in genetically epilepsy-prone rats (DeSarro and DeSarro, 1993; Smith and Chapman, 1993). The oral effective doses in these models range from 3 to 25 mg kg^{-1}, and intraperitoneal doses from 1 to 3 mg kg^{-1}. The maximal effects occur at 2–6 h and persist up to 24 h following a single oral administration. Repeated applications of D-CPPene did not result in statistically significant reduction of activity compared to acute administration demonstrating a lack of pronounced tachyphyllactic effect. In the so-called 'kindling model', where seizures are induced by the daily repetition of a subthreshold electrical shock to the amygdala, Loescher and Honack (1991) and Loescher *et al.* (1993) found that NMDA antagonists are relatively less potent than non-NMDA EAA receptor antagonists and that such animals are very sensitive to NMDA antagonist elicited side-effects. They therefore predicted that NMDA receptor antagonists might not be useful antiepileptic

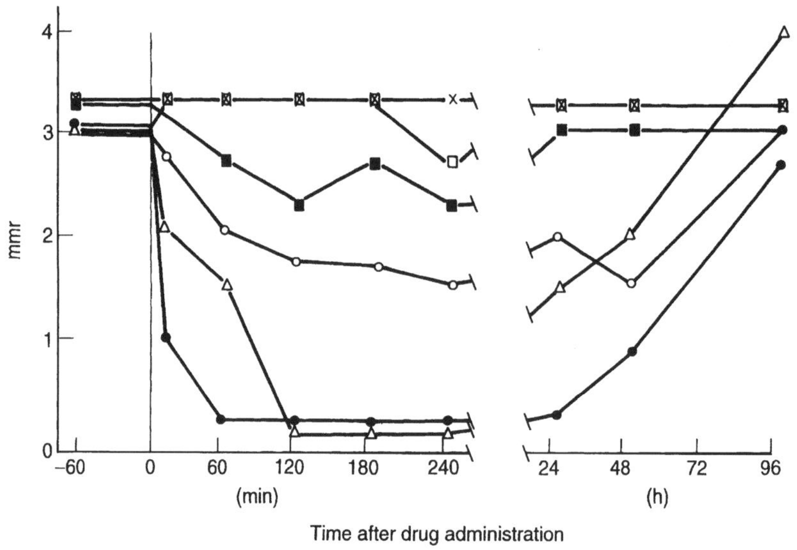

FIG. 5 The effect of intravenously applied D-CPPene on mean myoclonic responses (mmr) elicited by stroboscopic stimulation in three photosensitive baboons (mean scores). There is a dose-dependent inhibition of convulsions that can last over 24 h at doses of 4 mg kg^{-1} and above. ×, 0 mg kg^{-1}; □, 0.5 mg kg^{-1}; ■, 2 mg kg^{-1}; ○, 4 mg kg^{-1}; △, 8 mg kg^{-1}; ●, 16 mg kg^{-1}. (Modified, with permission, from Patel *et al.* (1990).)

agents in humans but should be rather used at low doses to potentiate, for example, AMPA antagonists. When D-CPPene was tested in kindled animals it was found to be effective, but at distinctly higher doses than those found in the electro-shock models described above (8 and 12 mg kg^{-1} i.p.; Duermueller *et al.*, 1994).

Chemically induced convulsions are equally inhibited by D-CPPene: metrazol (Stables, 1992); strychnine (McAllister, 1992); bicuculline and mercaptopropionic acid (Tutka *et al.*, 1992).

Primates An elegant primate study was performed by Meldrum and his colleagues in photosensitive baboons, where a stroboscopic light stimulus can induce seizures (Patel *et al.*, 1990). In this model, single applications of 8–16 mg kg^{-1} i.v. or >24 mg kg^{-1} p.o. cause strong inhibition of these light-induced convulsions. The inhibition can last for over 24 h, and it is noteworthy that in these experiments in which plasma concentrations were measured, significant protection was measurable at times when the plasma concentrations of D-CPPene had sunk below the detection level (Figs 5 and 6).

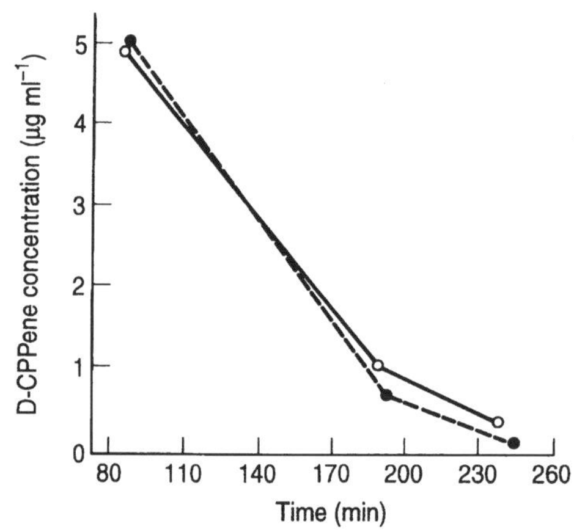

FIG. 6 Plasma concentrations of D-CPPene applied intravenously at 4 mg kg^{-1} to photoepileptic baboons in two animals (solid and broken lines). Note that plasma levels show rapid clearance ($t_{1/2}$ ~ 70 min) and they were undetectable at 24 h when this dose still afforded a significant protection from convulsions. (Modified, with permission, from Patel *et al.* (1990).)

Conclusion These results show that D-CPPene has clear inhibitory effects on convulsive phenomena after intraperitoneal, intravenous, and oral applications in different mammalian species. There is no evidence of a weakening effect of D-CPPene after repeated applications, indicating a lack of tachyphyllaxis. The kindling experiments indicate that D-CPPene is distinctly less active in convulsion models of this type, and this may have clinical consequences in epileptic syndromes where a kindling effect may be involved. The primate experiments suggest that D-CPPene is capable of inhibiting central NMDA receptors at a time where plasma levels of the drugs have sunk to the undetectable level. This may be due to a preferential retention of D-CPPene in brain tissue.

3.3.3 Effects of D-CPPene on the behavior of rodents

Gross behavior in mice The behavioral effects of D-CPPene have been studied both in mice and rats. In the so-called 'primary observation test', scored according to

Irwin (1968) and performed in OF-1 mice, the following threshold effects were observed at 1 h after intraperitoneal drug application (Lowe *et al.*, 1994): reduction of place motor activity and rearing at 1 mg kg^{-1}, full inhibition at 32 mg kg^{-1}; decrease in locomotion at 1 mg kg^{-1}, complete inhibition at 32 mg kg^{-1}; reduction in muscle tone at 1 mg kg^{-1}; ataxia at 10 mg kg^{-1} and above; reduction of pinna reflex at 3.2 mg kg^{-1} with full inhibition at 320 mg kg^{-1}; disturbance of respiratory rhythm and force at 3.2 mg kg^{-1} and above; reduction of defecation at 1 mg kg^{-1} and above; lacrimation at 100 mg kg^{-1} and with reddish coloration at 320 mg kg^{-1}.

Biphasic effects with increasing intravenous doses in rats An interesting experiment in rats is reported by Lowe *et al.* (1994) trying to assess the effects on gross behavior of increasing doses to very high levels where rats were given constant intravenous infusions of D-CPPene: 200 μg h^{-1} for up to 6 h induced distinct muscle relaxation and ataxia, these effects becoming apparent 1 h after the beginning of the infusion. An infusion of a 10-fold dose (2 mg h^{-1}) induced a strong muscle relaxation within 5–20 min of the beginning of the infusion accompanied by loss of righting and lying on their sides with splayed rear legs. Surprisingly, after 2–3 h of infusion at this rate, the animals started to show phases of episodic motor activity which continued for the remainder of the infusion (5–6 h). During these episodes the animals rose on their front legs and displayed side-to-side or up and down head movements and sniffing. Jaw movements and front paw treading were also observed, reminiscent of behavior seen when proconvulsive drugs are administered. Only the front legs had any muscle tone, while the hind legs remained inactive. All of these effects ceased 1–2 h after the end of the infusion, and the behavior appeared normal when observed 1 day later. In a separate experiment it was determined that steady plasma concentrations of D-CPPene were reached after about 3 h of infusion at a rate of 2 mg h^{-1}. The concentration reached was 17 μg ml^{-1}. One could speculate that at the lower doses where the animals show pure relaxation and ataxia, NMDA receptor-dependent pathways are mainly inhibited, which are responsible for behavioral activation, whereas at higher doses NMDA receptor-dependent pathways may additionally be affected, which themselves inhibit the inhibitory pathways resulting in partial reactivation of behavior.

Effects in a learning model D-CPPene was evaluated in a test for acquisition of passive avoidance in rats (Lowe *et al.*, 1994). At 2 mg kg^{-1} i.p. administered 20 min before testing (which may be too early in view of the other behavioral data described above) significantly shortened retention latency, suggesting an interference with memory formation in this assay. But other forms of learning seem to be affected in a more complex way by NMDA antagonists (Mondadori and Weiskrantz, 1993).

Conclusion D-CPPene elicits mainly inhibitory effects on the behavior of rodents, although this inhibition can be partially overcome by very high doses of continuously infused drug. Furthermore, an inhibitory effect on learning was seen in the passive avoidance assay.

3.3.4 *Neuroprotective effects of D-CPPene*

Quinolinic acid-induced excitotoxicity in rat striatum Quinolinic acid is an endogenous agonist of the mammalian NMDA receptor (Herrling *et al.*, 1983) and causes excitotoxic lesions (for a review, see Stone, 1993). This property of quinolinic acid has been used to develop a model of neuroexcitotoxicity (Foster *et al.*, 1987). D-CPPene was studied for neuroprotective properties in a similar assay but magnetic resonance imaging was used to determine the extent of neuronal damage in a noninvasive way (Sauter and Rudin, 1986, 1991). In these experiments, 250 nmol/2 μl of quinolinic acid was injected directly into the striatum, and D-CPPene was administered intravenously (immediately after lesioning), intraperitoneally (immediately after lesioning), or orally (p.o., 30 min prior to lesioning). D-CPPene 1.25 and 4 mg kg^{-1} i.v. reduced quinolinic acid-induced lesions in the striatum by 43 and 75%, respectively; 3,10, and 30 mg kg^{-1} i.p. by 35, 76 and 92%, respectively; and 15 and 50 mg kg^{-1} p.o. by 15 and 26%, respectively. All effects were statistically significant with the exception of the lowest intravenous and oral doses. Similar results were obtained by Massieu *et al.* (1993).

Cat middle cerebral artery occlusion D-CPPene was evaluated by Bullock *et al.* (1990 a, b) and Chen *et al.* (1991) in the cat middle cerebral artery occlusion (MCAO) model. The cat is a better species than the rat for this model as the anatomy of the medial cerebral artery is more similar to the human anatomy and it allows good control of blood pressure, temperature, arterial hematocrit, carbon dioxide tension, and metabolic acidosis as performed in the laboratory of McCulloch. D-CPPene was applied as a bolus of varying doses 15 min preocclusion followed by continuous infusion. There were five experimental groups: a low-dose pretreatment group (15 min pretreatment with a bolus of 1.5 mg kg^{-1} i.v., followed by an infusion at 0.017 mg kg^{-1} min^{-1}, three cats); a middle-dose pretreatment group (4.5 mg kg^{-1} bolus, followed by a 0.051 mg kg^{-1} min^{-1} infusion, seven cats); a high-dose pretreatment group (15 mg kg^{-1} bolus, followed by 0.17 mg kg^{-1} min^{-1}, six cats); a high-dose post-treatment group (15 mg kg^{-1} i.v. bolus 1 h postocclusion, followed by a 0.17 mg kg^{-1} min^{-1} infusion, six cats). The middle and high pretreatment doses of D-CPPene showed a significant reduction of the volume of ischemic damage assessed histologically (Chen *et al.*, 1991; Figs 7 and 8). Figure 8 shows the effect of the middle pretreatment dose of D-CPPene, with a clear reduction of the damaged area.

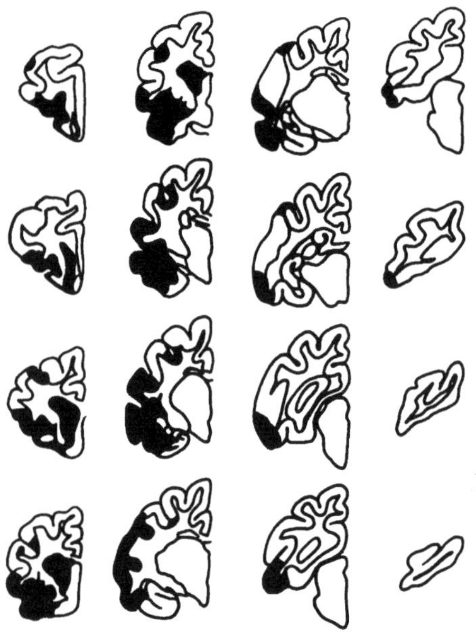

FIG. 7 Histological assessment ischemic damage following medial cerebral artery occlusion in 16 coronal sections of a cat brain. Dark areas indicate areas of damage. Such sections were used to determine the neuroprotective effects of D-CPPene. (Modified, with permission, from Chen *et al.* (1991).)

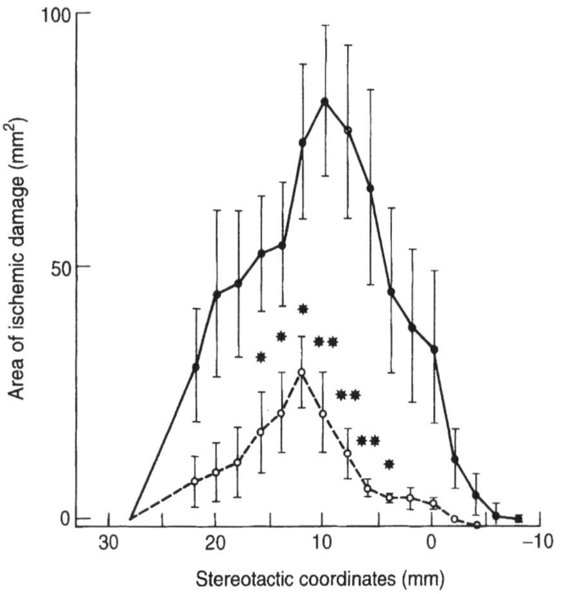

FIG. 8 Effect of D-CPPene (high pretreatment dose, 15 mg kg^{-1}) on the area of MCAO-induced ischemic damage in six control cats (solid circles) compared to treated animals (six cats, empty circles). The difference between controls and treated animals was statistically significant (*t*-test, Bonferroni correction). (Modified, with permission, from Bullock *et al.* (1990a and b).)

Effect of D-CPPene in a model for subdural hematoma in rats The model of Miller *et al.* (1990) was used for these studies where homologous venous blood was injected into the subdural space in halothane-anesthetised animals. D-CPPene 15 mg kg^{-1} was infused intravenously 15 min before the induction of the hematoma. The volume of damage induced by the hematoma was ~60 mm^3 in controls (n=8) and ~30 mm^3 (n=10) in treated animals, the difference being statistically different at p<0.019 in the Mann–Whitney test (Bullock *et al.*, 1990a, b).

Effect of D-CPPene in a model for brain edema in rats Edema can be a life-threatening consequence of ischemia. The effects of D-CPPene were therefore studied by Park (1994a,b). To calculate edema, the total volume of the non-ischemic hemisphere was subtracted from the total volume of the ischemic hemisphere (MCAO). D-CPPene was administered as a bolus of 4.5 mg kg^{-1} 15 min preocclusion followed by a continuous infusion at 3 mg kg^{-1} h^{-1}. Under these conditions D-CPPene reduced the mean volume of hemispheric edema by 34% (n = 6 in control and treatment groups, p <0.001).

Conclusion The experiments in cats and rats indicate that D-CPPene can distinctly reduce neuronal damage induced by ischemic injury and that best effect is obtained when the drug is given shortly before the injury. The time window for clinical effect cannot be easily determined from the present animal studies but it is reasonable to assume that the earlier the treatment begins after injury the better the chances of success.

3.3.5 Interactions with monoaminergic systems

D-CPPene was applied either subcutaneously or locally by microdialysis to different dopaminoceptive brain nuclei (nucleus accumbens and nucleus caudatus) in non-restrained animals (Imperato *et al.*, 1990). D-CPPene concentrations of 1, 10, and 100 μM in the transcranial dialysis tube stimulated the release of dopamine in a concentration-dependent fashion in both nuclei. At the same time, an increase in locomotion, sniffing, rearing, and grooming occurred. This behavioral stimulation followed the time-course of elicited dopamine release. The D-CPPene elicited behavioral elements could be prevented by the dopamine antagonist haloperidol. Dopamine release was observed in the same study after subcutaneous application of D-CPPene (15–20 mg kg^{-1}), but after this mode of application there was no behavioral stimulation, but rather muscle relaxation.

Conclusion D-CPPene interacts with midbrain dopaminergic systems by causing a release of dopamine. This could be related to the psychotomimetic effects seen after high doses in humans (see below).

3.3.6 *Pharmacokinetics and disposition of* D-*CPPene*

Absorption, distribution and elimination of D-CPPene were studied in rats and dogs using the ^{14}C-labeled compound.

Bioavailability, calculated by comparison of dose-normalized $AUC_{0-\infty}$ (unchanged drug) from oral versus intravenous applications, was only 2% in rats, but 7–8% in dogs, indicating significant species differences. Metabolite patterns in rats showed >90% unchanged drug both in brain and kidney; in dog plasma and urine the metabolite pattern showed almost exclusively unchanged drug. Elimination half-lives were 0.6 h in the rat and 0.8 h in the dog, a second half-life with a less than 10% AUC contribution was found in both species at 10–13 h. Total body clearance of unchanged drug was represented almost exclusively by renal clearance, as urinary recovery of unchanged drug was 84 and 99% in the rat and dog, respectively. This seems to be equally the case in humans (see below). Multiple dosing had no significant effects on pharmacokinetic parameters.

4 HUMAN STUDIES

Current human experience with EAA 494 is based on safety and tolerability studies in healthy volunteers, studies in patients with treatment resistant epilepsy and in patients with severe head trauma. Results of these studies are summarized below, except for the study in patients with severe head trauma, which is still under analysis.

4.1 Subjects and methods

Single oral doses of EAA 494 ranging from 10 to 900 mg were administered to nine cohorts of young, healthy, male volunteers in a double-blind, randomized, placebo-controlled dose escalation study. In each cohort, eight volunteers received a single dose of EAA 494, and four volunteers received a placebo; a total of 102 subjects were included, 68 receiving EAA 494 and 34 receiving placebo. In addition to the standard safety assessments, the effects of EAA 494 on selected neurological and cognitive functions and on the plasma profile of several hormones were examined.

After having established the safety and tolerability of single doses of EAA up to 900 mg, a multiple oral dose safety and tolerability study was conducted in young, healthy, male volunteers. In this double-blind, randomized, placebo-controlled study, 12 subjects (eight on the active drug, four on the placebo) received EAA 494 500 mg b.i.d. for 7 days, followed by 1000 mg b.i.d. for an additional 7 days. Standard safety evaluations were performed; in addition, neurological and neuropsychological functions and effects on plasma profile of selected hormones were assessed.

Subsequently in a single, ascending intravenous dose safety and tolerability study with a randomized, double-blind, placebo-controlled design, seven cohorts of healthy male volunteers were administered EAA 494 in the dose range of 1–100 mg. In each cohort six volunteers received EAA 494 and three received placebo, except for the 100 mg dose, at which only two volunteers were treated, and the 10 mg dose, at which 12 subjects received EAA 494 and six received placebo. The compound was administered as a short infusion over 5 min. As in the oral dose studies, standard safety parameters, neurological and neuropsychological functions, and effects of EAA 494 on plasma levels of selected hormones were examined. Subsequently, a multiple intravenous dose safety and tolerability study was performed in healthy male volunteers. In this randomized, double-blind, placebo-controlled study, subjects received either placebo or EAA 494 as a short intravenous infusion over 5 min. A total of 35 subjects were included, eight receiving the placebo, 12 receiving EAA 494 25 mg every 24 h for 7 days, and 15 receiving EAA 494 25 mg every 12 h for 7 days (except for the first and the last day, on which only a single dose of 25 mg was administered). In addition to the standard safety parameters, psychometric tests were performed and the EEG was recorded on selected study days.

An open label pilot study was conducted to assess the safety, tolerability, and the antiepileptic efficacy of oral EAA 494 in patients with frequent epileptic seizures resistant to available treatment. Ascending, oral, b.i.d. dosages of EAA 494 were administered as an add-on treatment to male and female patients with intractable complex partial seizures. Patients were given EAA 494 250 mg b.i.d. for 2 weeks; contingent on the tolerability and therapeutic need the dose was allowed to be increased to 500 mg b.i.d. for the next 4 weeks, and to 1000 mg b.i.d. for the following 4 weeks. A total of eight patients were included in this study, which was prematurely discontinued.

4.2 Results

4.2.1 *Behavioral and general CNS effects*

A single oral dose administration of EAA 494 up to 900 mg was not associated with any behavioral or other CNS effects which could reliably be attributed to the compound. There were no differences between the placebo and active compound nor a dose dependency in the severity and frequency of spontaneously reported behavioral effects, in a visual analog scale to assess mood or on a symptom questionnaire to assess behavioral and systemic adverse events.

Following multiple oral dose administration, six out of eight volunteers reported behavioral and other CNS effects which included lack of concentration, difficulty to fall asleep, sleeplessness, quickly changing dreams, bad dreams, hot flushes, feeling 'high', elevated mood, anxiety, restlessness, feeling of panic, loss of appetite, headache, and sweating. The duration and intensity of these effects varied across the subjects, they were more pronounced during the second week of treatment, i.e. at the higher dose level (1000 mg b.i.d.), and two volunteers discontinued the treatment prematurely in the first half of the second week because of these adverse effects. Three out of four subjects on the placebo also reported adverse events with similar qualities (lack of concentration, dizziness, tiredness, and vivid or bad dreams), but in general with milder intensity. The visual analog scale assessing mood did not reveal statistically significant differences between the groups.

The intravenous administration of EAA 494 to healthy young volunteers was not associated with any drug-related CNS effects up to a single dose of 10 mg. At the 10 mg dose level, five out of six subjects reported mild, non-specific adverse events such as headache, fatigue, and nausea. At 25 mg, four out of six subjects receiving EAA 494 reported CNS-related effects such as headache, difficulty to concentrate, numbness, and sleepiness. The severity was mild to moderate, the onset after dosing was 45 min to 4 h, and the duration was from 30 min to 2 h. The only adverse event reported in the placebo group was headache. The type, frequency, and severity of adverse events increased following the administration of 50 mg: five out of six volunteers reported adverse events, which included sleepiness, concentration difficulty, dizziness, distal paresthesia and tingling, headache, heavy-headedness, pressure in the head, light-headedness, accommodation difficulties, altered visual perception and visual hallucinations, spatial dissociation, taste disturbances, and mood disturbances. All these adverse events were noted as being mild to moderate. In general they appeared 1–4 h after administration and lasted 30 min to 4 h. Some of them, however, such as visual hallucinations, pressure in the head, spatial dissociation, distal paresthesias, and mood disturbances, appeared 24 h after administration and persisted for 10 min to 3 h. All of these adverse events resolved spontaneously. None of the subjects receiving the placebo reported adverse events at this dose level. The 100 mg dose of EAA 494 was administered to two subjects, and based on the adverse events observed further enrollment of subjects at this dose level was terminated. Both subjects reported a feeling of distal numbness and tingling, and numbness around the mouth at the end of the infusion. One hour after infusion they developed restlessness, hot flushes, temporal and spatial disorientation, euphoria, logorrhea alternating with apathy, enhanced sense of smell and taste, and altered perception of forms and colors. These effects appeared in clusters lasting 5–15 min and alternated with periods of relative normality, with cycles approximately every 30 min. The periods of recurrences followed by recovery phases lasted for 27 h after drug administration. The visual analog scale assessing mood did not show any systematic changes, and the symptom questionnaire revealed some spontaneously not reported effects such as light sensitivity, weakness, fatigue, drowsiness, feeling as if in a dream, altered taste, headache, and dizziness.

In the multiple intravenous dose administration study, eight out of 12 subjects receiving EAA 494 25 mg once a day reported adverse events related to the CNS. These included headache, concentration difficulty, excitement, drowsiness, fatigue, and giddiness. The intensity of these effects was mild in most cases, the onset was generally 1–5 h after infusion and the duration was 1–5 h. The frequency of adverse events decreased from day 1 to day 6. At this dose level, three out of four placebo subjects reported adverse events related to the CNS such as mild fatigue, headache, excitement, insomnia and chills. The type, frequency and intensity of CNS effects were clearly more in the group receiving EAA 494 25 mg twice a day. Fourteen subjects reported adverse events which included (in the order of frequency) concentration difficulty, pressure in the head, slurred speech, difficulty to fall asleep, headache, excitement, fitful sleep, giddiness, heaviness of the head, legs or body, colored images when the eyes were closed, fatigue, drowsiness, sleepiness, numbness of different body parts, hot flushes, euphoria, laughter, anxiety, unusual or colored dreams, lightheadedness, sadness, tingling/itching of the scalp, tears, bitter taste, buzzing in the ears, increased sensitivity, stimulation of intellectual functions, increase of ideas, and a sensation of general well-feeling. The majority of subjects started experiencing adverse events following the second administration on day 2, and the adverse events were observed most frequently on days 3 and 4. The majority of them were defined as moderate, but two subjects developed severe adverse events (giddiness and laughter). The time of onset was around 1 h after administration, and the duration was usually 1–5 h. Two subjects discontinued the study because of adverse events on day 3 (giddiness, nausea, heartburn, numbness in the lips, slurred speech, and fluctuations in mood) and on day 5 (anxiety). Three out of four placebo subjects reported mild CNS-related adverse events including sleepiness, headache, anxiety, buzzing in the ears, and heartburn. The visual analog scale assessing mood did not reveal any systematic changes in the mean values, and the symptom questionnaire reflected the above-described adverse events without providing additional information.

A number of CNS-related adverse events were reported by all patients participating in the pilot study in patients with treatment-resistant epilepsy (Sveinbjornsdottir *et al.*, 1993). These included impaired concentration, sedation, drowsiness, disorientation, confusion, depressed mood with emotional incontinence, dizziness, and amnesia. Some of these symptoms were severe and led to the discontinuation of the patients.

4.2.2 *Motor effects and effects on somatic neurological functions*

In the single and multiple oral dose studies no motor effects of treatment were reported spontaneously. Similarly, neurological functions which were assessed at preset intervals and which included assessment of balance (standing on one foot, and tandem walk) and quantified manual motor performance (aiming from right to left, steadiness, following a line, and tapping) did not reveal any statistically significant differences between the active treatment and placebo groups. No pathological nystagmus was observed at any dose level.

Following intravenous administration of single doses the neurological test battery described above did not reveal any differences between active treatment and placebo or any dose-dependent effects on stance, tandem gait, or eye movements. These functions were not systematically assessed in the multiple intravenous dose administration study. The spontaneously reported motor effects were non-specific and included slurred speech and general weakness described as fatigue, tiredness, or heaviness of extremities and body.

Ataxia was reported by six patients out of eight in the pilot study in treatment-resistant epilepsy. One patient was unable to walk for a few days because of severe ataxia. Motor speech was impaired in three patients, resulting in dysarthria, unintelligible speech, and, in one case, anathria. One patient reported intermittent diplopia, and one patient suffering from the Sturge–Weber syndrome with a right cerebral angiomatous malformation developed continuous choreo-athetoid movements of the left arm and leg together with increasing gait ataxia, which resolved several days after discontinuing the treatment.

The somatosensory functions were not systematically examined. In both single and multiple intravenous dose studies, however, distal numbness and tingling, numbness around the mouth, and itching and tingling in the scalp were spontaneously reported at high dose levels. In the epilepsy study, two patients complained of hemisensory symptoms without detectable focal sensory deficits.

4.2.3 *Effects on cognitive functions*

The most frequent spontaneously reported effects of EAA 494 on cognitive functions were on vigilance, concentration, and attention. Except for the single oral dose study, concentration difficulties, drowsiness, sedation, confusion, and disorientation were spontaneously reported in all other studies, especially at high dose levels. In the epilepsy study, two patients reported a total amnesia for a period of 10 days, commencing after an increase in the dose to 500 mg b.i.d.

In addition to the spontaneous reporting of effects on cognitive functions, a neuropsychological test battery was administered in all studies. This battery included tests to assess phasic alertness (i.e. attentional matrix, measured by a test of simple reaction time), focused attention (measured by a test of complex reaction time), cognitive processing, and psychomotor speed (digit symbol substitution test), immediate as well as delayed recall for verbal (selective reminding test or word recognition test), and nonverbal (Rey's 15 figures) and spatial memory (city map test, a subscale of the 'Lern and Gedaechtnis test').

In the single oral dose study there were no consistent effects of the compound on the attention and memory tests. In the multiple oral dose study, tests of attention and psychomotor speed remained unaffected. In the selective reminding test there was an impairment in the mean performance score in the active group; however, no consistent dose or time dependency was observed. No effects were seen in the Rey's 15 figures test.

In the single and multiple intravenous dose studies, verbal and nonverbal memory test performance was significantly impaired after administration of 50 mg single and 25 mg b.i.d. doses, without a concomitant, statistically significant and consistent (i.e. time- and dose-dependent) impairment in reaction time tests, although there was a trend toward impairment. In the multiple-dose study, psychomotor speed as measured by the digit symbol substitution test was also impaired. In contrast to the significant impairment in verbal and nonverbal memory tests, spatial memory test performance was not significantly affected in a consistent manner. The maximum effects occurred 2 h postmedication, lasted up to 8 h, and were more pronounced after repeated administration, i.e. on day 7 of the multiple-dose administration study. These results suggested differential and specific effects of NMDA receptor antagonism on memory processes.

A number of psychometric tests (including word list learning, information processing, semantic processing, and the Stroop test) were also employed in the epilepsy study. An impairment in all of these tests was observed during treatment with recovery after withdrawal of the compound. These results were, however, difficult to interpret because of the small number of patients and pronounced non-specific CNS effects.

4.2.4 *Effects on electroencephalography*

In the multiple intravenous dose study, computer-quantified EEG recordings were performed at the baseline, on day 4 (before and 2, 4, and 8 h after the morning dose), and, in addition, 24 h after the last dose in the 25 mg b.i.d. group.

The administration of EAA 494 was associated with an increase of power in the slow wave bands (δ and θ), a decrease of the power in the α band, a decrease of the maximum peak position, and a decrease of the total power. These changes were more pronounced in the b.i.d. group, lasted at least 12 h, and were still detectable to an extent already before the morning dose administration on day 4.

4.2.5 *Effects on treatment-resistant epileptic seizures*

There was no beneficial effect of treatment with EAA 494 up to oral doses of 500 mg b.i.d., for a period of up to 3 weeks at this dose level, in patients with frequent, treatment-resistant complex partial seizures. The trial had to be prematurely terminated because of pronounced adverse events occurring in all patients and lack of therapeutic effects. The results of this study are described in detail elsewhere (Sveinbjornsdottir *et al.*, 1993)

4.2.6 *Effects on plasma hormone profiles*

Because of findings in animal experiments suggesting that EAA 494 may effect plasma levels of certain hormones, daily plasma profiles of prolactin, cortisol, and luteinizing hormone were obtained in single and multiple oral and single intravenous dose studies. In the oral dose studies there were no effects of treatment with D-CPPene on the plasma levels of these hormones, and merely circadian fluctuations were detected. Similarly, in the single intravenous dose study, no noteworthy changes in plasma hormone levels were observed except for attenuation of normal daytime fluctuations in cortisone levels, 1–24 h after the administration of 100 mg dose level to two subjects.

4.2.7 *Systemic effects and effects on safety parameters*

Standard safety parameters were assessed in all studies, which included a complete physical examination, vital signs (blood pressure, radial pulse, body temperature), ECG, hematology, blood chemistry, and urine examination. There were no consistent or clinically relevant changes which were considered to be related to the test compound in any of these parameters in any of the studies. The two subjects who received 100 mg of EAA 494 intravenously had a slight increase in systolic and diastolic blood pressure (maximum absolute values 148 and 149 mmHg systolic, 103 and 107 mmHg diastolic), and one subject at this dose level had a slight tachycardia (104 bpm) 32 h after the administration. These subjects had pronounced psychotomimetic adverse events at the time of these changes and were visibly distressed, so that these changes may have been of secondary nature. One of the two subjects at the 100 mg dose level had a slight decrease in body temperature after drug administration (from 37.3 at the baseline to 35.8 at 1 h).

Spontaneously reported systemic effects which could not be directly linked to CNS effects were rare. The most frequent was dry mouth at high intravenous doses (50 and 100 mg single and 25 mg b.i.d. doses); the isolated events included sweating following multiple oral dose administration and facial/conjunctival hyperemia in one subject after the 50 mg i.v. dose.

4.2.8 *Drug exposure and pharmacokinetics*

Blood samples were obtained in all studies and plasma concentrations of EAA 494 were determined. The results confirmed that all subjects were exposed to the compound in all studies. In the single oral dose study, highest mean C_{max} values were obtained at the 900 mg level, and reached 223 ng ml^{-1}. In the multiple oral dose study the mean C_{max} value after the 1000 mg morning dose on day 14 was 199 ng ml^{-1}. After oral administration the t-$_{max}$ was roughly 2.5 h, and the terminal elimination half-life varied from 13 to 20 h, which may have been due to the variable absorption from large numbers of capsules administered. The bioavailability of the compound after oral administration was low.

In the single intravenous dose study the mean maximum plasma concentrations at the end of infusion ranged from 127 to 3711 ng ml^{-1} for the 1–50 mg dose range, and reached 17 225 ng ml^{-1} at 100 mg. The AUC for the 1–100 mg dose range was 186–15 301 ng h ml^{-1}. The increase in C_{max} was linear for the 1–50 mg dose range, and the increase in the AUC was linear for the 1–100 mg dose range. The nonlinearity in C_{max} at the 100 mg dose level was thought to be due to paucity of data (only two subjects were studied at this dose level). Total plasma clearance was in the range of 90–160 ml min^{-1} $t_{1/2}$ and V_z were approximately 1.6 h and 20 l, respectively. Essentially the entire dose of EAA 494 was recovered unchanged in urine over a 24 h period after the drug infusion.

In the multiple-dose intravenous study, C_{max} (5 min after the end of infusion) after 25 mg once a day was 2470 ng ml^{-1} on day 1 and 2701 ng ml^{-1} on day 7. These values were virtually the same in the group administered 25 mg b.i.d. (2531 ng ml^{-1} after the first dose on day 1 and 2584 ng ml^{-1} after the morning dose on day 7). Thus, there was no indication of drug accumulation during multiple once or twice daily administration. The mean terminal plasma elimination half-life was approximately 3.5–4 h. The plasma clearance and volume of distribution were similar to the results obtained in the single intravenous dose study (110–120 ml min^{-1} and 32–42 l, respectively). These results are consistent with the assumption that elimination of EAA 494 mainly occurs via glomerular filtration of the unchanged compound, and the compound is mainly distributed into the extracellular water, the volume of distribution being of the order of magnitude of total body water.

In summary, the pharmacokinetics of EAA 494 in humans is characterized by relatively rapid absorption after oral administration, relatively short plasma elimination half-life after intravenous administration, a small volume of distribution, virtual lack of metabolism, and linear kinetics both after oral and intravenous administration.

5 DISCUSSION

This chapter summarizes the development of a biologically active compound from the very beginning of the discovery of EAAs in the late 1950s and describes the steps in chemical thinking that led to the synthesis of the most potent pure competitive antagonist NMDA receptor antagonist (in 1986). Subsequent contributions from pharmacologists over several years revealed that D-CPPene was systemically active, and allowed for the first time an exhaustive characterization of the biological actions of such an NMDA antagonist in mammals at the cellular level, D-CPPene shows a highly selective and fully reversible competitive antagonism of endogenous and exogenous agonists at the NMDA receptors exclusively found in the CNS in mammals. To-date, no other direct effects on mammalian cells have been discovered, so that all tissue-, organ-, and systemic effects described in this chapter can be assumed to derive from the interaction of D-CPPene with the NMDA receptor in the brain and spinal cord. This is even more likely because of the observation that D-CPPene is practically not metabolized, so that none of the observed effects are expected to result from the action of metabolites interacting with other primary biological mechanisms. The major effects seen were muscle relaxation, inhibition of proconvulsive phenomena, neuroprotection following both ischemic or mechanical injury, modulation of learning mechanisms, and modulation of dopamine systems.

The results of studies with EAA 494 in humans confirmed that the compound penetrates the CNS and exerts potent CNS effects in the higher range of the doses administered. The increased frequency and intensity of CNS effects following multiple-dose administrations, despite the relatively short plasma half-life and lack of evidence for accumulation, suggests that the compound may have a longer residency time in the CNS as compared to plasma or that the pharmacodynamic effects of the compound in the CNS last longer than its actual residency time in the CNS.

The effects of EAA 494 in the CNS are complex, and suggest that the blockade of NMDA receptors are associated with both depressant and stimulant effects in the CNS. The CNS effects observed with EAA 494 were similar to those described with noncompetitive and other competitive NMDA receptor antagonists such as MK-801 and CNS1102, respectively (Troupin *et al.*, 1986; Muir *et al.*, 1994; Muir and Lees, 1995b). This observation suggests that, independent of the mechanism of receptor blockade, the inhibition of NMDA receptor function may result in a distinct pattern of CNS effects.

The use of EAA 494 in humans was not associated with any signs of systemic toxicity. The profile of psychotomimetic effects, however, may render the use of NMDA antagonists difficult in chronic indications involving ambulatory patients, should high doses be required to achieve a sufficient blockade of NMDA receptors for clinical efficacy. Acute indications in which NMDA antagonists are used for a short duration in a hospital setting may offer better opportunities for the therapeutic application of these compounds. These indications include severe head trauma, as a consequence of which the patients are comatose, artificially sedated, ventilated, and supervised in intensive care units. Under these circumstances, NMDA receptor antagonists could safely be administered at doses high enough to achieve sufficient blockade of NMDA receptors, and their psychotomimetic effects are not likely to be relevant. Similar considerations would apply to patients with severe stroke resulting in coma or patients with a severe enough stroke which would justify their treatment in an intensive care unit under sedation. Obviously, these considerations would not apply if neuroprotective doses in conditions involving CNS ischemia were shown to be lower than those inducing undue psychotomimetic effects or if there were novel NMDA antagonists devoid of psychotomimetic effects.

An interesting study with a closely related compound, D-CPP, appeared recently, showing that intrathecal administration of this similarly pure competitive NMDA antagonist was capable of inhibiting previously untreatable chronic pain resulting from a peripheral nerve injury (Kristensen *et al.*, 1992; Gordh *et al.*, 1995 and this book), suggesting that D-CPPene could also be useful in some conditions of severe chronic pain that might result from the NMDA 'wind-up' phenomenon (Urban *et al.*, 1994).

6 CONCLUSION

D-CPPene is one of the most extensively studied potent and pure competitive NMDA antagonists with systemic activity. It has been developed from the original discovery of neuroactive EAAs through rational chemical derivatization and structure–activity relationships. Over the course of more than 10 years (1986 to the present) it has been exhaustively characterized in animal studies, giving a very good picture of many physiological effects of the mammalian EAA NMDA receptor, as it is plausible that most biological effects seen are exclusively mediated by this receptor. More recently, human characterization has also begun to complete the picture. The major findings in intact organisms include muscle relaxation, inhibition of proconvulsive phenomena, modulation of learning mechanisms, neuroprotective effects, inhibition of some form of centrally mediated pain, and interaction with monoaminergic circuits probably leading to the psychotomimetic effects seen in humans at high doses. No organ-related toxicological effects have been seen up to high doses, so that the clinical potential of D-CPPene can be widely explored. The main clinical indication pursued at the time of writing is the reduction of damage following trauma to the CNS.

The scientific story of D-CPPene very clearly illustrates the inestimable value of a pure systemically active pharmacological tool: (i) scientifically it is essential in order to determine the physiology of a particular receptor in all species including humans; and (ii) medically it can potentially lead to better treatment in severe conditions such as traumatic injury of the CNS or some forms of chronic pain in the present case.

3 Intrathecal Administration of a Competitive NMDA Receptor Antagonist for Pain Treatment

JENS D. KRISTENSEN

Department of Anesthesiology and Intensive Care, University Hospital, DK-5000 Odense, Denmark

1 SUMMARY

This chapter discusses the role of the *N*-methyl-D-aspartate (NMDA) receptor in pain transmission. There is now substantial experimental evidence that increased nociceptive activity may result in neuronal plasticity of the spinal cord, and the possible role of the NMDA receptor in this process is discussed.

A case study where the competitive NMDA receptor antagonist 3-(2-carboxypiperazin-4-yl) propyl-1-phosphonic acid (DL-CPP) was given intrathecally at the lumbar level for the treatment of a severe, intractable, chronic pain state is summarized. DL-CPP, administered intrathecally in a dose of 200 nmol, completely eliminated the spread of pain sensation from the territory of the injured nerve to the left body half, and it abolished afterdischarge, i.e. the pronounced increase in pain level that outlasted the initial stimulus.

The primary information from this patient study supports the hypothesis that the NMDA receptor can be manipulated to reduce human pain. How this can be achieved in the clinical situation of pain remains to be seen. If controlled clinical studies can confirm that the clinical manifestations of central sensitization/neuronal plasticity can be reduced by interfering with the NMDA receptor function, several options for pharmacological intervention will be possible.

2 OVERVIEW OF THE PHARMACOLOGY

The NMDA receptor antagonist (DL-CPP) is a piperazine derivative that acts competitively at the glutamate recognition site. It is highly potent and selective to the NMDA receptor (Davies *et al.*, 1986; Lehmann *et al.*, 1987). Although DL-CPP is strongly hydrophilic, with a log *P* of –3.4 (Hays *et al.*, 1990), it is effective in the treatment of experimental convulsions after systemic administration in mice (Lehmann *et al.*, 1987), rats (Holmes *et al.*, 1990), and baboons (Patel *et al.*, 1990). After systemic

EXCITATORY AMINO ACIDS – CLINICAL RESULTS WITH ANTAGONISTS
ISBN 0-12-546820-2

administration in anesthetized pigs, the concentration of DL-CPP in the spinal fluid was estimated to be about 25% of that in serum, indicating relatively good penetration through the blood–brain barrier (Kristensen *et al.*, 1995). In this study, DL-CPP was also administered epidurally and intrathecally at the lower lumbar site, and it was found that only minute fractions of DL-CPP spread rostrally via cerebrospinal fluid bulk flow. The half-life of DL-CPP in the cerebrospinal fluid was between 1.5 and 3 h, and the main way of systemic elimination was renal. No effects on hemodynamics, arterial blood gases or acid–base balance could be correlated to epidural or intrathecal administered DL-CPP (Kristensen *et al.*, 1995).

Neurotoxicological studies have been performed on rat spinal cord after 2 weeks of intrathecal administration of 4 nmol of DL-CPP (13 nmol kg^{-1} of body weight (BW)) daily, divided in two doses. Light and electron microscopic examination and a morphometric method using an unbiased stereological estimator of cell number and cell volume were used as sensitive parameters of spinal cord neurotoxicity. The morphology and ultrastructure of the spinal cord showed no signs of neurotoxicity, and the cell number and volume were not different from saline-injected rats (Kristensen *et al.*, 1993b). At a dose of about 200-fold the lowest antinociceptive dose, *viz.* 150 nmol kg^{-1} BW, given intrathecally, the mitochondria were affected by swelling of the cristae in the spinal cord tissue near the tip of the intrathecal catheter (Kristensen, 1994).

However, when given systemically and in high doses, NMDA antagonists have been reported to produce vacuolization in specific areas of the rat brain (Allen and Iversen, 1990; Olney *et al.*, 1991; Duval *et al.*, 1992). It cannot be excluded that this effect is species-specific, as it has not been reproduced in other animals. Recently, it has been suggested that the vacuolization is an artefact resulting from aldehyde tissue fixation (Auer and Coulter, 1994).

Spinal cord blood flow (SCBF) has been studied in anesthetized rats after intrathecal injection of DL-CPP using two different methods. In the laser Doppler flowmetry technique, the spinal cord of anesthetized rats was exposed by laminectomy, and a laser Doppler probe was placed over the dorsal horn, allowing registration of spinal blood flow in a tissue hemisphere of 1–2 mm. Relative changes in spinal blood flow over time were then obtained after intrathecal administration of increasing doses of DL-CPP from 1 to 100 nmol (3–330 nmol kg^{-1} BW) dissolved in 10 µl of saline. No change in the relative blood flow of the spinal cord was seen at any dose level (Kristensen *et al.*, 1994a). Furthermore, SCBF was measured 30 min after intrathecal injection of 4 nmol of DL-CPP (13 nmol kg^{-1} BW) by quantitative autoradiography using [^{14}C]iodantipurine as a tracer. By this method, absolute values of SCBF could be obtained from a whole cross-section of the spinal cord, with a resolution that allows separation into white and gray matter blood flow. No difference was found in SCBF after intrathecal administration of DL-CPP, when compared to saline-injected rats (Kristensen *et al.*, 1993a).

3 RATIONALE FOR CLINICAL TESTING

The last decade of pain research has provided new insights into the physiology and pathophysiology of pain. An important single issue is the recognition of the dynamics of pain transmission. It has long been known that the central nervous system is capable of modifying its structure and function in relation to evolution, experience, and injury. What is new is the discovery that the NMDA receptor plays a fundamental role in the development of neuronal plasticity.

Our knowledge of the NMDA receptor role in pain processing derives primarily from basic scientific studies. From these experimental investigations it has been convincingly documented that the NMDA receptor complex is involved in the transmission and modulation of nociceptive information at the spinal cord level (Cahusac *et al.*, 1984; Davies and Lodge, 1987; Dickenson and Sullivan, 1987; Woolf and Thompson, 1991; Näsström *et al.*, 1992; Kristensen *et al.*, 1994b). Drugs that interfere with NMDA receptor function may constitute new classes of pharmaceuticals for the future treatment of pain. The few clinical studies that have been performed in this field suggest that this new principle for pain treatment may well be valid in humans.

3.1 Neurotransmitters for excitatory nociceptive signals

The main source of neurotransmitters of excitatory sensory information appears to be the excitatory amino acids (EAAs) and certain specific neuropeptides, such as substance P and calcitonin gene-related peptide (CGRP). EAAs act as agonists at both NMDA receptors and non-NMDA receptors, α-amino-3-hydroxy-5-methyl-4-isoxazole propionic acid (AMPA) and the metabotrope glutamate receptor (mGlu). The specific role of these receptors is not fully known. It is believed that AMPA and mGlu have a general effect by inducing short excitatory postsynaptic potentials, whereas the NMDA receptor amplifies and prolongs the response to depolarization. The EAAs and their corresponding receptors are widely distributed throughout the central nervous system, at both the spinal and the supraspinal levels.

3.2 Neuronal plasticity, the NMDA receptor, and pain

Neuronal plasticity is the ability of the central nervous system to modify its structure and function in response to evolution, experience, or injury. Hyperexcitability of the dorsal horn neurons may be referred to as an activity-dependent form of neuronal plasticity. There is now substantial evidence that these changes are caused by intracellular events following NMDA receptor activation or neurotoxic action of the EAAs.

A theoretical sequence of events has been proposed (Woolf, 1991; Dubner and Ruda, 1992) whereby C-fiber stimulation of the spinal cord results in the release of EAAs and neuropeptides, such as substance P and CGRP. The EAAs induce fast excitatory postsynaptic potentials via AMPA and mGlu receptors, whereas the neuropeptides produce slow synaptic potentials. If the stimulation is intense, these slow potentials may add, to produce a cumulative depolarization that can override the voltage-dependent Mg^{2+} block of the NMDA receptor, and hence result in amplification of the depolarization caused by the influx of calcium ions. The role of mGlu receptors may be to enhance the NMDA receptor activity. Glutamate binding to mGlu receptors will lead to activation of protein kinase C, which is believed to catalyze the phosphorylation of subunits of the NMDA receptor, resulting in enhanced NMDA receptor-mediated Ca^{2+} entry (Scoepp and Conn, 1993). This NMDA receptor-mediated increase in intracellular calcium initiates a cascade of intracellular events, responsible for the development of neuronal plasticity, i.e. changes in the function or structure of the nerve cell. Thus, while neuropeptides and EAAs acting at the non-NMDA receptors lower the threshold for neuron activation, the NMDA receptor activation amplifies and prolongs the response to depolarization.

3.3 Implications for clinical pain and its treatment

Possible clinical correlates to this hyperexcitability of the dorsal horn neurons may be such hyperexcitability states as allodynia and secondary hyperalgesia that may occur in acute nociceptive pain, e.g. postoperative pain. As hyperexcitability is most likely a constituent of acute and postoperative pain (Woolf, 1991), pharmacological intervention in the effects of NMDA receptor activation is a theoretical possibility, in combination with drugs that reduce afferent input to the spinal cord. However, the clinical situation of acute and postoperative pain is relatively easy to handle with currently available pharmacological agents, such as local anesthetics, NSAIDs (non-steroidal anti-inflammatory drugs), and opioids.

In contrast, patients suffering from neuropathic pain often describe types of pain sensation that are refractory to conventional pain treatment, and it is for these patients that the need for new treatment strategies is most urgent. Research into the role of NMDA receptors in nociceptive perception has given valuable knowledge about the pathophysiology underlying neuropathic pain, and gave rise to the idea that intervention in the NMDA receptor function may represent a new pharmacological tool for its treatment. Neuropathic pain states are often followed by signs of spinal cord hyperexcitability, such as persistent pain, hyperpathia, spread of pain outside the territory of the injured nerve, and aftersensation. Excessive and/or long-lasting activity in nociceptive afferents may by different forms of synaptic plasticity, resulting in neuronal dysfunction and even neurotoxicity, thus implying a contribution of the NMDA receptor system to the development of the abnormal sensations experienced in chronic pain (Dubner and Ruda, 1992). Activation of the NMDA receptor triggers the expression of gene transcription factors, such as c-*fos* and related immediate-early genes, that can produce long-term changes in synaptic function, lasting from a few hours to several months. Loss of inhibitory neurons may result, as excessive depolarization induces neurotoxicity, either from high concentrations of EAAs (Rothman and Olney, 1986) or by the action of, for example, dynorphin (Caudle and Isaac, 1988). These mechanisms may cause permanent changes in nociceptive processing, hence correlating to chronic pain states.

4 HUMAN RESULTS

We have previously published a report on the clinical use of a competitive NMDA receptor antagonist (Kristensen *et al.*, 1992). The patient suffered from a severe neuropathic pain condition, caused by a nerve injury. Her pain syndrome had the following four characteristics: (1) at rest she suffered from a continuous deep pain sensation mediolaterally on her left thigh, i.e. matching the innervation area of the anterior cutaneous branches of the femoral nerve; (2) low threshold and brief mechanical or thermal stimulation of her left thigh in the territory of the injured nerve caused an immediate, severe pain sensation (allodynia), followed by; (3) a pronounced increase in the basic pain level lasting from minutes to hours after termination of the stimulation ('afterdischarge', 'wind-up'); (4) this evoked pain sensation spread beyond the territory of the injured nerve, extending downward to the foot and upward to the left side of the abdomen and thorax up to the level of Th_2 (Fig. 1). This pain, which embraced most of the left side of the body, lasted from minutes to hours. A main problem was sleeplessness due to pain triggered by movements and mechanical stimulation of her leg during the night. The condition frequently caused the patient to become totally exhausted, requiring hospitalization and deep sedation with benzodiazepine for a day or two to allow her to obtain some rest.

Intrathecal administration of 90 mg of hyperbaric 5% lidocaine produced an apparent sensory and motor block extending from Th_{10} to S_5. However, the continuous deep pain persisted unchanged. The allodynia to mechanical and thermal stimulation, as well as the afterdischarge and spread of pain, was markedly alleviated, but none of the components was completely abolished.

After intrathecal injection of 200 nmol of DL-CPP, the afterdischarge and the spread of pain beyond the area of the injured nerve was completely abolished following brief, low-threshold, mechanical, and thermal stimulation. The immediate allodynia and the continuous deep pain sensation were still present with the same severity as before the injection of DL-CPP, but these pain sensations

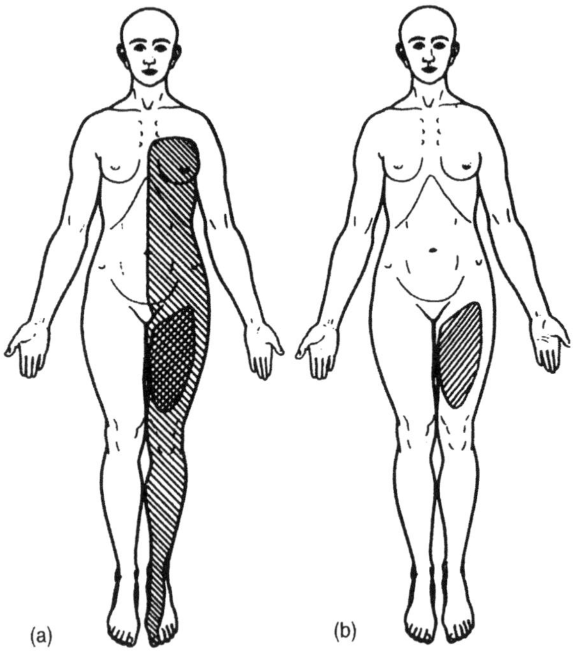

FIG. 1 Anatomical mapping of the area of pain sensation following low-threshold mechanical stimuli of the left thigh (a) before and (b) after DL-CPP treatment. The area of pain sensation after DL-CPP treatment was restricted to the area of the injured nerve (double-hatched area). Intrathecal administration of DL-CPP completely abolished the afterdischarge and the spread of pain sensation. (Reproduced, with permission, from Kristensen *et al.* (1992a).)

were now strictly limited to the area of the injured nerve (Fig. 1).*

The normal sensory and motor function was not affected by spinal NMDA receptor inhibition. The patient could walk, and experienced normal sensations to nociceptive stimulation. The thermal sensory and pain thresholds were unchanged, when compared with pre-CPP values, over both the affected and the unaffected skin areas. No changes were observed in the response to low- and high-threshold mechanical stimulation, proprioception, or sensitivity to vibration. No changes could be observed in motor function, coordination, or reflexes, and the patient could micturate normally. Heart rate and blood pressure were unaffected.

About 4 h after the administration of DL-CPP the patient developed a state of anxiety, uneasiness, and hyperacusis, similar to the side-effects that can be produced by ketamine. She responded well to small doses of diazepam. Due to the psychotomimetic side-effects, we stopped monitoring the effects of DL-CPP on her pain. At regular follow-ups, she seemed not to have suffered any permanent harm from these side-effects, with the exception of memories of an unpleasant experience. Subjectively, the spread of pain as well as the afterdischarge was reduced during the subsequent 2 weeks.

*An additional five doses each of CPP 100 nmol were administered over the next 2 h without further improvement in pain relief.

5 DISCUSSION

The present patient study implies a fundamental role of the NMDA receptor system in clinical pain types involving pathological pain, such as postoperative pain and neurogenic pain. If the result can be confirmed in controlled clinical studies, interference with the NMDA receptor system may represent an important new principle for clinical pain treatment.

DL-CPP, administered intrathecally in a dose of 200 nmol, completely eliminated the spread of pain sensation from the territory of the injured nerve to the left body half, and it abolished afterdischarge, i.e. the pronounced increase in pain level that outlasted the initial stimulus. Antagonism of the spinal NMDA receptor did not, however, affect allodynia in this patient. Experimentally, stimulation of the NMDA receptor by strychnine can elicit allodynia, which could be reduced by both competitive and noncompetitive NMDA antagonists (Yaksh, 1989). However, in a patient where the allodynia has been present for several years, the mechanism may be different from that producing acute allodynia in experimental animals. In the chronic situation of allodynia, excessive neuroplasticity that includes degenerative processes may have resulted from protracted hyperexcitability. Recently, it was proposed that neurotoxicity caused by EAAs or dynorphine may result in dysfunction of small local circuit neurons and loss of inhibitory function (Dubner and Ruda, 1992), and such a mechanism could theoretically explain both the allodynia and the lack of effect of antagonism on the NMDA receptor.

The finding that intrathecal administration of lidocaine markedly reduced the allodynia may have been due to reduced excitability of the primary afferent, hence reducing afferent input, or to the hypersensitive dorsal horn neurons themselves. Neither DL-CPP nor lidocaine affected the constant deep pain, indicating that this sensation was 'fixed' at the spinal or supraspinal level.

An important finding in this human study was that normal sensory and motor functions appeared to be completely unaffected by intrathecal DL-CPP. Hence, only pathological pain, i.e. pain sensations resulting from spinal cord hyperexcitability due to neuronal plasticity, was affected. This is consistent with the results of several experimental studies in 'neuropathic pain' models where NMDA antagonists could alleviate hyperalgesia without affecting latency to noxious stimulation in normal tissue (Ren *et al.*, 1992; Yamamoto and Yaksh, 1992). This stands in contrast to the effects of opioids, which in the clinical situation are effective in the treatment of nociceptive pain, whereas they are often ineffective against neuropathic pain with hyperpathia (Arnér and Meyerson, 1988). This characteristic of NMDA receptor antagonists, *viz.* their capacity to dampen hyperexcitability of the spinal cord without affecting normal sensory transmission, is of fundamental importance to its potential clinical use. It implies the possibility of avoiding the drawbacks associated with other spinal analgesics, such as motor dysfunction (following local anesthetics) resulting in immobilization, and reduced sensibility (following both local anesthetics and morphine). Sensibility fulfills an important physiological function in warning of potential tissue damage. Furthermore, the finding supports the theory proposed by Woolf (1989), based on experimental studies, that pain can be divided into two distinct and qualitatively different categories: physiological pain caused by transient non-tissue injurious noxious stimuli, and pathological pain caused by peripheral and central sensitization of the somatosensory system following tissue injury.

The ketamine-like side-effects that developed about 4 h after the last dose of DL-CPP was given were probably a result of the rostral spread of DL-CPP to supraspinal levels. The results from a pharmacokinetic study following intrathecal injection in the anesthetized pig indicated that supraspinal appearance of DL-CPP was caused mainly by rostral transportation, and that the concentration peaked after about 2 h at the cervical level (Kristensen *et al.*, 1995). The total dose given was obviously too large in this patient. However, as the therapeutic effect was obtained following only 25% of the total given dose, it may still be possible to obtain pharmacologically active doses at the lumbar spinal cord without eliciting supraspinal side-effects. Furthermore, hydrophilic drugs, such as DL-CPP, penetrate rather slowly into the spinal cord tissue (Kessler *et al.*, 1976), and the onset time of analgesia may have been longer than the interval we allowed between dose iteration. Hence, an analgesic effect may have been found after the first 100 nmol, or even a lower dose, if the observation period had been longer than 20 min.

The primary information from this patient study is in support of the hypothesis that the NMDA receptor can be manipulated to reduce human pain. How the NMDA receptor or effects from its activation can be manipulated in the clinical situation of pain remains to be seen. If controlled clinical studies can confirm that the clinical manifestations of central sensitization/neuronal plasticity can be reduced by interfering with the NMDA receptor function, several options for pharmacological intervention will be possible.

5.1 Inhibition of glutamate release

The straightforward approach is to prevent the release of EAAs and neuropeptides from the primary afferent. The obvious clinical correlate to this is to prevent afferent signalling to the spinal cord, e.g. by using local and regional anesthetic techniques as well as systemic analgesics. The concept of pre-emptive analgesia, i.e. to administer analgesics before the pain occurs, was re-actualized by Woolf (1989), and its importance is generally acknowledged (McQuay and Dickenson, 1990; Woolf, 1991). Specific pharmacological treatment can reduce presynaptic glutamate release. It has been suggested that stimulation of the adenosine A_1 receptor situated at the presynaptic end-terminal inhibits glutamate release in experimental studies (Sosnowski and Yaksh, 1989). When adenosine was given intravenously to volunteers, it was found to alleviate ischemic tornique pain (Segerdahl *et al.*, 1994), and, recently, intrathecal administration of the adenosine analog R-PIA (R-phenylisopropyladenosine) was shown to reduce allodynia in a patient with neuropathic pain (Karlsten and Gordh, 1995). Recently, a new class of drugs has been introduced which is claimed to inhibit presynaptic glutamate release. One of these drugs, lamotrigine, has been introduced as a new anti-epileptic drug, acting by inhibition of glutamate release through use-dependent blockade of voltage-sensitive sodium channels and stabilization of neuronal membranes (Leach *et al.*, 1986). The clinical role of this class of drugs for the treatment of pain has not yet been established. Experimentally, systemic lamotrigine raised the pain threshold in a rat model of hyperalgesia (Nakamura-Craig and Follenfant, 1994). According to clinical trials for antiepileptic therapy in over 4000 patients, lamotrigine appears to be relatively safe from a toxicological point of view (Vajda, 1992).

5.2 Antagonizing the NMDA receptor sites

There are several ways to block the effect of released glutamate at the NMDA receptor complex, but much research remains to be accomplished on this issue before the potentials of blocking each of the different receptor sites can be mapped. NMDA receptor antagonists, acting at the transmitter recognition site, the polyamine site, the glycine site and at the ion channel, are tested in clinical trials for the treatment of epilepsy and stroke (Lipton, 1993). The variety of ways for the NMDA receptor to regulate nociceptive processing implies that there may also be several ways to inhibit its functioning, thus encouraging the development of new drugs having specific functional effects on the NMDA receptor complex.

5.2.1 Glutamate site

As indicated by the present study, DL-CPP or related drugs may provide the possibility of clinical pain treatment, if they can be proven beneficial in controlled studies. Furthermore, it has to be established whether therapeutic concentrations can be achieved by spinal application without eliciting supraspinal side-effects. However, there are newer, more potent and specific, competitive antagonists acting at the glutamate-binding site which are of interest for clinical pain treatment. Most of these have been introduced for the treatment of stroke and other kinds of ischemic neuronal injury (e.g. *cis*-4-phosphonomethyl-2-piperidine carboxylic acid (CGS19755), D-3-(2-carboxy-piperazin-4-yl)-1-propenyl-1-phosphonic acid (CPPene)). The major drawback of these drugs is their psychotomimetic side-effects, which can result from systemic administration. As NMDA receptors are widely distributed in the central nervous system and are involved in a large number of neuronal processes, the competitive property of these blocking agents probably makes it difficult to achieve a therapeutic drug concentration at the target organ with a high glutamate concentration without interfering with other NMDA receptor-mediated activities, where the concentration of glutamate may be within the normal physiological range. To avoid supraspinal effects, spinal administration may be the only route by which to administer competitive NMDA receptor antagonists to conscious patients. However, gene cloning of NMDA receptors has revealed differences in the configuration of the receptor complex, depending on its location in the central nervous system (Monyer *et al.*, 1992). This may lead to second- or third-generation antagonists which are active on spinal, but not supraspinal, NMDA receptors.

5.2.2 *Channel blockers*

Repetitive, high-frequency synaptic activity which produces prolonged depolarization of the cell membrane to levels above –50 mV can reduce the Mg^{2+} blocking of the channel, hence allowing Ca^{2+} to enter the cell when the NMDA receptor is activated by recognition site agonists. Pharmacological blocking of the Ca^{2+} channel is use-dependent, i.e. the channel must be open before the blockers can enter the channel. Even unblocking of the channel is agonist-dependent, suggesting that the blockers can become trapped in the channel. Theoretically, this type of NMDA antagonist ought to have the advantage of being most effective at sites with high NMDA receptor activity.

The dissociative anesthetic ketamine represent the NMDA antagonist that has been most extensively used to test the influence of NMDA receptors in pain transmission in humans. Of the two enantiomers, the affinity of (*S*)-ketamine for the phencyclidine (PCP) site is about four times as high as that of (*R*)-ketamine (Klepstad *et al.*, 1990). Ketamine also interacts with the μ and nonopioid σ binding site but at a lower affinity than that for the PCP site. Ketamine in high doses, resulting in plasma concentrations of about 1 μg ml^{-1}, produces a dissociative anesthesia. When comparing the effects of ketamine *in vivo* with their receptor affinities obtained from *in vitro* studies, a positive correlation was observed between the analgesic effect and binding to the PCP site of the NMDA receptor (Klepstad *et al.*, 1990). Some 50% NMDA receptor occupancy corresponded to anesthetic levels of (*S*)-ketamine, whereas a 20–30% receptor occupancy was related to analgesia (Øye *et al.*, 1991). Low-dose ketamine administered to volunteers showed analgesic properties that were independent of opioid receptors (Maurset *et al.*, 1989). Using (*S*)-[*N*-methyl-^{11}C]ketamine as a tracer in combination with 'cold' (*S*)-ketamine in positron emission tomography (PET), a direct relationship was shown between specific (*S*)-ketamine binding in the human central nervous system and its analgesic and psychopharmacological effects (Hartvig *et al.*, 1995). In this study it was also shown that the analgesic effect on ischemic tornique pain could be reduced in a dose-dependent manner, without affecting warm and cold thresholds or heat and cold pain, measured with the Marstock Thermotest. Furthermore it was found that, even at the lowest dose of 0.1 mg kg^{-1} of (*S*)-ketamine, the analgesic effect was accompanied by cognitive disturbances, such as changes in hearing and visual functions, altered body image, and feelings of unreality and insobriety.

The analgesic effect of low-dose ketamine has also been demonstrated in several studies, using clinical postoperative pain as a model (Hagelin and Lundberg, 1981; Joachimsson *et al.*, 1986; Dich *et al.*, 1992; Jahangir *et al.*, 1993; Roytblat *et al.*, 1993). Postoperative pain includes acute nociceptive pain with peripheral and central sensitization, resulting in, for example, NMDA-mediated secondary hyperalgesia. An important advantage of ketamine is that it produces less respiratory depression and sedation than opioids. A few small studies describe the beneficial effect of ketamine in patients with neuropathic pain where conventional pain treatment was without effect (Mankowitz *et al.*, 1982; Stannard and Porter, 1993; Backonja *et al.*, 1994; Eide *et al.*, 1994; Hoffmann *et al.*, 1994; Persson *et al.*, 1995).

These studies demonstrate an analgesic effect of low-dose ketamine in experimental ischemic pain, postoperative pain, and neuropathic pain in doses where 'normal' physiological pain transmission seems unaffected. It is also quite evident that this analgesic effect was difficult to separate from disturbances in cognitive functions, such as hearing, vision, and mood. Other clinically available channel blockers are dextrorphan, dextromethorphan, and memantine, all weak channel blockers with a low affinity for the PCP site and a short duration of action.

Felbamate is a promising new antiepileptic drug whose mechanism of action is unknown. In whole-cell voltage clamp recordings from cultured rat hippocampal neurons, clinically relevant concentrations of felbamate (0.1–3 mM) inhibited NMDA responses and potentiated γ-aminobutyric acid (GABA) responses (Rho *et al.*, 1994). Single-channel recordings indicated that the effect on NMDA responses occurred via a channel-blocking mechanism, though it has also been suggested that felbamate is active at the glycine site (McCabe *et al.*, 1993). Felbamate is the first anticonvulsant drug with a dual action on both excitatory (NMDA) and inhibitory (GABA) brain mechanisms.

5.2.3 *Glycine site*

The glycine site is a modulator site at the NMDA receptor complex present in the same protein subunit as the NMDA recognition site (Moriyoshi *et al.*, 1991). Occupancy at the glycine site is an absolute prerequisite for NMDA receptor activation, and the role of the glycine site seems to be modulation of the NMDA-mediated synaptic response (Mayer *et al.*, 1989). Thus the release of glutamate from the primary afferent mediates the synaptic transmission, whereas changes in extracellular glycine modulate the role of NMDA receptors in the transmission. Cloning of NMDA receptor subunits suggests the existence of subtypes with differentiated affinity for glycine (Kutsuwada *et al.*, 1992; Monyer *et al.*, 1992). Kynurenic acid was among the first-developed glycine site antagonists. It is nonselective, being similarly potent at NMDA and non-NMDA receptors (Perkins and Stone, 1985). Newer antagonists, such as L687414, are more selective and potent at the glycine site. L687414 is effective in experimental studies on stroke and convulsion in doses that do not produce vacuolization of neurons or alterations in brain glucose metabolism (Hargreaves *et al.*, 1993a).

5.2.4 *Polyamine site*

The polyamine site is an intracellular binding site that can modulate the affinity of other agonists and antagonists. The modulatory effect seems complex, and dependent on residual activity at the other binding sites (Oblin and Schoemaker, 1994). Spermidine and spermine, which are agonists to the polyamine site, enhance binding of MK-801 and glycine to the NMDA receptor. In the formalin test in rats, a synergistic effect was found for the combined administration of the channel blocker MK-801 and spermine, whereas the polyamine site antagonist ifenprodil failed to produce any analgesic effects (Coderre, 1993). Ilfenprodil and its analog SL 82.0715 exert a dose-dependent, but incomplete, inhibition of the NMDA receptor by antagonist action at the polyamine site (Carter *et al.*, 1989). More information is needed on the role of the polyamine site in regulation of the NMDA receptor.

5.3 Antagonizing the effects of NMDA receptor activation

A third possibility is to intervene in the secondary events that result from activation of the NMDA receptor complex. This principle may already be in use clinically, as it has been suggested that NSAIDs and acetaminophen are effective at the spinal level (Jurna *et al.*, 1992, Björkman *et al.*, 1994), probably by interacting with the nitric oxide (NO) system. NO seems to play a central role in the secondary events that result from activation of the NMDA receptor (Meller and Gebhart, 1993). NO is catalyzed by the enzyme NO synthase (NOS). This enzyme converts arginine into free NO and citruline. Only certain types of neurons contain NOS. In the spinal cord, NOS-like immunoreactivity has been identified in the dorsal horn and around the central canal, loci important for sensory processing. NOS is a calmodulin-sensitive enzyme. The binding of Ca^{2+}–calmodulin complexes to the enzyme results in activation of catalytic activity. It is through the generation of NO that several neurotransmitters, including glutamate, acetylcholine, substance P, histamine, and bradykinin, are thought to activate guanylyl cyclase and increase cellular concentrations of cGMP in the central nervous system. Thus, neurotransmitters that increase intracellular calcium levels could be expected to stimulate guanylyl cyclase activity in neurons containing NOS (Mocanda *et al.*, 1991).

NMDA activation enhances NO synthesis from arginine, implicating NO as an important mediator of NMDA effects, such as nociception and neurotoxicity. This also implies that antagonism of NO production could have antinociceptive effects, especially in pain states with signs of central sensitization, hypothetically produced by NMDA receptor activation. Some structural analogs to arginine function as NOS antagonists, and block the production of NO. Experimentally, it has been shown that intrathecal antagonism of the NOS by L-N^G-monomethyl-arginine (L-NMMA) produces a dose-dependent antinociceptive effect (Hedner *et al.*, 1991), suggesting a theoretically new mechanism to induce spinal analgesia.

Protein kinase C (PKC) activity seems to play a central role in the events resulting from NMDA receptor activation (Ben-Ari *et al.*, 1992). Inhibitors of protein kinase C, of which staurosporin is one of the most potent, inhibit ischemic neuronal damage, and may also be a tool for the study of the role of PKC in nociceptive transmission. However, clinical inhibition of PKC is probably not feasible, as PKC plays a crucial role in signal transduction for the activation of many cellular functions including the control of cell proliferation.

5.4 Interaction with other receptor systems

Though the NMDA receptor may have a fundamental part to play in the development of pathological pain, it should be remembered that pain processing results from the integrated action of a number of receptor systems. Experimental studies on interaction have shown synergistic effects from the combination of NMDA antagonists acting at different sites in the receptor complex (Coderre, 1993) or by combination of NMDA antagonists with substances acting on other receptor systems. Reports that tolerance to μ-opioids can be attenuated by NOS inhibitors (Kolesnikov *et al.*, 1992) as well as by NMDA antagonists (Elliott *et al.*, 1994) emphasize the importance of the study of interaction. Hence, polypharmacy may be a key to achieving effective pain control.

In conclusion, there is much evidence to suggest that different types of pain can be relieved by manipulation of the NMDA receptor function. Unique properties of NMDA receptor such as the numerous mutually differing binding sites at the NMDA receptor, the existence of NMDA receptor subunits with variable configuration, variability in the combination of NMDA receptor subunits depending on anatomical location in the central nervous system, and interference from other neurotransmitters acting both on non-NMDA receptors such as AMPA and mGlu and on other receptors, such as neuropeptide receptors and intrinsic inhibitory receptors, constitute mechanisms by which NMDA receptor function can be regulated, thus giving the NMDA receptor a central role for modulation of not only afferent nociceptive input but also the function and structure of the entire synapse. This also implies a multitude of possibilities for pharmacological manipulation of receptor function. The major concern is the psychotomimetic side-effect produced by drugs that interfere with the NMDA receptor complex. As the NMDA receptor is widely distributed in the central nervous system, and is involved in a number of neuronal processes, it is difficult to use this class of drugs systemically. Other routes of administration, e.g. spinal, may circumvent the problem, though in so doing will limit its clinical use. Molecular cloning of the genes for the NMDA receptor as well as for other EAA receptors has revealed that the receptors consist of a number of subunits, and the variability in the way these subunits combine, at different anatomical locations, may represent a molecular basis for the functional diversity of the receptors (Sprengel and Seeburg, 1993), and hence a way to identify antagonists acting specifically on receptors involved in pain processing. Furthermore, the roles of the different modulatory sites at the NMDA receptor have not yet been explored, and the development of drugs exerting a specific action on these sites may also be a way to separate therapeutic effects from side-effects. Finally, even though each of these potential approaches, *per se*, may not be sufficient to relieve pain, it is possible that synergistic effects can be obtained by their combination or when combined with drugs acting at other receptor systems involved in pain transmission.

ACKNOWLEDGMENTS

This work is supported by The Swedish Medical Research Council, grant No. 9077, and the Lions Foundation for Cancer Research.

4 Clinical Experience with the NMDA Ion Channel Blocker, Aptiganel Hydrochloride (CERESTAT®)

ANDREW G. KNAPP, LAIMA I. MATHEWS AND ELKAN R. GAMZU

Cambridge NeuroScience, Inc. 1 Kendall Square, Building 700, Cambridge, MA 02139, USA*

1 SUMMARY

Aptiganel hydrochloride (CERESTAT, CNS 1102), is a novel *N*-methyl-D-aspartate (NMDA) ion channel blocker (Reddy *et al.*, 1994) being developed for the treatment of stroke and severe traumatic brain injury (TBI). Originally identified in radioligand binding experiments, aptiganel is a potent (28 nM) and selective ligand for the ion channel site of the NMDA receptor (Kirk *et al.*, 1994). It exhibits the *in vivo* pharmacology expected for compounds with this mechanism of action: the principal effects in animals are signs of central nervous system excitation (increased spontaneous activity, stereotypies) and depression (ataxia, sedation), as well as modest increases in blood pressure in conscious animals. These effects have a rapid onset, indicating that aptiganel enters the brain readily, a conclusion confirmed by studies with radiolabeled aptiganel. The elimination half-life of the compound is approximately 1 h in nonhuman species. Aptiganel is a potent anticonvulsant and has been shown to be neuroprotective in several *in vitro* and *in vivo* models of cerebral ischemia. When aptiganel HCl is administered following middle cerebral artery occlusion (MCAO) in rats, neurological improvement and reductions in infarct volume of up to 70% have been reported (Minematsu *et al.*, 1993b). In such animal models of stroke, plasma levels of $\geq$10 ng ml^{-1} were associated with neuroprotection. The compound has a favorable safety profile in animals: single-dose, repeated-dose (up to 90 days), and reproductive toxicity studies have been completed.

Over 300 volunteers and patients have been exposed to aptiganel HCl in a series of phase I (Muir *et al.*, 1994) and phase II studies (Gamzu, 1995). These studies have demonstrated that aptiganel HCl can be safely given in doses that produce plasma levels and plasma 'areas under the curve' (AUCs) that have been associated with

* Aptiganel HCl is being developed jointly by Cambridge NeuroScience, Inc. and Boehringer Ingelheim GmbH. CERESTAT is a registered trademark of Boehringer Ingelheim International GmbH.

EXCITATORY AMINO ACIDS – CLINICAL RESULTS WITH ANTAGONISTS
ISBN 0-12-546820-2

neuroprotection in animal models. In conscious individuals, the principal dose-limiting effects have been central nervous system complaints, including disorientation, confusion, sedation, and nausea. Agitation and hallucinations occur infrequently. Increases in systolic blood pressure and heart rate have also been observed. The pharmacokinetics of this compound in man are highly favorable for parenteral administration in an acute-care setting and do not change with prolonged administration. The plasma elimination half-life in man is about 4 h. Continuous intravenous treatment durations of 72 h have been achieved in severe TBI patients; 12 h treatments have been achieved in stroke patients (Fayad *et al.*, 1996). Although not designed to evaluate efficacy, the early studies have shown some nonsignificant beneficial effects. Pivotal efficacy studies in stroke and TBI began in 1996.

2 PHARMACOLOGY

2.1 Site and mechanism of action

The NMDA receptor is a ligand-gated ion channel whose normal function is to mediate synaptic transmission. In addition to the agonist recognition site at which glutamate binds, the NMDA receptor contains several other pharmacological binding sites that can modulate the activity of the receptor (Wong and Kemp, 1991). Of particular significance is a site, located within the ion channel pore, to which several anticonvulsants and anesthetics bind. Ligands for this site occlude the ion channel of the NMDA receptor, preventing ions from entering the cell, hence noncompetitively antagonizing the actions of glutamate. Such noncompetitive NMDA antagonists have neuroprotective actions in *in vitro* and *in vivo* models of excitotoxic neuronal injury and cerebral ischemia (Albers *et al.*, 1989; Choi, 1990; McBurney, 1994; McCulloch *et al.*, 1991).

Aptiganel HCl (*N*-(1-naphthyl)-*N*′-(3-ethylphenyl)-*N*′-methylguanidine hydrochloride; Fig. 1) is a novel, potent, and selective ligand for the ion channel site of the NMDA receptor.

In radioligand binding assays on rat brain membranes, aptiganel HCl had a K_i of 28 nM for the NMDA receptor-associated ion channel site (as defined by displacement of [^{3}H]dizocilpine binding; in our hands, the K_d for dizocilpine was ~3 nM). Interestingly, the 10-fold difference in potency between aptiganel and dizocilpine has held true in most of the *in vitro* and *in vivo* situations where we

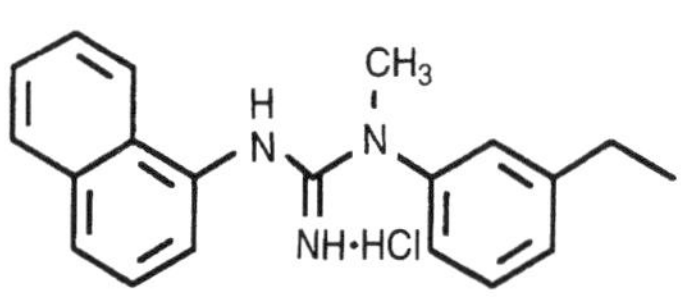

FIG. 1 Chemical structure of aptiganel HCl.

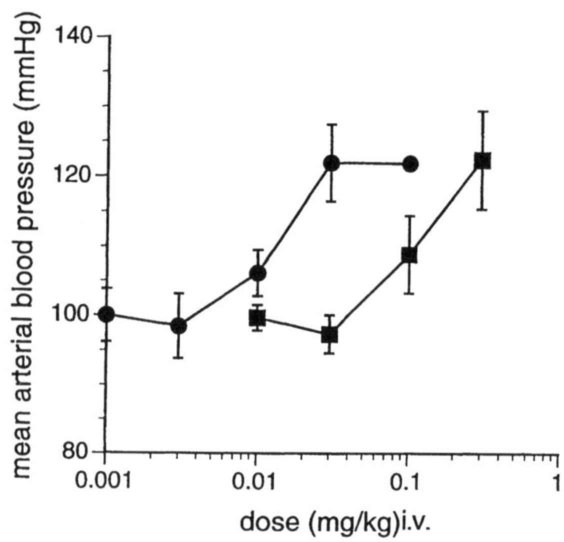

FIG. 2 The effects of intravenously administered aptiganel and MK-801 on mean arterial blood pressure in conscious instrumented baboons. ●, MK-801 (n = 2); ■, aptiganel (n = 4).

have compared the two molecules (see Fig. 2 for one such example). Aptiganel had negligible affinity for other binding sites associated with NMDA or non-NMDA glutamate receptors, nor for adrenergic, noradrenergic, cholinergic, serotoninergic, or opiate receptors, nor for voltage-gated cation channels (Kirk *et al.*, 1994).

2.2 General *in vivo* pharmacology

The major pharmacological effects of aptiganel are on the central nervous system and the cardiovascular system. Aptiganel HCl has been evaluated in a standard battery of neuropharmacological screening tests in mice and rats. The most prominent effects were protection against electroconvulsive seizures, and effects indicative of central nervous system depression and sedation. Following intravenous administration in rats, behavioral effects were evident at 0.25 mg kg^{-1} and became pronounced at 1 mg kg^{-1} and above. These included excitation, stereotypies, ataxia, and sedation. Following a single dose of 4 mg kg^{-1}, the effects lasted about 4 h.

The effects of aptiganel HCl on cardiovascular parameters have been examined in rats, guinea-pigs, Yorkshire pigs, and baboons. In general, intravenous administration of aptiganel HCl produced only modest changes in cardiovascular performance. In unanesthetized animals, the predominant effect was an approximately 20% increase in blood pressure. In conscious, instrumented baboons (Fig. 2), this pressor response was evident following intravenous administration of 0.1 mg kg^{-1} and was maximal (increase of 20 mmHg in mean arterial pressure) following administration of 0.3 mg kg^{-1}, a dose which also caused marked sedation.

2.3 *In vitro* and *in vivo* neuroprotection

Aptiganel protected cultured brain neurons against exposure to toxic concentrations of glutamate (ED_{50} = 0.38 μM) at concentrations consistent with its affinity for the NMDA receptor ion channel site (Kirk *et al.*, 1994). In the absence of added glutamate, aptiganel (10 μM) alone did not injure neurons *in vitro*.

Intraperitoneal administration of aptiganel HCl to neonatal rats dose-dependently reduced (by up to 86%) the brain damage caused by direct intracerebral injection of NMDA. The duration of action of aptiganel in this model was estimated by administering the compound at varying intervals before the NMDA challenge. Neuroprotection was achieved when aptiganel HCl was administered up to 4.5 h before the NMDA. In contrast, the duration of action of an equieffective dose of MK-801 was at least 12 h. Aptiganel was also able to protect neonatal rats against brain damage induced by ligation of the common carotid artery followed by hypoxia (Wang *et al.*, 1995b). A dose of 6 mg kg^{-1} i.p. provided >95% protection in this model.

In rat models of focal cerebral ischemia, which are considered to be good models of human ischemic stroke, intravenous administration of aptiganel HCl substantially reduced (by 40–70%) the amount of brain damage (and the concomitant neurological dysfunction) caused by unilateral occlusion of the middle cerebral artery. Aptiganel was active in both permanent (Minematsu *et al.*, 1993a; Park *et al.*, 1993) and reversible (Minematsu *et al.*, 1993b) occlusion models in which a variety of different intravenous dosing regimens were used. Protection was observed when aptiganel was administered up to 1 h postocclusion (Meadows *et al.*, 1994). In rats, the lowest maintained plasma concentration of aptiganel associated with neuroprotection was ~10 ng ml^{-1}, achieved by a bolus dose of 0.25 mg kg^{-1} followed by continuous infusion of 0.17 mg kg^{-1} h^{-1}. This dose was associated with a mild degree of ataxia and/or sedation. With higher doses, increased sedation occurred, but the degree of neuroprotection was maintained.

2.4 Toxicology

Toxicology studies performed to date indicate that aptiganel HCl can be administered safely for the projected clinical usage.

Aptiganel HCl was found to be nonmutagenic in a standard battery of genotoxicity tests (Ames test, mouse lymphoma forward mutation assay, mouse micronucleus assay, and chromosome aberrations assay in human lymphocytes). The aqueous formulation of the compound did not produce vascular or perivascular irritation, and was not hemolytic.

Single-dose, repeated-dose, and reproductive toxicology studies have been performed in rodents. The intravenous LD_{50} was 17.6 mg kg^{-1} for rats and 53.6 mg kg^{-1} for mice. Studies of 14 and 90 days in rats were conducted to determine the potential cumulative toxic effects associated with repeated, daily intravenous injection of aptiganel HCl (0.1–6.0 mg kg^{-1} day^{-1}). During and after administration of the compound, dose-dependent behavioral signs and symptoms lasting up to several hours were observed. These included increased locomotor activity, ataxia, abnormal gait, prostration, and tremors. Dose-dependent reductions in the rate of increase of body weight were observed in the 90 day study, perhaps as a consequence of the increased spontaneous activity of the animals. However, no dose-dependent, treatment-related histopathological abnormalities were observed. No potential for reproductive toxicity was seen in a segment I/II study in rats, although maternal and fetal weights were reduced.

In 14 and 90 day repeated-dose toxicity studies in cynomolgus monkeys, aptiganel HCl (0.05–2.1 mg kg^{-1} i.v. administered once daily) produced dose-related clinical signs of central nervous system depression, including lethargy, ataxia, and prostration. The only dose-dependent, treatment-related histopathology finding was lymphoid depletion in the thymus, which occurred in the 14 day study, but not the 90 day study.

In specialized toxicology studies, the propensity of aptiganel HCl to induce neuronal vacuolization (Olney *et al.*, 1989) in the cingulate gyrus and retrosplenial cortex of the female rat brain was assessed. The vacuolization phenomenon has been associated with high-dose administration of NMDA antagonists to rodents – its relevance to the clinical use of such agents is unclear. For subcutaneous administration, the threshold dose for inducing vacuoles is approximately 1 mg kg^{-1}. Vacuolization was reversible within 12 h, even following doses as high as 10 mg kg^{-1} s.c. No vacuoles were observed after 0.25 mg kg^{-1}, nor 2.5 mg kg^{-1}, administered intravenously.

2.5 Pharmacokinetics and metabolism

The pharmacokinetics of aptiganel have been examined across species. Following intravenous administration, aptiganel distributes rapidly (<10 min) into a large volume. Distribution is also rapid (~30 min) following subcutaneous administration. The plasma half-life of aptiganel is approximately 60 min in the rat, and 90 min in primates. The half-life in man is longer: approximately 4 h, with a mean clearance of 18 ml min^{-1} kg^{-1} (Muir *et al.*, 1994). Pharmacokinetic parameters (C_{max}, AUC) are linear with dose in the rat, monkey, baboon, and human. The pharmacokinetics of aptiganel in stroke patients (majority over 70 years old) is similar to that in normal volunteers (Block (for the CNS 1102-003 Study Group), 1995). In patients with severe traumatic brain injury, drug clearance was sometimes observed to be more rapid than in normal volunteers or stroke patients (Gamzu (for the CNS 1102-002 Study Group), 1994). No gender differences with respect to plasma pharmacokinetics have been observed in man.

Aptiganel is approximately 88% bound to human plasma proteins. Binding occurs to both serum albumin and α-1-acid-glycoprotein. Binding of aptiganel to plasma proteins is unaffected by the presence of warfarin, and vice versa.

The tissue distribution of [^{14}C]aptiganel following intravenous administration in albino rats is rapid: the observed t_{max} for most tissues (including the central nervous system) was 2 min. Radioactivity was completely eliminated by 20 h. The majority of radioactivity appeared in the liver, lung, kidney, intestinal contents, and carcass (radioactivity in the kidney and intestinal contents reflects urinary and fecal elimination). Six minutes after administration of [^{14}C]aptiganel HCl, the level of radioactivity in the brain was approximately 12 times that in plasma. High-performance liquid chromatography analysis has confirmed that the radioactivity in brain results exclusively from the parent compound. Radioactivity in the brain and in plasma decline over the same time-scale, indicating that aptiganel is not retained excessively in the brain.

The major metabolic pathway for aptiganel appears to be oxidation of the ethyl substituent, with *N*-dealkylation occurring as a minor metabolic pathway. Rat and monkey livers metabolize the compound more rapidly and more extensively than does human liver. Except for one minor metabolite (*N*-demethylated aptiganel), the metabolic products are not neuroactive.

Table 1 Side-effects and the dose at which they were first reported after exposure of volunteers to ascending intravenous doses of aptiganel HCI

Dose (μg kg^{-1})	Side-effects first occurring at this dose
10	Tired, weak; perioral numbness
30	Lightheaded/dizzy; numbness in extremities, tingling/increased sensation, flushing/ sweating; slowed speech; increased heart rate and blood pressure
45	Detached feeling, abnormal gait, frank sedation, numbness (general); cold extremities, thirst/dry mouth; dulled hearing, nystagmus, blurred/altered vision; nausea/ vomiting, anxiety, insomnia
60	'Catatonia'/deep respiration/limb plasticity; pallor, trembling; paranoia; urgency, loss of appetite
100	Choreiform movements

3 CLINICAL STUDIES IN NORMAL MALE VOLUNTEERS

Three separate studies examining the safety and tolerability of different intravenous doses of aptiganel HCl have been conducted in normal male volunteers.

The first study examined the effects of single doses administered over a 15 min period in an ascending manner from 3 to 100 μg kg^{-1}. Beginning at 30 μg kg^{-1} of aptiganel HCl, transitory increases in heart rate and blood pressure were observed. The increases were dose-dependent, reaching a maximum of approximately 25–30% over the baseline by 1 h after administration. The effects lasted for 2–3 h, with the highest dose group (100 μg kg^{-1}) exhibiting the most protracted response. All parameters returned to normal by 12 h (Muir *et al.*, 1994).

Other side-effects noted were primarily central nervous system complaints with a dose-dependent increase in intensity and duration (Table 1). The most severe reactions occurred in the 60 and 100 μg kg^{-1} groups. A condition termed 'catatonia' was observed in four subjects. The term was used to describe a state of open-eyed sedation in which subjects were not responsive to verbal commands. This was accompanied by limb plasticity, and lasted 2–3 h. All subjects recovered without medical intervention.

In the second study, volunteers received an initial bolus delivered over 15 min followed by a 4 h infusion designed to achieve and maintain a set plasma drug level. The total dose ranged from 15 to 73 μg kg^{-1} over 4 h and 15 min. The results from this study did not differ greatly from those observed after a single-bolus administration. The increases in heart rate and pressure were clinically meaningful only at initial doses of 45 μg kg^{-1} resulting in a sustained increase relative to the baseline. The overt clinical symptomology was also similar. An 'open-eye sedation' was not observed. However, one subject experienced some olfactory hallucinations and increased sensory awareness. These effects were transitory.

The third study examined, in a cross-over manner, the effects of a single 2 mg fixed dose of aptiganel HCl administered over 15 min versus 1–2 min. This dose was equivalent to approximately 30 μg kg^{-1} for a 70 kg person. Subjects were selected based upon weight, both light and heavy, in order to test for variability in responses. No significant differences on vital signs, side-effects, or response times were noted between the mode of administration or between individuals of differing weights, suggesting that a weight-adjusted dose may not be necessary.

In all of the studies of normal volunteers, the plasma pharmacokinetics of the drug were linear with dose; the clearance was about 18 ml min^{-1} kg^{-1} with a plasma elimination half-life of the parent drug of approximately 4 h.

In summary, the effects of intravenously administered aptiganel HCl in normal male volunteers were primarily related to the central nervous system. The most common complaints were dizziness, numbness and blurred vision. At higher doses, increases in both heart rate and blood pressure are observed in a dose-dependent manner in normotensive individuals. All effects dissipated without medical intervention, and there were no long-term effects.

4 CLINICAL STUDIES IN PATIENTS

4.1 Stroke

4.1.1 Initial safety study

The first study in stroke patients (Study No. 003) was a double-blind, dose-escalating evaluation (Block (for the

CNS 1102-003 Study Group), 1995). Because the trial was intended to assess safety and not efficacy, patients presenting within 18 h of an ischemic stroke in the carotid or vertebrobasilar artery territories were screened for entry. Patients were required to have a minimal neurological deficit score of at least 4 on the NIH Stroke Scale (NIHSS), with the exception of those with isolated hemianopia or aphasia. A computed tomography/magnetic resonance imaging scan was required for confirmation of clinical diagnosis. The patient was randomized (central randomization) to either an active or placebo (saline) treatment administered first as a bolus dose over 15 min followed immediately by a second dose infused over 4 or 6 h.

Patients were continuously monitored for changes in vital signs and emerging clinical signs and symptoms. Standard clinical laboratory testing was performed along with neurological assessments at 12 h, weekly during hospitalization, at discharge, and again at 30–60 days post-treatment. The Barthel Index was assessed at discharge and at follow-up. Blood samples were taken periodically to determine plasma drug concentrations. Patients were treated with either the drug or placebo administered first as a loading dose over 15 min followed immediately by a continuous 4 h infusion. The dose was administered in an escalating fashion after review of each dose level for effects on vital signs, side-effects and general observations by the attending physicians as to tolerability. The dose escalation was halted during administration of the drug at 30 μg kg^{-1} over 15 min followed by an infusion of 20 μg kg^{-1} h^{-1} for 6 h (total dose 150 μg kg^{-1}) because of the preponderance of severely compromised patients enrolled in that group.

No patient's treatment was discontinued, and all patients effectively completed the protocol with the exception of three patients lost to follow-up. The number of patients by group are summarized in Table 2. There was an equal distribution of sexes, with the majority of patients being over 70 years old (within-group mean ages ranged from 69 to 76 years). The population was typical in that concurrent diseases such as hypertension, obesity, coronary artery disease, and diabetes mellitus were present. The within-group mean times from stroke to the beginning of treatment ranged from 9.1 to 12.0 h.

Table 2 Exposure and mortality in stroke patients exposed to ascending intravenous doses of aptiganel HCl or placebo in Study No. 003

Aptiganel HCl total dose (μg kg^{-1} × 15 min^{-1} + μg kg^{-1} h^{-1} × 4 h	Number enrolled		Number of deaths	
	Drug	Placebo	Drug	Placebo
30 (10 + 5)	7	2		
32 (20 + 3)	8	2		
30 (30 + 0)	5	1	1	
50 (30 + 5)	14	3	1	
70 (30 + 10)	9	3	1	1
90 (30 + 15)	13	3	1	
110 (30 + 20)	13	4	1	
150 (30 + 20/6 hrs)	5	2	3	
Total	74	20	8	1

The stroke etiology was classified based upon diagnostic and clinical assessments prior to discharge from hospital (Table 3).

Vital signs The majority of patients were hypertensive on entry, and had been determined to be stable during the baseline evaluation. Isolated instances of increasing blood pressure were treated with labetalol HCl or other anti-hypertensive medications during the test article administration. Within-group averages for mean arterial pressures (MAP) are presented in Fig. 3. Elevations in heart rate were observed, but there was no clear dose–response relationship. In the highest dose group, increases in heart rate of ~10 beats min^{-1} were recorded during the fifth and sixth hours of the infusion.

Deaths A total of 94 patients were enrolled, with 74 receiving the active drug at differing doses and 20 receiving saline as the placebo. The death rate up to 2 months post-stroke was not atypical for this population; 1/20 (5%) for placebo and 8/74 (11%) for patients that received the active drug. The primary causes of death included cardiac and respiratory arrest, pneumonia, and recurrent cerebrovascular events. The deaths occurred from 2 to 38 days

Table 3 Percentage of patients with a given stroke etiology by TOAST classification within each dose group of Study No. 003

	Total dose (μg kg^{-1})*						
Stroke etiology	Placebo (*n* = 20)	30–32 (*n* = 20)	50 (*n* = 14)	70 (*n* = 9)	90 (*n* = 12)	110 (*n* = 12)	150 (*n* = 4)
Large artery atherosclerosis (%)	25	25	43	33	33	17	25
Cardioembolism (%)	30	15	21	33	17	42	50
Small artery occlusion (%)	20	30	7	0	8	17	0
Other determined (%)	0	0	7	0	8	0	0
Undetermined (%)	25	30	21	33	33	25	25

* Some totals may not equal 100% due to rounding error.

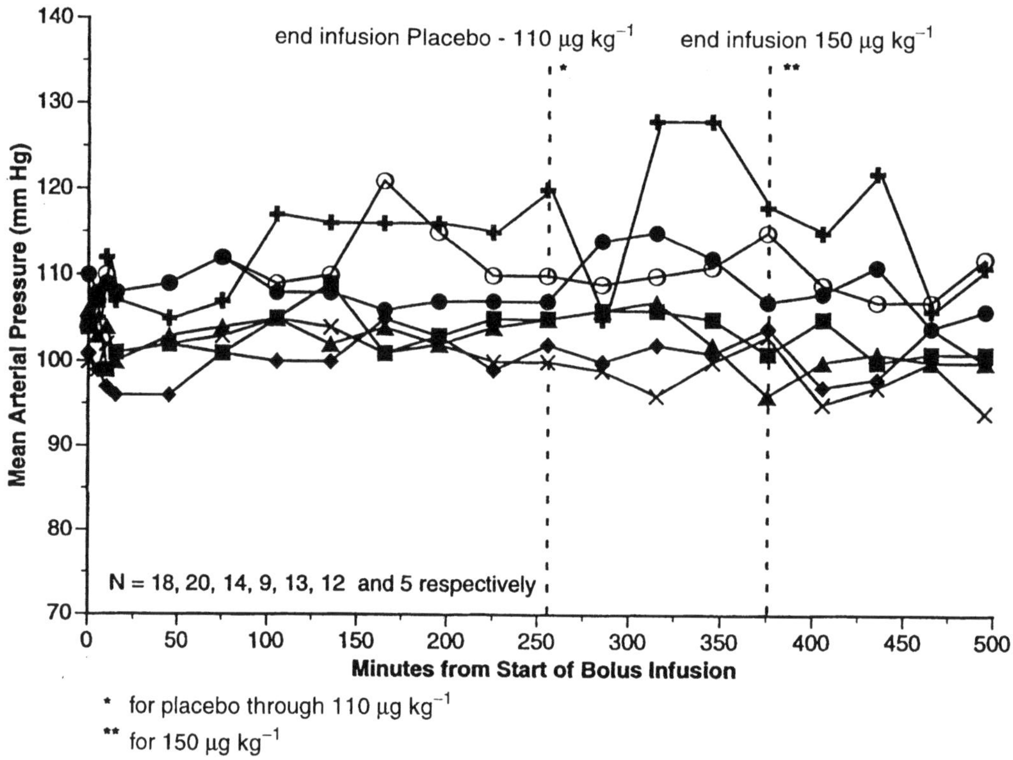

FIG. 3 Mean arterial blood pressure over time in each dose group of stroke patients in Study No. 003. ×, placebo; ●, 30–32 μg kg^{-1}; ▲, 50 μg kg^{-1}; ◆, 70 μg kg^{-1}; ■, 90 μg kg^{-1}; ○, 110 μg kg^{-1}; +, 150 μg kg^{-1}.

following treatment, with all but two occurring at least 1 week post-treatment.

All deaths but one were ascribed to the underlying diseases or ischemic stroke. The exception involved a 68-year-old male treated with 30 μg kg^{-1} over 15 min followed by 20 μg kg^{-1} h^{-1} of aptiganel HCl for 6 h. The patient had presented with a right middle cerebral artery infarct and normal blood pressure. The patient's blood pressure was elevated prior to treatment (190/100), and increased during the infusion. This latter increase was successfully treated with labetalol HCl. The patient was treated for nausea, vomiting, and agitation following the end of the infusion. A second computed tomography scan the next day showed a large left-hemisphere parenchymal hemorrhage contralateral to the right infarct, the cause of which could not be definitively determined because no autopsy was performed.

A safety committee of three noninvestigator physicians felt that it was unlikely that the hemorrhage was directly related to the administration of aptiganel HCl. However, the rise in blood pressure could have precipitated the process and the drug may have contributed to the hypertensive episode.

Adverse events All nonfatal serious adverse events through the entire follow-up period (60 days) were recorded. None was attributed to drug administration. These included the occurrence of pneumonia, extension of stroke, and other cardiovascular events such as a myocardial infarction or sporadic episodes of dysrhythmias. In this population, these were expected events with the incidence of 7/20 (35%) placebo and 12/74 (16%) drug patients reporting. Expected side-effects (based on the signs and symptoms reported by normal volunteers) were recorded during the infusion and up to 12 h thereafter. Table 4 presents these signs and symptoms as a percentage of all placebo or drug patients taken together.

During the course of the study, all adverse events were recorded. Table 5 lists the most common of all other adverse events that occurred with an incidence of ≥5% of all patients that received active drug at any dose.

The overall incidence of any one event is low and associated with an elderly population with concurrent diseases (i.e. diabetes or hypertension) and medications.

Neurological and functional outcome The mean change from the baseline to follow-up (approximately 6 weeks) on the NIHSS by dose group is presented in Table 6.

Between-group comparisons were not attempted since the trial had not been designed to assess efficacy. Not only was entry permitted up to 18 h post-stroke, but because of

Table 4 Percentage of expected side-effects reported up to 12 h postinfusion in Study No. 003

Side-effect	Placebo (n = 20)	Drug (n = 74)
Blurred/double vision (%)	0	4
Catatonia/nonresponsive (%)	0	4
Cold extremities (%)	0	1
Disorientation/detached feeling (%)	5	10
Dry mouth/thirst (%)	0	5
Flushing (%)	0	3
Lighteaded/dizzy (%)	0	7
Nausea (%)	0	11
Nystagmus (%)	0	5
Paresthesia (%)	5	5
Sedation/decreased responses (%)	5	12
Speech/dysarthric or sluggish (%)	0	7
Sweating (%)	0	3
Tired/weak/malaise (%)	0	10
Vomiting (%)	0	8

Table 5 Percentage of adverse effects reported by ≥ 5% of aptiganel-treated patients for the entire Study No. 003 period

Side-effect	Placebo (n = 20)	Drug (n = 74)
None (%)	10	20
Hypertension	5	15
Tachycardia (%)	5	10
Agitation (%)	5	11
Abnormal ECG (i.e. PVCs, sagging S-T) (%)	15	10
Chest pain (%)	0	5
Headache (%)	20	5
Increased blood glucose (%)	0	5
Pnemonia (%)	0	5

Table 6 Mean NIHSS scores[a] in stroke patients exposed to various doses of aptiganel or placebo in Study No. 003

Total dose	Baseline		Follow-up	
(μg kg^{-1})	n	Mean	n	Mean
Placebo	20	9.8	17	5.8
30–32	20	9.5	19	3.7
50	13	12.1	12	8.0
70	8	13.1	8	6.1
90	12	10.8	10	8.3
110	12	14.2	11	11.0
150	5	11.8	2	1.5

[a] Score range: 0–42. Lowered score indicates improvement.

the escalating dose and non-parallel entry design there were differences in severity of the neurological impairment between dose groups at entry into the study (as reflected in the NIHSS scores). The Barthel Index was used as an assessment of the quality of daily living. As might have been expected, the mean Barthel values at follow-up were correlated with the mean neurological impairment at the baseline, and ranged from 50 to 83, with the placebo having a value of 74.

Summary The drug has been well tolerated in patients experiencing an acute ischemic event. The incidence of central nervous system complaints was low and did not approach the severity observed in the normal male volunteers. Isolated cardiovascular effects, such as increasing blood pressure and heart rate, were manageable with treatment.

The pharmacokinetics of the drug were not affected by the age, sex, or disease state of the patients. The levels achieved and the clearances were identical to those observed in normal healthy volunteers. Neurological outcome in the aptiganel-treated patients, as measured by the NIHSS, was equivalent to, or better than, that observed in the placebo-treated patients.

4.1.2 *Ongoing studies*

Two additional studies in stroke patients have completed enrollment, but neither database is finalized at the time of writing. The first (Study No. 008), a parallel, double-blind comparison of three doses of aptiganel (30–110 μg kg^{-1}) versus placebo will examine the dose–response relationship by utilizing the NIHSS at 7 days, 1 month and 3 months after treatment. In addition, the Barthel Index and a modified Rankin will be examined at 3 months. A total of 132 patients with an ischemic stroke within 6 h of the event were entered into the trial. Final blinded observations are expected to occur in January 1996.

In order to determine an appropriate tolerable non-weight-adjusted dose for phase III trials, an additional dose-escalating study in patients within 24 h of an ischemic stroke was undertaken (Study No. 010). The primary measure for safety evaluations was an assessment of any dose-dependent significant rises in blood pressure or any other untoward cardiovascular event (*e.g.* hypotension) (Fayad *et al.*, 1996).

The first phase of the study included only a single dose administered over 5 min in increments of 1.5 mg to groups of at least four patients each (three drug, one placebo). This phase administered doses of 3.0–7.5 mg. It was determined that a dose of 7.5 mg would probably be unacceptable to a majority of physicians and patients due to a consistent significant increase in systolic blood pressure (in some cases greater than 30 mmHg). In addition, the frequency and severity of central nervous system side-effects were deemed not to be adequately tolerable for large-scale studies.

The second phase of the study examined several bolus doses in conjunction with an infusion selected to maintain a given plasma level for up to 12 h. After the first 6 h, the duration of the infusion was extended at the discretion of the investigator and research staff, with input in some cases from the patient's family members.

The highest dose studied was 6 mg initially followed by a 1 mg h^{-1} infusion. At this dose level, most of the patients exhibited some degree of sedation, including a nonresponsiveness to commands. Of the eight patients receiving the drug, only three received the complete 12 h.

Subsequently, all but two of 12 patients tolerated 12 h of a dosing regimen that comprised a 4.5 mg bolus followed by 0.75 mg h^{-1}. One of the two patients was treated for 6 h, at which point he was required to be transferred to another hospital. The second patient's treatment was stopped after 5 h due to hypotension secondary to the use of morphine sulfate for pain. This dose regimen achieved and maintained a drug plasma level of over 10 ng ml^{-1} (Fig. 4)

There were three deaths in the 30 day follow-up period: one of 10 in the placebo-treated patients (10%) and two of 36 in aptiganel-treated patients (5%).

Although not designed to assess efficacy, neurological function was evaluated by the NIHSS and the Scandinavian Stroke Scale (SSS) at the baseline and at day 7 (or discharge, if earlier). Summary descriptive statistics are shown in Table 7, and indicate that, on average, patients receiving aptiganel exhibited greater neurological improvement than patients receiving the placebo. Based on considerations similar to those discussed for Study No. 003, no statistical comparisons were made.

4.2 Severe traumatic brain injury

4.1.1 Initial safety study

The first study in patients with severe TBI (Study No. 002) was an ascending single-dose safety and tolerance study of 4 h infusions of aptiganel on a weight-adjusted basis (Gamzu (for the CNS 1102-002 Study Group), 1994).

Drug dose levels were administered in an escalating fashion to patients who had experienced a severe traumatic head injury within 72 h (see Table 8). The administration consisted of an initial loading dose over 15 min followed

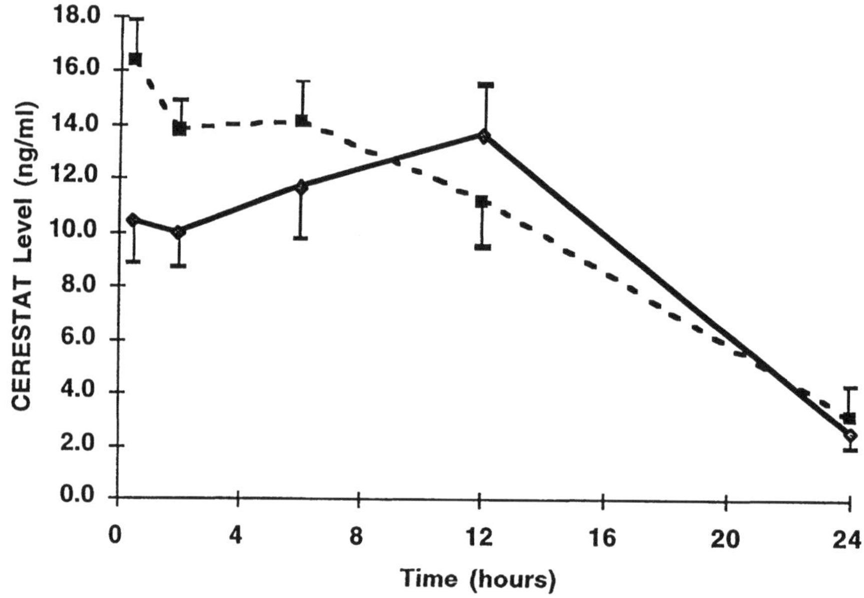

FIG. 4 Mean (± 1 SEM) plasma aptiganel levels over time for two different 12 h intravenous dosing regimens of aptiganel HCl (Study No. 010). ■--■, 6 mg + 1 mg h^{-1} (n = 8); ◆—◆, 4.5 mg + 0.75 mg h^{-1} (n = 12).

Table 7 Neurological status as indicated by mean NIHSS scores and percentage improved on the motor components of the SSS for placebo and aptiganel-treated groups in Study No. 010

Group	n	Mean NIHSS scores			SSS% improved[a] motor score
		Baseline	~7 days	Change	
Placebo	10	9.8	8.6	1.2	50
All bolus	15	10.0	5.9	4.0	80
4.5 mg + 0.75 mg h^{-1}	12	12.5	7.8	4.7	75
6.0 mg + 1.0 mg h^{-1}	8	8.5	4.6	3.8	75

[a] Gait improved by ≥ 3 points, or all three motor components improved by 1 grade (2 points).

Table 8 Dosing regimens, sample size, and patient status in Study No. 002 on the safety of intravenous aptiganel HCl in TBI patients

Dose group total dose (μg kg^{-1} (15 min)$^{-1}$ + μg kg^{-1} h × 4 h)	Number of patients treated	Number of deaths	Number completed at ~6 months	Number lost to follow-up at ~ 6 months
70 (30 + 10)	6	0	4	2
121 (45 + 19)	6	1	5	–
160 (60 + 25)	7	4	3	–
140 (100 + 10)	1	0	1	–
200 (100 + 25)	1	0	1	–
268 (100 + 42)	8	0	6	2
Total	29	5	20	4

immediately by a continuous infusion over 4 h. The objectives of the study were safety assessments including monitoring of vital signs, intracranial pressure; arteriovenous oxygen difference, and general clinical laboratory chemistries.

The doses were chosen to attain and maintain specified plasma drug levels, and the dose was increased only after an evaluation of the previous patients' physiological results. The drug was administered as an adjunctive treatment, and all other standard measures were permitted including the use of mannitol, pentobarbital, barbiturates, and surgical evacuations.

Twenty-eight males and one female were treated 17–73 h postinjury. Of the 29 people treated, 22 had initial (postresuscitation, presedation) Glasgow Coma Scores (GCSs) scores of 3–6.

There were no deleterious effects during the drug infusion on the parameters measured. Within the dose groups with sufficient patients ($n \geq 6$) for analysis, the mean changes from the baseline by hour were combined. A decrease in intracranial pressure (ICP) with a combined increase in MAP resulted in maintaining or increasing cerebral perfusion pressure (CPP, Fig. 5).

Outcome data (including physical examinations, memory assessment, the Galveston orientation test, and the Glasgow Outcome Score (GOS)) were obtained at 1 and 6 months post-treatment. These were used for safety evaluations only, because the patient numbers were not sufficient for assessing a dose–response relationship.

The GOS ratings at 1 and 6 months are summarized in Table 9 by admission GCSs. The 6 month data exclude patients lost to follow-up ($n = 4$).

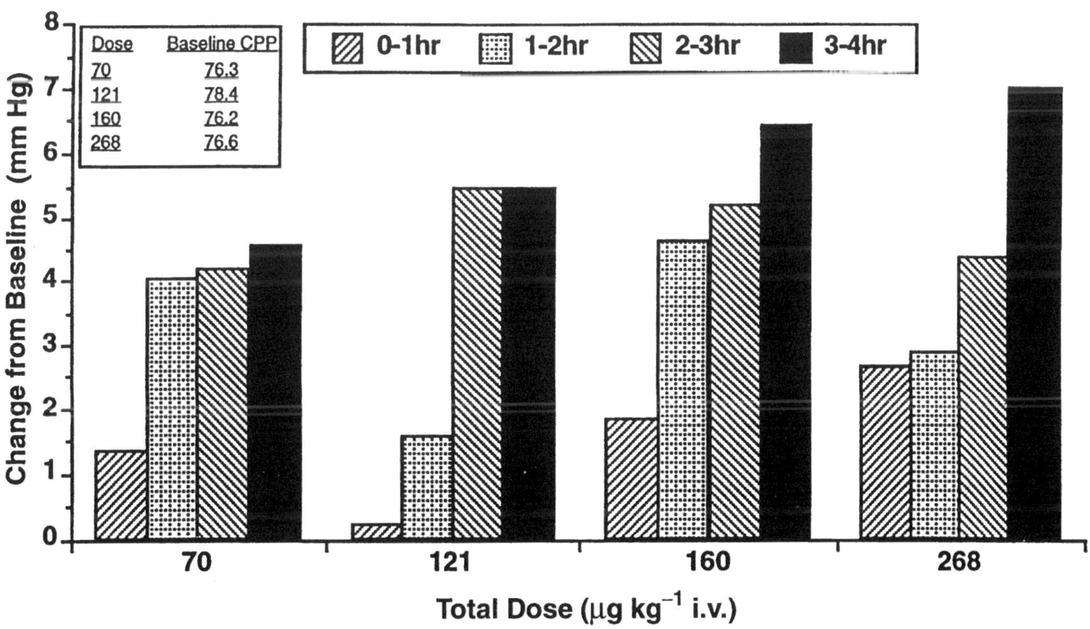

FIG. 5 Mean increase in CPP from the baseline by hour for four different intravenous dosing regimens in severe TBI patients in Study No. 002.

Table 9 GOS for available TBI patients at 1 and 6 months as a function of entry GCS in Study No. 002

	GOS at 1 month (n = 29)			GOS at 6 months (n = 25)		
GCS at entry	Good/moderate	Severe	Vegetative/dead	Good/moderate	Severe	Vegetative/dead
3–4	0	7	4	5	1	5
5–6	3	3	5	2	5	1
7–8	1	4	0	3	0	1
>9	2	0	0	1	1	0
% of total	21	48	31	44	28	28

Deaths All deaths (5/29, 17%) were considered related to or a consequence of the initial injuries, and occurred 2–42 days following treatment. Mortality after severe traumatic head injury is usually high. The death rate expected for this study was about 25%.

Adverse events During hospitalization, one report of each of the following was recorded: agitation due to alcohol withdrawal; bacteremia; lung infection; and staphylococcus septicemia. Additionally, observations at 1 and 6 months following treatment included both neurological and psychological changes expected after a severe head injury. These included personality changes such as mood swings, depression, and anxious agitation. Two patients had continuing seizure activity that was directly attributed to the injury. None of these or any changes in clinical laboratory measures were directly associated with aptiganel HCl.

4.2.2 *Extended infusion study*

A second study (Study No. 005) has been recently completed. The objectives of the study were to assess the safety of extended (12–72 h) infusions of aptiganel HCl administered within 8 h of a severe traumatic head injury (GCS 4–8). Assessments of safety were based on vital signs, ICP, CPP, and clinical laboratories. The drug dose was the highest examined in the previous study – 100 μg kg^{-1} followed by an infusion of 40 μg kg^{-1} h^{-1} (approximately 1 mg kg^{-1} per 24 h). Six patients were treated in each of the four ascending duration groups of 12, 24, 48, and 72 h. Nine patients received a placebo in this phase. In addition, 14 patients were treated in a second, open-label phase of the study using a fixed dose calculated to deliver and maintain a plasma drug level of at least 30 ng ml^{-1}.

The first patient enrolled in the study was discontinued due to the occurrence of seizures first observed within the first hour of infusion. An error by the pharmacist resulted in the patient receiving 10 times the prescribed amount of drug. The patient received 1 mg kg^{-1} over 15 min followed by 400 μg kg^{-1} h^{-1} for approximately 2 h. The case history is as follows:

Patient 101, a 48-year-old male, fell from a scaffolding, resulting in a subarachnoid hemorrhage with bilateral brain contusions. No other serious injuries were noted. His medical history included alcohol abuse, for which he had been recently treated for delirium tremors. Less than 7 h after the injury he was dosed with aptiganel HCl in conjunction with the standard therapy of vecuronium bromide and midazolam for sedation. At approximately 1 h into test drug administration, tonic/clonic seizures were observed and treated with intravenous Valium (diazepam). During the second hour, the seizures recurred. The test drug infusion was stopped and treatment consisted of Valium, Dilantin (phenytoin), and Ativan (lorazepam). During this time, the patient did not experience any other abnormalities. There were no clinically significant changes in vital signs, including blood pressure.

The next day, a repeat computed tomography scan showed an increase in cerebral edema. The standard therapy was continued with no recurrence of seizures. Within a week, the patient was greatly improved, with a GCS of 14 from an initial GCS of 6. He had occasional episodes of confusion but was fully oriented and alert with a discharge GOS of 2 (moderately disabled) at 1 week. There did not appear to be any short-term sequelae. On day 90, the patient's GOS was 2 and his Disability Rating Scale score was 4.

Plasma drug concentrations for this patient were monitored over the 24 h after administration of the drug was halted. At the end of the infusion, the level was 148.9 ng ml^{-1}. After 24 h, the level had declined to 16.8 ng ml^{-1}. The clearance was estimated to be 16.88 ml min^{-1} kg^{-1}, which is within the range observed for normal volunteers and elderly stroke patients.

Over the entire study, there were no adverse drug effects observed on any of the physiological parameters measured (ICP, MAP, and CPP). With the exception of the first patient described above, discontinuation of the drug infusion only occurred when a patient was determined to have died.

The target plasma level of 30 ng ml^{-1} was achieved for all infusion durations. The plasma kinetics of the drug were unchanged by the length of the infusion. In addition, a non-weight-adjusted dose yielded a consistent plasma level independent of body weight. The clearance was approximately 22 ml min^{-1} kg^{-1}, which does not differ from that observed in elderly stroke patients or normal volunteers.

The open-label portion of the study enrolled a number of patients with a poor prognosis, including patients with no activity on EEGs due to extended periods of increased ICP (>30 mmHg). Seven of the 14 patients died or were removed from life support within the first week after the injury. Over the entire study, the death rate was within the expected range (30%) for this severity of injury.

The number of placebo patients (nine) was too small to permit a comparative analysis of outcome measures. Five of these patients were admitted with a GCS of 7 or 8 were under 35 years of age. These variables are indicators of a better outcome. In contrast, the drug-treated groups included an older population. As of this date, outcome data are available for 34 patients. Even though increasing age is considered a risk factor, the older patients (>36 years) who had received aptiganel had outcomes at 3 months that were better than expected from historical data. According to the Traumatic Data Coma Bank (TDCB) statistics for 6 month outcome, the Good and Moderate disability outcome is only expected for approximately 30% of patients in the 36–55-year-old age range (Choi *et al.*, 1994). In contrast, 55% of the patients in this age range who received aptiganel in Study No. 005 had a Good or Moderate disability outcome. Moreover, based on the TDCB, one could expect further improvement from 3 to 6 months.

5 DISCUSSION

Over the last decade, an impressive body of evidence has implicated the excitatory amino acids – particularly glutamate – and their receptors in the etiology of various neurological disorders (Lipton and Rosenberg, 1994; Ginsberg, 1995a,b). A large number of *in vitro* and animal experiments has identified the NMDA receptor as being singularly important in the cascade of events that leads to neuronal death following an ischemic or traumatic injury to the brain. Calls for the clinical evaluation of NMDA antagonists (Albers *et al.*, 1989) have been answered by the identification of several different classes of molecules directed toward various sites within the NMDA receptor–ion channel complex. Arguments in favor of particular pharmacological approaches to NMDA antagonism (e.g. competitive versus noncompetitive antagonists, full versus partial antagonists, etc.) can and have been made on theoretical grounds or on the basis of preclinical findings (Fisher, 1995; Muir and Lees, 1995). However, all such arguments must ultimately be tested in the clinic, and so far the clinic has produced some surprises.

The clinical development of NMDA antagonists has been dominated by two questions: (1) Can such agents be administered safely at doses that will confer measurable benefit? (2) How does one go about extrapolating from results obtained in animal models to the more complicated clinical situation? In the development of aptiganel HCl, we have had to wrestle with both of these questions.

With regard to the question of safety, we have considered both the diseases to be treated and the type of molecule to be used. We have concentrated initially on stroke and traumatic brain injury because these are life-threatening disorders with no existing treatment. Thus, a certain degree of adverse effects can be justified, if benefit to the patient can be demonstrated. In addition, both stroke and TBI present an opportunity for short-term treatment in an acute-care setting, where patients can be monitored closely.

With respect to the choice of molecules, the field has undergone a change in outlook. Initially, it was felt that high-affinity noncompetitive NMDA antagonists would be prone to cause unacceptable 'psychotomimetic' side-effects, such as agitation and hallucinations. As it happens, aptiganel HCl has proved to be relatively free of such effects, particularly in patients, with the main symptoms being ataxia and sedation. In contrast, studies of the competitive NMDA antagonist selfotel and the low-affinity noncompetitive antagonist dextrorphan have reported a much higher incidence of problematic neuropsychological effects (Albers *et al.*, 1995; Grotta *et al.*, 1995).

In addition to its central nervous system side-effect profile, aptiganel HCl has several other attributes that make it especially suitable for use in the proposed clinical indications. It enters the brain rapidly, which is an advantage in disorders where the time from injury to treatment is presumed to be crucial. It also leaves the brain quickly, such that adverse effects resolve quickly upon stopping administration. The cardiovascular effects of aptiganel are manageable, and are in a direction (pressor rather than depressor) that does not worsen outcome either in stoke or in TBI and may be of benefit (Drummond *et al.*, 1991; Rosner *et al.*, 1995). Finally, the pharmacokinetics of the drug are extremely well behaved, even over prolonged administration periods, and dosing can be chosen to maintain a consistent plasma drug level.

The second question noted above, that of extrapolating from animal models to the clinical situation, represents a formidable challenge. In both stroke and traumatic brain injury, the degree of neurological and functional deficits in patients is extremely heterogeneous, and is complicated by uncontrolled variables such as time to treatment, age, concurrent disease, concomitant medications, other residual deficits, and extent of rehabilitation. In stroke assessment, a number of well-documented neurological scales have been used, but the recently published results of the tPA study (NINDS-rtPA Stroke Study Group, 1995) suggest that neurological assessment may not be the most sensitive measurement in the acute phases following the treatment. Rather, benefit can only be assessed later (3 months), as is the case with TBI (6 months). Moreover, there is a general consensus that any benefit incurred during treatment should be substantiated by a long-term benefit in the functional outcome of the patient. In order to convincingly demonstrate that such clinically meaningful effects have been achieved, the numbers of patients per group in a clinical study must range from 300 to 500. Faced with the likelihood that traditionally designed phase II studies are unlikely to yield statistically significant evidence of efficacy, we have concentrated on satisfying ourselves of the safety of aptiganel in both patient populations, and on optimizing the dosing regimens to be used in large-scale efficacy studies.

In any development program there is always a search for reasonable alternative surrogate measures that will expedite the process. Recent work with new neurological imaging techniques, such as diffusion-weighted magnetic resonance imaging and magnetic resonance spectrosopy hold some promise in possibly evaluating the effects of neuroprotective treatments (Warach *et al.*, 1996). However, it is unlikely that a validation across a large heterogeneous population will be undertaken in the short term. In TBI patients, the markers of decreased CPP (<70 mmHg) and increased ICP (>20 mmHg) have been associated with poor outcomes (Rosner *et al.*, 1995). Nevertheless, confirmation in large therapeutic trials using only these measures has not been accepted.

Our approach to the clinical development of aptiganel HCl has taken the pragmatic approach of determining the maximally tolerated dose and trying to attain and maintain the plasma drug level associated with neuroprotection over the projected period of vulnerability. Thus, aptiganel has been studied using escalating doses and plasma drug levels in placebo-controlled studies in relatively small patient populations. The importance of placebo groups is crucial in these studies, due to the heterogeneity of these disease states. The results, to date, have permitted us to delineate adequately the safety profile of the drug, while observing some encouraging results in neurological and functional parameters. However, the sample sizes and the study designs were not conducive to any definitive interpretation. The most important result of these early clinical studies has been the ability to achieve the presumed neuroprotective plasma drug levels with an adequate safety margin and tolerable side-effect profile. In this respect, aptiganel appears to have an advantage over other NMDA antagonists that have been studied in comparable populations.

The next challenge for the development of aptiganel HCl will be to assess the neuroprotective effects of the drug in a large number of patients. Pivotal trials of aptiganel in 700 TBI patients (one active dose versus placebo) and in 900 ischemic stroke patients (two active doses versus placebo) started in the first half of 1996.

ACKNOWLEDGMENTS

The authors would like to thank all their colleagues at Cambridge NeuroScience and Boehringer Ingelheim who participated in various stages of the development of aptiganel, but especially Robert N. McBurney, who was the original project leader and has been a continual, important contributor to the advancement of CERESTAT. James B. Fischer also deserves special recognition for his time and devotion to establishing, monitoring, and analyzing the pharmacokinetic data. We are grateful to all our preclinical and clinical collaborators. Special thanks are due to Jane Wagner and Mario Pita for their help in preparing this chapter.

5 Development of ACPC, A Partial Agonist of the Glycine Site on the NMDA Receptor

MARIA-LUISA MACCECCHINI

Symphony Pharmaceuticals, Inc., 3624 Market Street, Philadelphia, PA 19104, USA

1 INTRODUCTION

Alterations of ion flow at the *N*-methyl-D-aspartate (NMDA) receptor have been linked to a variety of disorders. Overactivation of the NMDA receptor has been associated with the damage that occurs following stroke and other ischemic insults to the central nervous system (CNS) (Monaghan *et al.*, 1989; Collingridge and Singer, 1990; Meldrum and Garthwaite, 1990; Nakanishi, 1992; Von Lubitz *et al.*, 1992; Szatkowski and Attwell, 1994; Greenamyre and Porter, 1994), while underactivation of the receptor may be involved with memory loss (Monaghan *et al.*, 1989). There is growing evidence that compounds active at this receptor may also be useful as antidepressants (Trullas and Skolnick, 1990; Nowak *et al.*, 1993; Paul *et al.*, 1993), anxiolytics (Trullas *et al.*, 1991), anticonvulsants (Chapman *et al.*, 1987; Chapman and Meldrum, 1993, Gean and Shinnick-Gallaher, 1988), and inhibitors of morphine and opiate addiction (Trujillo and Akil, 1991; Kolesnikov *et al.*, 1992). However, in animal models as well as clinical trials, many of the ion channel blockers or antagonists active at the NMDA receptor produce severe side-effects (Willets, 1990; Diemer, 1990; Buchan, 1990; Faden and Salzman, 1992). To address the unmet need for safe and effective CNS therapeutics, we are developing a novel group of receptor modulators, called partial agonists. We have identified partial agonists that can act at three different sites on the NMDA receptor. The most advanced of these compounds, 1-aminocyclopropanecarboxylic acid (ACPC), is in clinical trials for treatment of ischemic injury and depression.

2 PARTIAL AGONISM AS A THERAPEUTIC APPROACH

As early as 1956 it was noted that some agents can simultaneously act as both agonists and antagonists, and the term 'partial agonist' was coined to describe this phenomenon (reviewed by Jasper, 1992). Partial agonists are unique in that they have a bimodal action. In regions of the brain where levels of endogenous neurotransmitters are reduced, partial agonists may serve to increase receptor stimulation. Alternatively, in regions where there are elevated concentrations of neurotransmitters, partial agonists can act as antagonists to reduce receptor stimulation. Because this reduction is not complete and neurotransmission can still take place, partial agonists may be referred to as 'regulators'. Partial agonists have been developed as

EXCITATORY AMINO ACIDS – CLINICAL RESULTS WITH ANTAGONISTS
ISBN 0-12-546820-2

drugs in a number of disorders (e.g. acebutalol in hypertension, ciladopa in Parkinson's disease, and bretazenil in anxiety) where they offer all of the efficacy of antagonists or full agonists but have superior side-effect profiles (Jasper, 1992).

To develop partial agonists, Symphony has utilized its proprietary system of computer-assisted design to construct models of the binding pockets on glutamate receptors and to define the conformational transition from agonist to antagonist activity. These models, combined with structure–activity relationship (SAR) data, enable the design of compounds with high specificity and with agonist, antagonist, or partial agonist properties. An example of the results of such an effort is the SAR data shown in Table 1. The parent compound acts as a noncompetitive antagonist of polyamine responses on a novel site of the NMDA receptor. By introducing selective changes we obtained compounds with reduced ability to inhibit polyamine responses at the NMDA receptor, here assessed by the ability of polyamines to enhance the binding of [^{3}H]dizocilpine ([^{3}H]MK-801) to the receptor complex. The first two compounds shown, SYM 1010 and SYM 1035, are antagonists: they close the channel and cut off ion flow completely. The next compound, SYM 1007, closes the channel 70% in the presence of stimulation or opens the channel 30% in the absence of stimulation. Therefore, it gives a constant ion flow of about 30%, independent of the original channel state. The ability to progressively alter the degree of function as a partial agonist and a partial antagonist is seen in each of the next three compounds, SYM 1007, SYM 1022, and SYM 1037, that permit 50, 65 or 80% stimulation, respectively. The two last compounds, SYM 1001 and SYM 1027, show the full agonist state. Basically, we have established the capability to regulate ion flow through a channel to the desired degree.

Table 1 Spectrum of pharmacological activity for selected Symphony compounds active at the NMDA receptor[a]

Compound	Antagonist activity	Agonist activity
SYM 1010	100	–
SYM 1035	100	–
SYM 1007	70	31
SYM 1022	51	53
SYM 1037	35	68
SYM 1008	22	79
SYM 1001	–	100
SYM 1027	–	100

[a]The activity of this series of compounds at the NMDA receptor was determined by measuring their effect on the binding of [^{3}H]dizocilpine ([^{3}H]MK-801, 20 μM) to rat forebrain membranes. Antagonist activity was determined as the percentage inhibition of [^{3}H]dizocilpine binding that was enhanced by the addition of spermine (12.5 μM). Agonist activity was measured as the percentage enhancement of [^{3}H]MK-801 binding in the absence of other additives (Zhou *et al.*, 1996).

3 ACPC – PHARMACOLOGY AND MECHANISM OF ACTION

ACPC acts as a partial agonist at the glycine-binding site of the NMDA receptor. Because ACPC has the properties of an agonist, it permits normal function to occur under physiological conditions, but since it is not a full agonist, it is able to limit the influx of calcium and the cascade of events otherwise initiated in the presence of pathologically high concentrations of glutamate. In the absence of glycine, ACPC potentiates NMDA-induced currents but only to 80% of the amplitude obtained by glycine. However, ACPC also reduces the glycine-induced potentiation of the NMDA receptor to 80% of the maximal response (Fig. 1; Watson and Lanthorn, 1990).

Therefore, the partial agonist ACPC offers an approach to neuroprotection that is distinct from channel blockers, competitive antagonists, and glycine antagonists. In addition, ACPC has the advantage of a superior side-effect profile in both preclinical and clinical studies, as will be shown later in this chapter.

4 EFFICACY OF ACPC IN ANIMAL MODELS OF NEUROPROTECTION

The neuroprotective action of ACPC has been demonstrated in animal models of stroke (focal ischemia), myocardial infarct (global ischemia), and spinal cord injury.

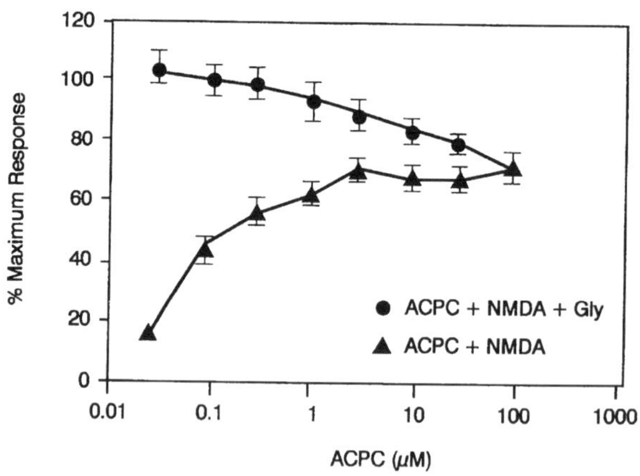

FIG. 1 ACPC is a partial agonist of the glycine site on the NMDA receptor. Oocytes from *Xenopus laevis* were injected with rat brain mRNA and subjected to single-electrode voltage clamp after 2–6 days. The data shown are normalized inward currents induced by 100 μM NMDA in the absence or presence of 10 μM glycine and increasing amounts of ACPC as shown. Each point represents the mean percentage of the maximum response for 6–7 oocytes; the maximum response is induced by the application of 100 μM NMDA and 10 μM glycine. (Adapted with permission, from Watson and Lanthorn (1990).)

4.1 Global and focal ischemia

Pretreatment with ACPC was neuroprotective in an animal model of global ischemia. When global ischemia was induced in gerbils by a 20 min bilateral occlusion of the carotid arteries, 80% of the saline-treated animals died within 7 days. However, if the animals were treated with daily administration of ACPC for 6 days prior to surgery, with or without an additional dose 30 min before the surgery, 85% of the animals survived the full 7 days post-surgery (Fig. 2; Von Lubitz *et al.*, 1992).

ACPC is also an effective neuroprotectant when administered acutely after the induction of global ischemia (Fig. 3; Von Lubitz *et al.*, 1992). Again, over 80% of the saline-treated control animals died. However, when ACPC was administered 5 min after the 20 min ischemic insult, only one animal died. When animals were treated with the NMDA channel blocker MK-801 before the ischemic event, there was no improvement in survival following surgery (D.K.J.E. Von Lubitz, unpublished data). Taken together, these findings suggest that ACPC can protect animals from death following severe global ischemia more effectively than MK-801.

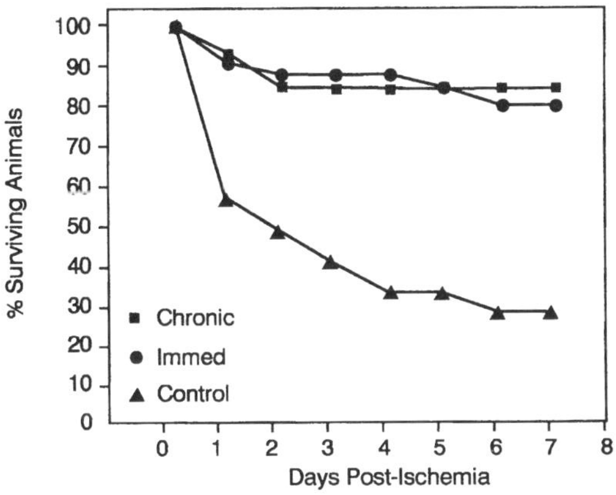

FIG. 2 ACPC administered prior to surgery enhances survival of gerbils following severe global ischemia. Animals (6–20 per group) received daily injections of saline (vehicle) for 7 days, ACPC for 6 days with a 1 day wash-out period (chronic), or ACPC for 7 days (immediate) with the last injections in all cases administered 30 min prior to a 20 min period of bilateral carotid occlusion. Blood flow was then restored, and animals were allowed to survive for 7 days. (Adapted, with permission, from Von Lubitz *et al.* (1992).)

Both ACPC and MK-801 prevented neuronal cytotoxicity in the gerbil global ischemia model. Figure 4 shows the percentage of viable neurons in the animals that survived the ischemic insult. In the vehicle-treated animals about 60% of the neurons died, whereas MK-801 and ACPC protected the neurons in the CA1 region of the hippocampus. In this study, ACPC also reduced the incidence of seizures and improved the neurological score following the onset of severe global ischemia (Fig. 5). Therefore, MK-801 protected the neurons in surviving animals but did not increase animal survival, whereas ACPC protected the neurons and improved the outcome for the animals as well (Fossom *et al.*, 1995).

These studies show that ACPC is neuroprotective in animal models of global ischemia as determined by increased animal survival, reduction of neuronal cytotoxicity, and reduction of neurological defects and seizures. This compound is equally effective when administered before, immediately before, or after the onset of ischemia.

The ability of ACPC to reduce the damage resulting from permanent focal ischemia was recently reported by Lopez and Lanthorn (1996). Six daily injections of ACPC, 300 mg kg^{-1} i.p., followed by a 24 h wash-out prior to permanent middle cerebral artery occlusion in mice caused a significant reduction in infarct volume compared to vehicle-treated controls (Fig. 6).

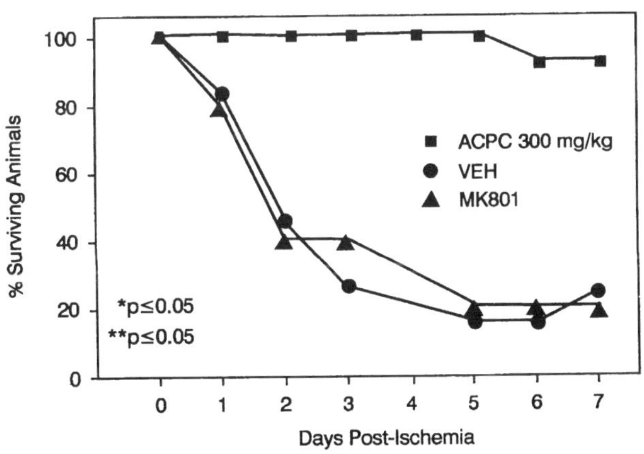

FIG. 3 ACPC administered after surgery enhances survival of gerbils following severe global ischemia induced by a 20 min period of bilateral carotid occlusion followed by reperfusion. Animals (10–12 per group) received saline (vehicle), ACPC (300 mg kg^{-1} i.p.,) 5 min after reperfusion, or MK-801 (3 mg kg^{-1} i.p.) 30 min before occlusion. Animals were allowed to survive for 7 days. (Adapted, with permission, from Von Lubitz *et al.* (1992) and D.K.J.E Von Lubitz (unpublished data).)

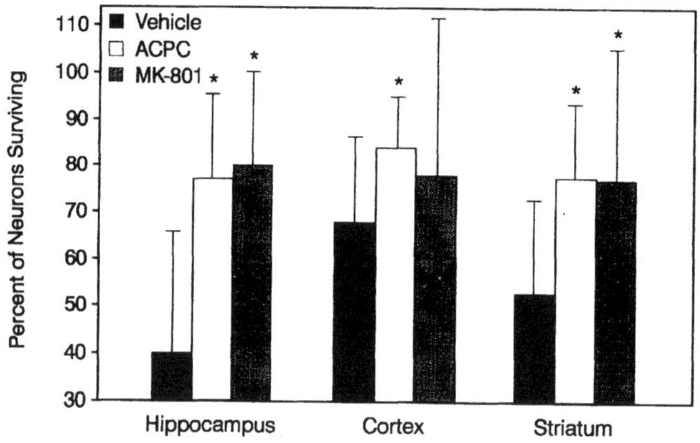

FIG. 4 ACPC administered after surgery enhances survival of neurons in selected brain regions of gerbils following severe global ischemia. Animals were treated as described in the legend to Fig. 3. *p <0.05 by the Student–Newman–Keuls test. (Adapted, with permission, from Von Lubitz *et al.* (1992) and Fossom *et al.* (1995).)

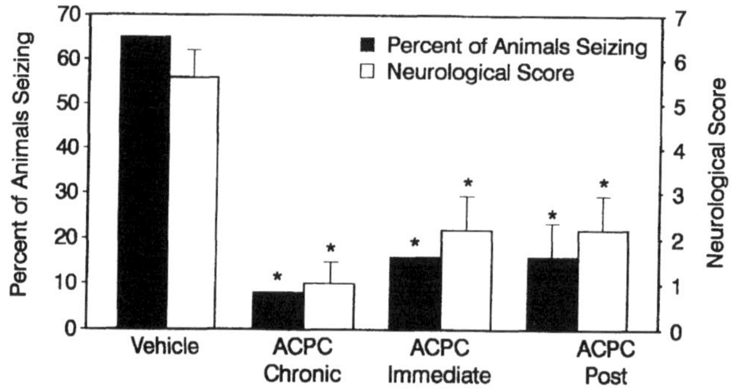

FIG. 5 ACPC administration prior to or following surgery improves the neurological outcome measures following global ischemia in gerbils. Animals were treated as described in the legend to Figs 2 and 3. The incidence of seizures was monitored continuously during the first 8 h after surgery and for 30 min periods at four 12 h intervals thereafter. The neurological scale represents the following: 0, normal; 1, sleepy/lethargic; 2, hyperactive; 3, circling/ptosis; 4, jumping; 5, rolling seizures/opisthotonus; 6, tonic convulsions; 7, coma, weak pain response; 8, coma, no pain response; 9, death. *p <0.05 versus vehicle. (Reproduced, with permission, from Von Lubitz *et al.* (1992) and D.K.J.E. Von Lubitz (unpublished data).)

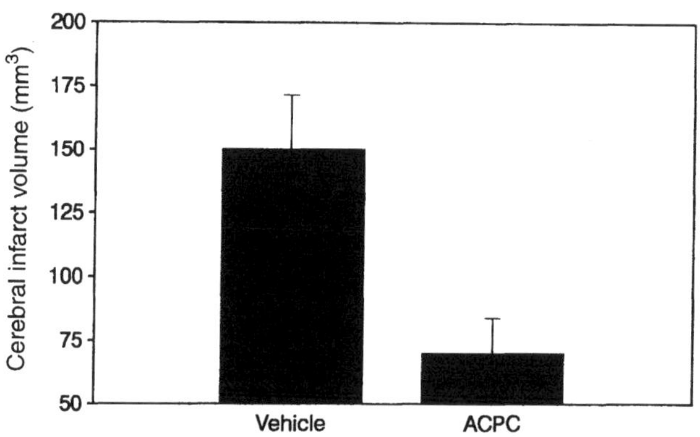

FIG. 6 ACPC is neuroprotective in a model of focal ischemia. Mice received ACPC, 300 mg kg^{-1} i.p, for 6 days followed by a 1 day wash-out prior to induction of permanent focal ischemia. The volume of cerebral infarct was assessed by standard techniques; the reduction in infarct volume seen in the ACPC-treated animals is statistically significant compared to vehicle controls. (Lopez and Lanthorn, 1996, in press.)

4.2 Spinal cord injury

ACPC also shows neuroprotective properties in an animal model of spinal cord injury. Injection of dynorphin A into the spinal cord of rats leads to hind limb paralysis, which can be prevented by NMDA antagonists. In this model, ACPC inhibited dynorphin A-induced paralysis in a dose-dependent fashion (Fig. 7) (Long and Skolnick, 1994). ACPC was as effective as other NMDA receptor inhibitors, such as the glycine site antagonist 7-chlorokynurenic acid (7 Cl K) and the NMDA site antagonist 5-fluoroindole-2-carboxylic acid (FICA). Additionally, ACPC reduced the number of necrotic neuronal cells following the spinal trauma (Fig. 8). In these studies, ACPC was effective when administered intraperitoneally, whereas other NMDA receptor antagonists (e.g. 7 Cl K and FICA) were injected intracerebroventricular, suggesting that ACPC is a superior therapeutic agent because it is readily available to the CNS following systemic administration.

These animal data indicate that the neuroprotective activity of ACPC may prevent the paralysis and other consequences of cell death following global ischemic injury (e.g. myocardial infarction), focal ischemia (e.g. stroke), spinal cord injury, or head trauma. Additionally, ACPC may be useful as a prophylactic therapy for individuals who are at a high risk of stroke due to a history of elevated blood pressure or transient ischemic attacks.

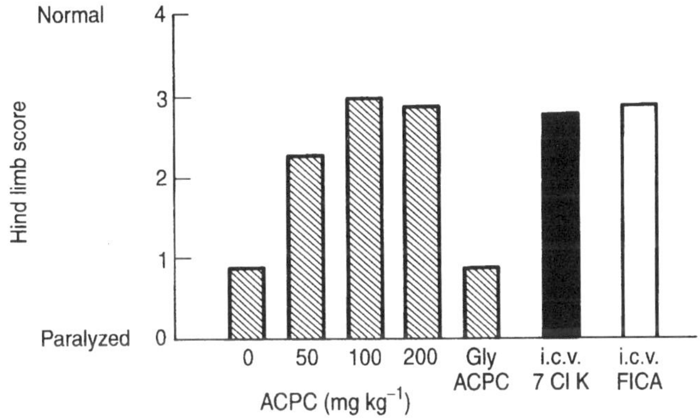

FIG. 7 ACPC reduction of hind limb paralysis following dynorphin A administration. Spinal subarachnoid injection of 20 nmol of dynorphin A produces severe hind limb paralysis 24 h later. Pretreatment with intraperitoneal ACPC (30 min prior to administration of dynorphin A), or with intracerebroventricular administration of 7 Cl K or FICA, prevented some or all signs of paralysis. The prior administration of glycine (800 mg kg^{-1} i.p.) blocked the protective effect of ACPC (100 mg kg^{-1} i.p.). Neurological function was evaluated by a blinded observer using a 5 point scale where 4 = normal motor function and 0 = paralysis, with complete absence of hind limb movement. (Long and Skolnick, 1994, and J.B. Long *et al.*, in preparation.)

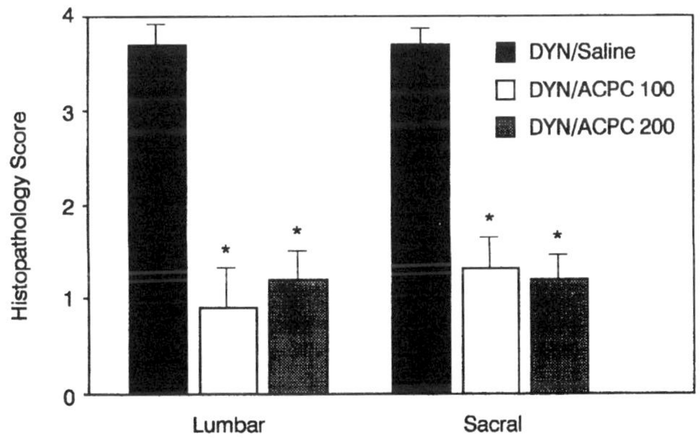

FIG. 8 Neuroprotection of ACPC following spinal cord injury produced by a spinal subarachnoid injection of 20 nmol of dynorphin A. Histopathology scores of tissue taken from the lumbar (L_6) and sacral regions 72 h after injection were determined by a blinded observer. The scores ranged from 1 (normal) to 4 (severe necrosis involving >60% of gray matter and adjacent white matter). Pretreatment with ACPC, 100 or 200 mg kg^{-1} i.p., 30 min prior to dynorphin A prevented the neurotoxic effects. (Long and Skolnick, 1994.)

5 EFFICACY OF ACPC ANIMAL MODELS OF DEPRESSION AND ANXIETY

5.1 Antidepressant activity

Recently, the NMDA receptor was found to be the locus of antidepressant action (reviewed by Paul *et al.*, 1994). Chronic administration of each of the four types of antidepressant therapies (tricyclic antidepressants, monamine oxidase inhibitors (MAOIs), selective serotonin reuptake inhibitors (SSRIs), and electroconvulsive shock therapy (ECT)) alters the ligand-binding properties of the glycine recognition site on this receptor (Nowak *et al.*, 1993; Paul *et al.*, 1993, 1994). The time-course of these changes parallels the onset of action, with ECT requiring about 5 days and existing antidepressants about 14 days of treatment (Paul *et al.*, 1994). In the antidepressant area there has been no discovery of a novel mechanism of action since the synthesis of the first SSRI 20 years ago. Therefore, the discovery of a novel mechanism of action for the treatment of depression would facilitate development of a drug with a different profile from existing antidepressants, such as a faster onset of activity, a safer side-effect profile, a different group of responders, and dual antidepressant and anxiolytic activity.

Comparison of various antidepressants for changes in glycine binding shows that the lag time for the changes is: ACPC ≥3 days; ECT ≥5 days; SSRIs ≥10 days; and tricyclics and MAOIs ≥14 days (Paul *et al.*, 1994; G. Nowak *et al.*, unpublished data). Thus, ACPC shows a faster onset of action than ECT and existing antidepressants. If alteration of the glycine site on the NMDA receptor is the final common pathway for therapeutic intervention in depression, then these findings suggest that ACPC may be a fast-onset antidepressant.

ACPC is active in behavioral models of antidepressant efficacy such as the forced swim test and stress-induced anhedonia. Following oral or parenteral administration, ACPC produced a statistically significant reduction in immobility in the forced swim test (Trullas and Skolnick, 1990; Trullas *et al.*, 1991; Capdevilla *et al.*, 1995; Fig. 9). The effects of orally administered ACPC persisted for at least 6 h, which is compatible with convenient therapeutic administration. Additionally, chronic, but not acute, treatment with either ACPC or imipramine altered the binding of glycine and other ligands to NMDA receptors in mice (Nowak *et al.*, 1993).

Chronic exposure to very mild, unpredictable stress has been found to depress the consumption of, and preference for, highly palatable sweet solutions (Willner *et al.*, 1987; Papp *et al.*, 1991). This animal model of anhedonia has been shown to be predictive for antidepressant efficacy with both typical and atypical antidepressants (Muscat *et al.*, 1992; Papp and Moryl, 1993; Papp *et al.*, 1993). Chronic administration of antidepressants, NMDA receptor channel blockers, or NMDA antagonists reversed the stress-induced anhedonia after 3–4 weeks (Fig. 10; Papp *et al.*, 1992; Papp and Moryl, 1994). When this model was used to measure the onset of antidepressant activity after chronic administration of ACPC, the stress-induced anhedonia was reversed after only 1 week of treatment (Fig. 11; Papp and Moryl, 1996). Therefore, in this model of antidepressant activity, ACPC reversed the anhedonia about four times faster than existing antidepressants or NMDA channel blockers and antagonists.

Since multiple studies have demonstrated the fast onset of ACPC activity *in vitro* (glycine-binding adaptive changes) and *in vivo* (using two different animal models of antidepressant activity), ACPC may well be a fast-onset antidepressant in man.

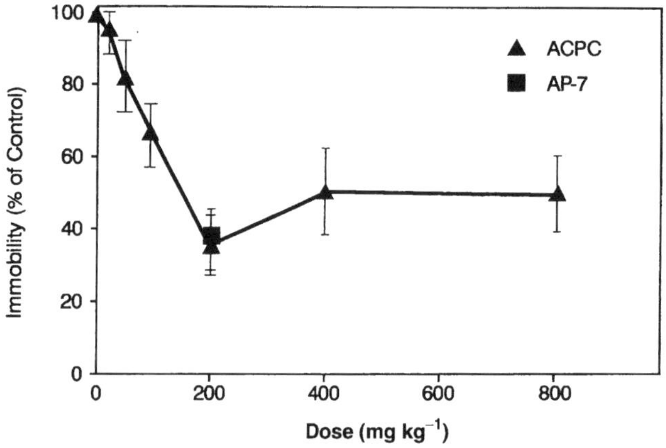

FIG. 9 Effects of ACPC treatment on immobility in the forced swim test. Mice received ACPC (25–800 mg kg^{-1} i.p.), and the forced swim test was administered 1 h later. The duration of immobility in the forced swim test, measured in seconds, was scored during the last 4 min of a 6 min period in which mice were placed in a water-filled cylinder. Each animal was tested only once. Values represent mean ± SEM of the percentage reduction in immobility; $n = 8$ mice per group. (Capdevilla *et al.*, 1996.)

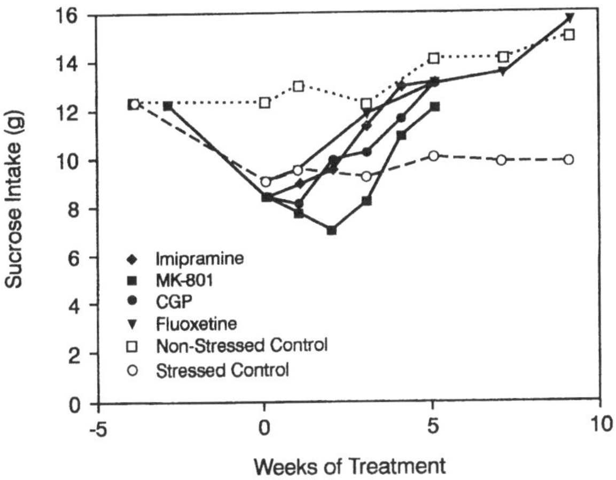

FIG. 10 Effects of various antidepressants on recovery from anhedonia in a chronic mild stress model of depression. Animals were trained to consume a 1% sucrose solution. Then test animals were subjected to chronic stress for 7–8 weeks from a variety of sources as described by Papp *et al.* (1991) and Papp and Moryl (1994), which caused a reduction of sucrose intake (anhedonia). After 3–5 weeks of treatment the antidepressants were able to reverse the stress-induced anhedonia. Reproduced with permission. (Muscat *et al.*, 1992; Papp and Moryl, 1994.)

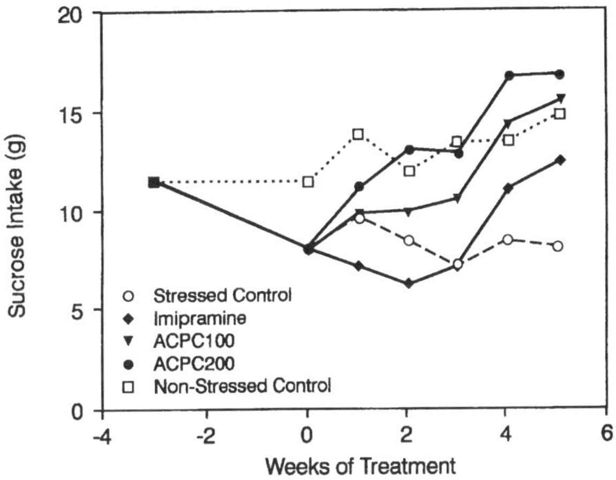

FIG. 11 Reversal of stress-induced anhedonia by ACPC. Experimental conditioning of sucrose intake was as described in Fig. 11. ACPC, 100 or 200 mg kg^{-1} i.p., reversed the anhedonia in as little as 2 weeks while imipramine, 10 mg kg^{-1} i.p. showed no effect until week 4. (Papp and Moryl, 1996.)

5.2 Anxiolytic activity

ACPC is also active in several animal models of anxiety such as the elevated plus maze and pup vocalizations following isolation from the mother. However it does not exhibit the muscle relaxant, sedative, and addictive properties common to benzodiazepines. In the plus maze test, administration of ACPC significantly increased both the number of times the animals entered the open arms of the maze and the percentage of time spent in the open arms (Fig. 12; Trullas *et al.*, 1991). When the effects of ACPC on vocalizations of rat pups maintained under conditions of social isolation were measured, ACPC at doses as low as 12.5 mg kg^{-1} significantly reduced the vocalizations (Fig. 13; Winslow *et al.*, 1990). Under these conditions, ACPC did not affect the animals' motor activity or core body temperature.

6 PREVENTION OF OPIATE TOLERANCE AND TOXICITY BY ACPC

ACPC can prevent or reverse morphine tolerance in mice. Since the importance of NMDA receptors in morphine tolerance is well established, ACPC was evaluated for its

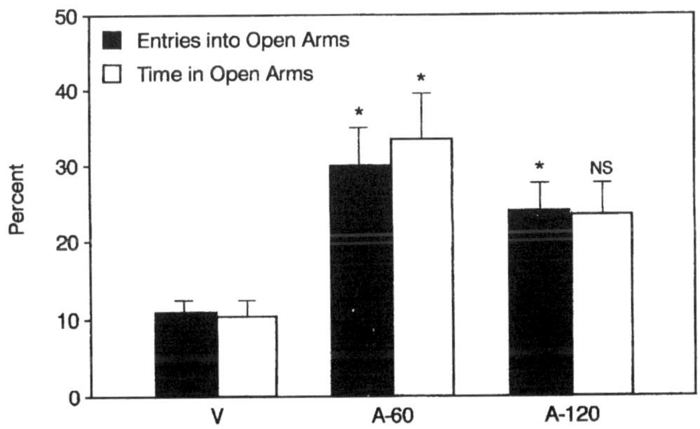

FIG. 12 Effects of ACPC, 400 mg kg^{-1} i.p., on plus-maze performance. Solid bars represent the percentage of entries into the open arms of the plus maze (number of open arm entries ÷ total entries × 100%), and the open bars represent the percentage of time in the maze that was spent in the open arms. Conditions tested were vehicle control (V), 60 min after ACPC (A-60), and 120 min after ACPC (A-120). *p <0.05 by the Newman–Keuls test. (Trullas *et al.*, 1991.)

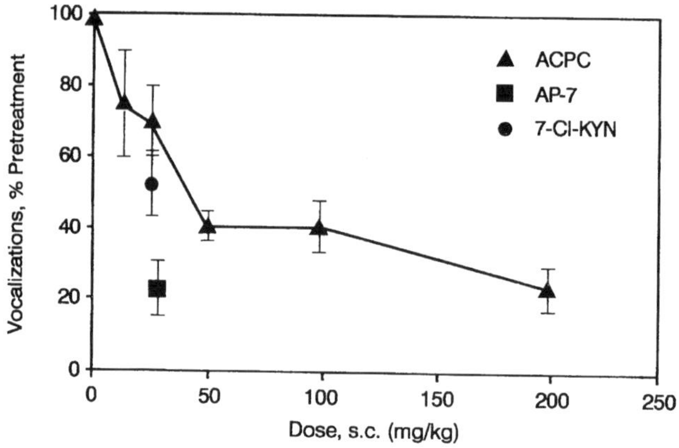

FIG. 13 Acute effects of ACPC (12.5–200 mg kg^{-1} s.c.), AP-7 (30 mg kg^{-1}), and 7 -Cl -K (25 mg kg^{-1}) on ultrasonic vocalizations of 10-day-old rat pups during social isolation. The amount of vocalization, determined by the number of calls, was measured over a 2 min period beginning 20 min after administration of the anxiolytic. Data are expressed as percentage changes from a 2 min observation period immediately before the injection. No change in vocalizations after injection was observed in the vehicle control group. (Reproduced, with permission, from Winslow *et al.* (1990).)

ability to prevent or reverse morphine tolerance. Morphine-induced analgesia diminished with repeated daily injections until, after 5 days, no analgesia was observed in any of the mice tested (Fig. 14; Kolesnikov *et al.*, 1994). However, administration of ACPC (50 or 150 mg kg^{-1}) prevented the development of tolerance to the analgesic effect of morphine. ACPC was effective in both preventing tolerance (Fig. 14(A)) and reversing the animals' tolerance (Fig. 14(B)) to morphine.

ACPC can prevent the convulsive effects of cocaine; similar activity has been shown for several NMDA receptor antagonists. Administration of ACPC, intraperitoneally, at 10 or 30 min before administration of a convulsive dose of cocaine (60 mg kg^{-1}), prevented the occurrence of convulsions (Witkin and Tortella, 1991). In this model, the efficacies of ACPC, diazepam, and the glycine-site NMDA antagonist 7 CI K were dose-dependent, with EC_{50} values of 200, 0.9, and 60 mg kg^{-1}, respectively.

7 PHARMACOKINETICS

ACPC is readily absorbed into the blood following intravenous, intraperitoneal, and oral administration. The bioavailability of the compound is 90–95% following oral administration (Fig. 15; Cherkofsky, 1995). Additionally, ACPC readily enters the CNS following administration by various routes. For example, ACPC can be detected in the cerebrospinal fluid of monkeys within 1 h of oral administration (Fig. 16). While the half-life of ACPC elimination from the blood is 2.5 h, ACPC is cleared much more slowly from the CNS, where the half-life is at least 6 h. These data from primates are similar to the pharmacokinetics of ACPC in rodents. When ACPC was adminis-

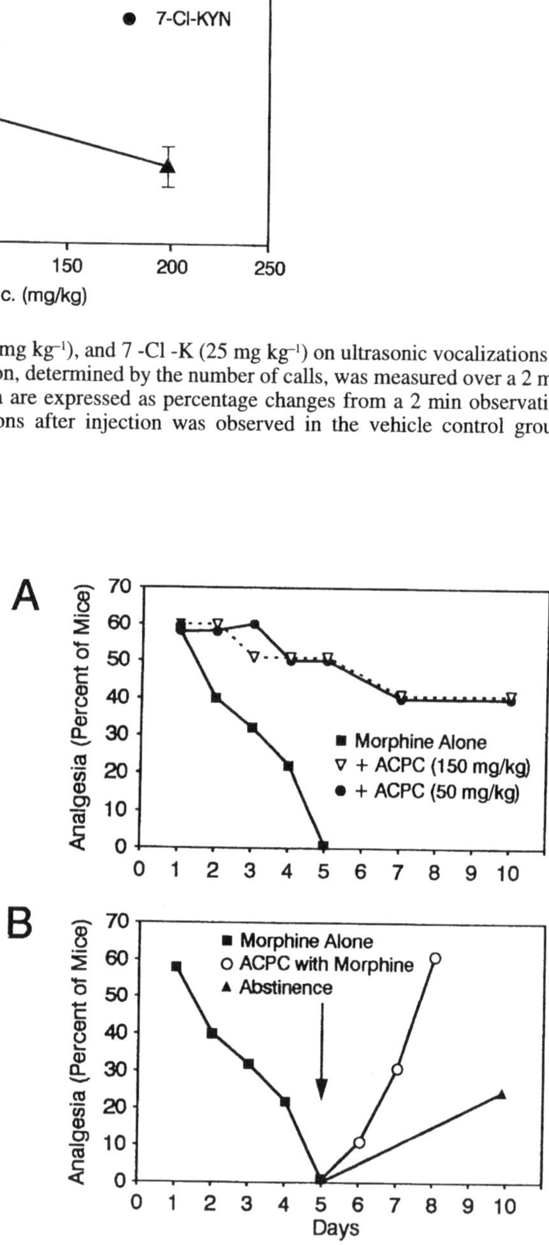

FIG. 14 Effects of ACPC on morphine tolerance. (A) Prevention of tolerance by ACPC. Groups of mice ($n \geq 10$) received 5 mg kg^{-1} of morphine alone or in combination with ACPC (50 or 150 mg kg^{-1} i.p.) daily. Analgesia was assessed using the tail-flick test on the days indicated. (B) Reversal of tolerance by ACPC. Mice received daily injections of morphine for 5 days and then were divided into two groups of 10. One group continued to receive morphine along with ACPC (150 mg kg^{-1}); in this group the tolerance to morphine was reversed after 3 days' treatment with ACPC. The other group received no drugs until day 10 (abstinence), when they were challenged with morphine. (Reproduced, with permission, from Kolesnikov *et al.*, 1994.)

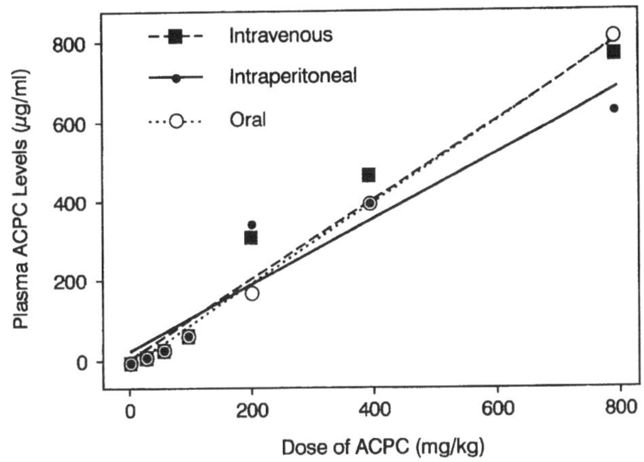

FIG. 15 Plasma levels of ACPC in mice after different routes of administration. ACPC was administered orally, intravenously, or intraperitoneally, and blood samples were obtained 1 h after administration. ACPC concentration was determined by gas chromatography–mass spectroscopy analysis (Cherkofsky, 1995). The data were analyzed using least-squares linear regression.

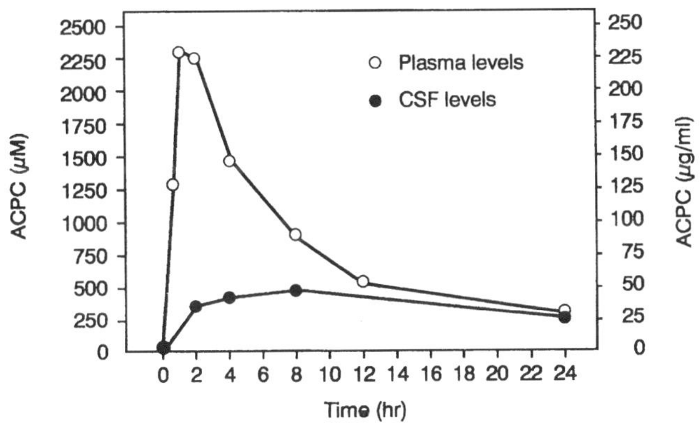

FIG. 16 Concentrations of ACPC in plasma and cerebrospinal fluid (CSF) following oral administration. [^{3}H]ACPC, 300 mg kg^{-1}, was administered nasogastrically to cynomolgus monkeys. Blood and CSF were collected at selected intervals for 24 h. Since thin-layer chromatography analysis of samples showed no formation of ACPC metabolites, ACPC levels were determined by direct liquid scintillation counting of samples.

tered to mice, 200 mg kg^{-1} i.p., it was quickly absorbed into the blood and then cleared from the blood plasma with a half-life of 1.5 h (Fig.17; Miller *et al.*, 1992). In the same study, ACPC reached maximum levels in the brain 1 h after administration, and a significant amount remained in this tissue after 8 h. The pharmacokinetics of ACPC in humans was assessed by Symphony in phase I clinical trials, which are described later in this chapter.

The absorption and pharmacokinetics of ACPC are consistent across the species tested, with the exception of an unusually slow elimination in the dog (Table 2). ACPC has a long retention time in the CNS. In all three species in which CNS pharmacokinetics were measured, the half-life of ACPC in the brain or cerebrospinal fluid was greater than 6 h, which is 2–3 times the plasma half-life of the compound.

In metabolic studies, ACPC was excreted intact by the kidneys; no evidence of metabolites was found in any of the species tested: rat, dog, monkey or human (Cherkofsky, 1995). For the animal studies, radiolabeled ACPC was extracted from tissue and/or body fluids at various time intervals and then analyzed by thin-layer chromatography. After 24–240 h (depending on the species) 93, 80 or 71% of the administered dose was recovered in the urine of rats, dogs, and monkeys, respectively.

The minimally effective dose of ACPC in animal models has been determined by several investigators. Winslow *et al.* (1990) found that 12.5 mg kg^{-1} of ACPC was an effective anxiolytic that significantly reduced vocalizations in stressed rat pups. Further reductions in the number of vocalizations were observed at higher doses

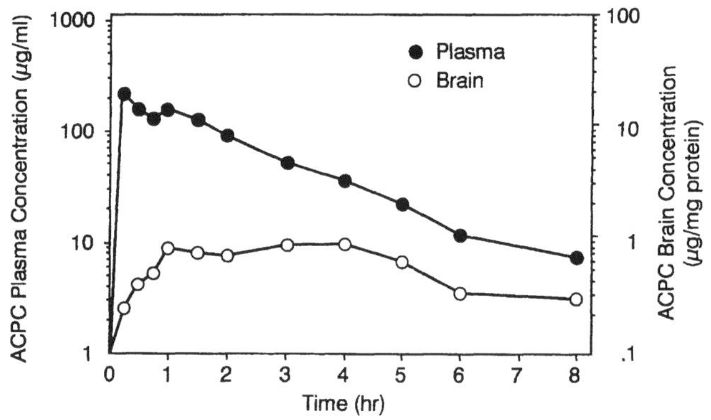

FIG. 17 Concentrations of ACPC in plasma and brain following intraperitoneal injection of 200 mg kg^{-1} of ACPC in mice. Blood and brain were collected from animals sacrificed at the indicated time intervals. ACPC was extracted from the tissues and analyzed by high-performance liquid chromatography. (Miller *et al.*, 1992.)

Table 2 Minimally effective doses of ACPC in models for CNS therapeutics

Species	Condition and model	Minimally effective dose (mg kg^{-1})	Dose range tested (mg kg^{-1})[a]	Reference
Rat pups	Anxiolytic activity – vocalization in isolated animals	12.5	12.5–200	Winslow *et al.* (1990)
Rats	Dynorphin A induced spinal injury	50	50–200	Long and Skolnick (1994)
Mice	Antidepressant activity – forced swim test	100	100–400 (oral)	Trullas *et al.* (1991)
		50 (i.v., i.p., p.o.)	25–800	Capdevila *et al.* submitted
Mice	Prevention of morphine tolerance	50	50, 150	Kolesnikov *et al.* (1994)

[a]ACPC was administered parenterally, unless otherwise noted.

(25–200 mg kg^{-1}). Comparable dose-dependent effects of ACPC, with minimally effective doses of 50–100 mg ml^{-1}, were observed in animal models of neurotoxocity, depression, and morphine tolerance (Trullas *et al.*, 1991; Kolesnikov *et al.*, 1994; Long and Skolnick, 1994). In each of these studies, which are summarized in Table 2, ACPC was effective at the lowest dose tested.

From the pharmacokinetic data in Table 3 and the efficacy data in animal models in Table 2, the minimally effective doses of ACPC in monkeys, dogs, and humans can be calculated. These determinations were made using the allometric relationships between data measured in all of the species tested (plasma half-life, area under the curve (AUC), and clearance rate) and the minimally effective dose of ACPC in mice and rats. From these calculations, we expect the minimally effective doses of ACPC will be 22 mg kg^{-1} in monkeys, 2.8 mg kg^{-1} in dogs (due to the slow clearance of ACPC in this species), and 11 mg kg^{-1} in humans (Table 3).

8 SAFETY PROFILES OF ACPC IN ANIMAL MODELS

In multiple-dose studies of ACPC, lasting up to 90 days, the pharmacokinetics of the last daily dose administered to rats and monkeys was comparable to that of the first dose, and no accumulation of the compound was observed. However, when ACPC was administered to dogs over seven daily doses, a significant (2–3-fold) accumulation of ACPC in plasma was observed (Symphony, unpublished data). These findings are consistent with the long clearance time of ACPC and large AUC for dogs (Table 3). Hemodynamic studies of ACPC in dogs revealed no observable alterations of cardiac function. Specifically, no changes in the blood pressure, heart rate, cardiac output, electrocardiogram (ECG), left ventricular pressure, or contractile force were observed after a cumulative dose of 440 mg kg^{-1} i.v.

The accumulation of ACPC in dog plasma resulted in toxicity in multiple-dose studies (7 days or longer); however,

Table 3 Pharmacokinetics and bioavailability of ACPC following a single dose

Species	Plasma half-life (h)	Estimated CNS half-life (h)	AUC (μg h ml^{-1})	Clearance (ml min^{-1} kg^{-1})	Minimally effective dose (mg kg^{-1})
Mouse	1.5	6	47[a]	7.3	50
Rat	2.5	> 6	105[a]	3.2	12.5–50
Monkey	2.9	> 6	108[a]	3.1	22[b]
Human	5.9	NA	216	1.5	11[b]
Dog	20.3	NA	833[a]	0.40	2.8[b]

Data adapted from Cherkofsky, 1995. NA, data not available.
[a]Extrapolated for a 20 mg kg^{-1} dose.
[b]Calculated.

the accumulation of ACPC and concomitant toxicity was not observed in either rats or monkeys, whose pharmacokinetics of ACPC better mimic those of humans (Table 3).

In both rats and monkeys, ACPC was tolerated at very high doses for up to 90 days (Table 4). No adverse effects were seen at doses up to 900 mg kg^{-1}. Among animals receiving the highest doses of ACPC, the only adverse effects observed were reduced growth rates relative to control animals, lethargy, and some abnormal pathology on necropsy. These data show that, for both acute and chronic administration of ACPC, there is a very large window (approximately 100-fold) between the maximum safe dose and the minimal effective dose.

ACPC has shown few side-effects in various animal studies, particularly in comparison to other channel blockers and NMDA receptor antagonists. For example, the channel blocker MK-801 has been reported to produce intraneuronal vacuoles, a proposed sign of neurotoxicity (Olney *et al.*, 1990). This compound, and two glycine site antagonists, 7 Cl K and HA 966, induced the production of a 70 kDa heat shock protein (HSP-70), which is a marker for intraneuronal vacuoles in at least one neuronal region (Berger *et al.*, 1994). In comparison, no production of vacuoles or HSP-70 was observed in any regions following administration of ACPC (Table 5).

Table 4 Minimum lethal doses of ACPC (LD_{30}, mg kg^{-1})

Species	Acute (i.v.)	7 days (i.v.)	28 days (i.v.)	90 days (p.o.)	Minimal effective dose (mg kg^{-1})
Rat	>5000	>5000	2700	1800	12.5–50[a]
Monkey	>8000	>3500	1800	1800	22[b]

[a]Depending on indication evaluated.
[b]Estimated.

9 PCP-LIKE EFFECTS

A number of experiments have been performed to evaluate the propensity of ACPC to produce phencyclidine (PCP)-like and discriminating stimulus effects (Table 6). The potential of ACPC to induce euphoria or addictive behavior was tested in rats trained to discriminate either MK-801 or phencyclidine from saline by drug lever selection. In one study using Sprague–Dawley rats, ACPC did not mimic the discriminative stimulus effect of either MK-801 or PCP, suggesting that ACPC may be devoid of the subjective side-effects characteristic of NMDA channel ligands (Witkin and Steele, 1992). In another study using Wistar rats (Koek and Colpaert, 1992), the ability of ACPC to block NMDA-induced lethality in mice and NMDA-induced discriminative stimulus in rats was compared with its ability to produce PCP-like behavioral effects in mice and PCP-like discriminative stimulus effects in rats. While

Table 5 Measures of neuronal injury and their induction by neuroprotective agents

Compound and concentration tested	Intraneuronal vacuolization		Induction of HSP-70 protein			
	Cingulate cortex	Retrosplenal cortex	Cingulate cortex	Retrosplenal cortex	Piriform cortex	Dentate hilus
MK-801 (2 mg kg^{-1})	+++	+++	+++	+++	None	None
7 Cl K (50 mg kg^{-1})	NA	NA	None	None	++	None
HA-966 (180 mg kg^{-1})	NA	NA	None	None	++	+++
ACPC (2700 mg kg^{-1})	None	None	None	None	None	None

Adapted from Berger *et al.* (1994).
++, modest induction; +++, induction noted in all animals; NA, data not available.

Table 6 Functional evaluations of ACPC in producing PCP-like and discriminating stimulus effects

Parameter Tested	Results	Reference
Production of PCP-like behavioral effects in Wistar rats as measured by drug lever selection	PCP-like drug lever selection was observed at 630 mg kg^{-1}	Koek and Colpaert (1992)
Production of MK801- and PCP-like behavioral effects in Sprague–Dawley rats as measured by drug lever selection	No PCP-like drug lever selection observed up to the limit tested, 2000 mg kg^{-1}	Witkin and Steele (1992)
Production of PCP-like effects in Wistar rats as measured by locomotion and falling	PCP-like increase of locomotor activity and falling was observed at 630 mg kg^{-1}	Koek and Colpaert (1992)
Locomotor performance in genetically epilepsy-prone (GEP) rats	No effects seen up to the limit tested, 1000 mg kg^{-1}	Smith and Meldrum (1995)
Increase in locomotion in mice	Increase in locomotion was observed above 300 mg kg^{-1}	Trullas and Skolnick (1990)
Production of PCP-like effects in Sprague–Dawley rats and mice as measured by locomotion and falling	No PCP-like increase of locomotor activity and falling was observed up to the limit tested, 600 mg kg^{-1}	Evoniuk *et al.* (1991)
Induction of PCP- and pentobarbital discriminative stimulus (DS) in Wistar rats	No DS was induced for both drugs up to 630 mg kg^{-1}	Koek and Colpaert (1992)
Induction of HA-966 DS in Swiss mice	ACPC and D-cycloserine fully substitute for the DS effect of HA-966	Witkin *et al.* (1995)
Blockade of NMDA-induced DS in Wistar rats	Dose-dependent blockade of NMDA-induced DS	Koek and Colpaert (1992)
Induction of conditioned place preference in mice	No induction of conditioned place preference up to the limit tested, 400 mg kg^{-1}	M. Papp *et al.* (1996)
Prevention of NMDA-induced lethality in mice	Dose-dependent prevention of NMDA toxicity	Koek and Colpaert (1992)

ACPC antagonized the lethal effects of NMDA in a dose-dependent manner, it produced little or no grossly observable PCP-like behavioral effects. At a dose of 630 mg kg^{-1}, ACPC produced drug lever selection in 70% of the rats discriminating PCP from saline (Koek and Colpaert, 1992). This level of ACPC is significantly higher than the minimally effective dose of 25 mg kg^{-1} in this species.

ACPC does not produce ataxia, loss of motor coordination, impairment or muscle relaxation in mice at doses of up to 560 mg kg^{-1} i.p. (Koek and Colpaert, 1992). Frequently, as in the Rotarod test, no effect is seen at 2000 mg kg^{-1} i.p., which is 40 times the minimally effective dose in this species (Symphony, unpublished data). In a study by Trullas and Skolnick (1990), ACPC did produce a small, but significant, increase in locomotor activity in mice at a dose of 300 mg kg^{-1}. However, in another study in mice, no changes in locomotor activity and falling were observed at the highest concentration tested of 600 mg kg^{-1} (Evoniuk *et al.*, 1991). In GEP rats, no increase was observed up to the limit tested of 1000 mg kg^{-1} (Smith and Meldrum, personal communication).

10 PHASE I CLINICAL TRIALS OF ACPC

A phase I clinical trial was conducted to determine the safety, tolerance, and preliminary pharmacokinetic properties of ACPC. In this study, a single intravenous dose of 10 or 20 mg kg^{-1} was administered to a total of 14 healthy male volunteers. Each volunteer entered into the study had no adverse findings from physical examination and history, clinical laboratory tests, and electrocardiography prior to the study, and had no abuse or dependence on psychoactive drugs for at least 2 years prior to the study.

The clearance of ACPC from the plasma following a single intravenous administration is shown in Fig. 18. As was found in rats and monkeys, ACPC readily distributed throughout the body and was excreted with a half-life of elimination of 5.9 h. This half-life is long enough to be compatible with the therapeutic administration of ACPC as either a single-dose injection (for acute treatment) or as a daily pill (for chronic therapy). At the same time this half-life is short enough to avoid accumulation of the drug in

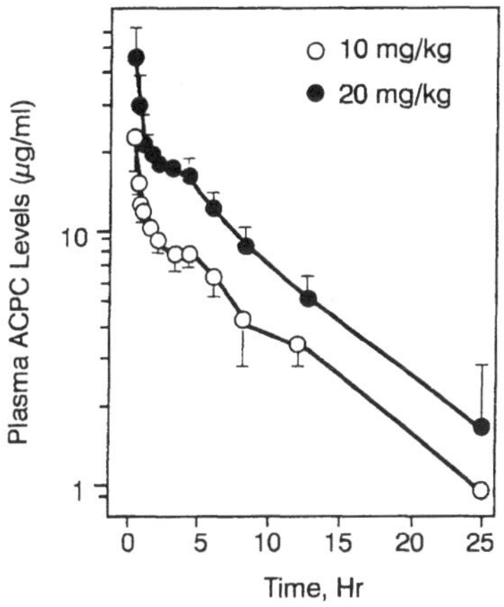

FIG. 18 Clearance of ACPC from human plasma. Normal male volunteers received single intravenous doses of either 10 mg kg^{-1} ($n = 5$) or 20 mg kg^{-1} ($n = 4$) over periods of approximately 7 and 14 min. Blood plasma was collected at selected intervals for 24 h, and ACPC was determined by gas chromatography–mass spectroscopy (Howell *et al.*, 1995). The plasma half-life of ACPC in the studies was calculated to be 5.9 h.

the body. ACPC was secreted intact into the urine; no metabolites were detected. These findings are consistent with the absence of ACPC catabolism in the animal studies.

Each subject received a physical examination, including vital signs measurement, clinical laboratory tests, ECG, electroencephalogram (EEG), psychomotor function, and cognitive function at 24 h after dosing. Throughout the study, no adverse hemodynamic, neurological, or psychiatric signs were noted in any of the subjects, regardless of treatment group. In particular, the subjects exhibited no changes in ECG or EEG recordings, blood chemistry and physiology, and no signs of sedation, agitation, or euphoria. This study demonstrated that the administration of ACPC at dose levels calculated to be efficacious in man, 10 and 20 mg kg^{-1} was safe and well tolerated.

These clinical findings and the supporting animal data suggest that, under therapeutic conditions, ACPC will have a favorable side-effect profile relative to currently available antidepressants and neuroprotective agents.

6 Ifenprodil and Eliprodil: Neuroprotective NMDA Receptor Antagonists and Calcium Channel Blockers

CHRIS CARTER[1], P. AVENET[2], J. BENAVIDES[2], F. BESNARD[1], B. BITON[2], A. CUDENNEC[2], D. DUVERGER[2], J. FROST[2], C. GIROUX[2], D. GRAHAM[1], S. Z. LANGER[2], J. P. NOWICKI[2], A. OBLIN[2], G. PERRAULT[2], S. PIGASSE[2], P. ROSEN[2], D. SANGER[2], H. SCHOEMAKER[2], J. P. THÉNOT[2] AND B. SCATTON[2]

[1]Synthélabo Recherche, 10, rue des Carrières, BP248, 92405 Rueil Malmaison, France
[2]Synthélabo Recherche, 31 avenue Paul Vaillant-Couturier, 92220 Bagneux, France

1 INTRODUCTION

Ifenprodil was originally developed as a moderately potent α_1 adrenoceptor antagonist (Carron *et al.*, 1971) intended for use as a cerebral vasodilator to improve cerebral blood flow. In fact, it was found to constrict cerebral arteries (via a still uncharacterized mechanism) (Young *et al.*, 1981; MacKenzie *et al.*, 1984) and its neuroprotective efficacy,

EXCITATORY AMINO ACIDS – CLINICAL RESULTS WITH ANTAGONISTS
ISBN 0-12-546820-2

initially demonstrated in a cat model of focal ischemia (Gotti *et al.*, 1984) was then interpreted in relation to its potential effects on the cerebral circulation. These were hypothesized as vasoconstrition in healthy brain areas and vasodilatation (or lack of constriction) in the ischemic penumbra, the combined effects of which would tend to increase irrigation in areas of hypoperfusion (MacKenzie *et al.*, 1984). This hypothesis has not in fact been rigorously tested, although ifenprodil does to some extent restore cerebral blood flow in hypoperfused areas of the infarcted cat brain (Delage *et al.*, 1983).

Ifenprodil is rapidly metabolized by the liver and a halogenated back-up (SL82.0715) was synthesized, essentially to limit this hepatic metabolism. This drug, now known as eliprodil (Fig. 1), is in phase II/III clinical trials for stroke and cerebral trauma. Ifenprodil and eliprodil have been revealed as potent noncompetitive *N*-methyl-D-aspartate (NMDA) receptor antagonists with a unique mechanism of action. They act via a polyamine-sensitive modulatory site and exert selective antagonistic effects at a particular NMDA receptor subtype (containing NR2B subunits) and, unlike other nonselective NMDA antagonists, have relatively few side-effects.

In the early 1980s Meldrum's group showed that the competitive NMDA antagonist 2-amino phosphonopentanoic acid, when injected into the hippocampus, markedly reduced the cell death produced by reversible global cerebral ischemia in the rat (Simon *et al.*, 1984), and others showed that glutamate receptors were also involved in hypoxic or hypoglycemic neuronal death (Benveniste *et al.*, 1984, 1989; Wieloch *et al.*, 1985; Weiss *et al.*, 1986; Goldberg *et al.*, 1987a,b, 1988a,b; Monyer and Choi, 1988; Monyer *et al.*, 1989; Tecoma *et al.*, 1989; Choi, 1988a,b, 1992). The neurotoxic effects of glutamate and of the overstimulation of glutamate receptors were well known from the pioneering studies of Olney and others, and a role for glutamate generally suspected in neurodegenerative processes (Olney, 1969, 1981, Olney *et al.*, 1974; Olney and de Gubareff, 1978; Coyle *et al.*, 1981). Nevertheless, the rather startling idea that ischemic neuronal death could be blocked by receptor antagonism overturned a great deal of dogma concerning the inevitability of postischemic neuronal death and, not unnaturally, excited an unprecedented interest in NMDA receptors and drug development to prevent neuronal death. During our screening program, we found that ifenprodil and eliprodil (both of which had been previously shown to be neuroprotective in focal ischemia) were NMDA antagonists with a very atypical *in vitro* and *in vivo* profile (Carter *et al.*, 1988). Very generally, these compounds are potent noncompetitive NMDA antagonists and, like others, neuroprotective in animal models of stroke and cerebral trauma (Gotti *et al.*, 1988, 1990; Toulmond *et al.*, 1993b). However, at neuroprotective doses, they lack the side-effect profile of other types of NMDA antagonists (psychostimulant effects, motor incoordination, amnesia, neurotoxicity, tachycardia, and hypertension) (Perrault *et al.*, 1989; Carter *et al.*, 1991; Sanger and Joly, 1991; Sanger and Jackson, 1989; Balster *et al.*, 1994; Giroux *et al.*, 1994; Scatton *et al.*, 1994a,b). Over the years, we have learnt more about these compounds, which interact with a polyamine-sensitive modulatory site on the NMDA receptor (Reynolds and Miller, 1989; Carter *et al.*, 1990) and which are, furthermore, selective antagonists of NMDA receptors containing the NR2B subunit (Williams, 1993b). They are also potent σ ligands (Schoemaker *et al.*, 1990; Contreras *et al.*, 1990; Karbon *et al.*, 1990; Hashimoto and London, 1995) and relatively potent blockers of voltage-operated L-, N-, and P-type neuronal calcium channels (Biton *et al.*, 1994, 1996; Church *et al.*, 1994).

The neuroprotective effects of these compounds are very likely due to a fortuitous blend of NMDA receptor and calcium channel antagonism that shuts down the potentially disastrous excessive influx of calcium ions into the ischemic neuron. The fact that these compounds do not produce the side-effects associated with NMDA receptor blockade likely reflects their selective antagonism of particular NMDA receptor subtypes (predominantly those containing the NR2B subunit).

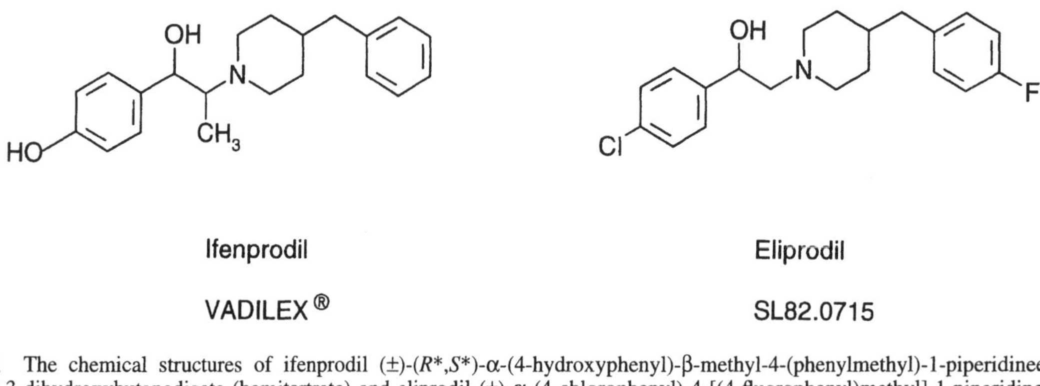

FIG. 1 The chemical structures of ifenprodil (±)-(*R**,*S**)-α-(4-hydroxyphenyl)-β-methyl-4-(phenylmethyl)-1-piperidineethanol-(*R*,*R*)-2,3-dihydroxybutanedioate (hemitartrate) and eliprodil (±)-α-(4-chlorophenyl)-4-[(4-fluorophenyl)methyl]-1-piperidineethanol (hydrochloride).

The pharmacological profiles of ifenprodil and eliprodil and their neuroprotective effects are reviewed below.

2 THE NMDA RECEPTOR COMPLEX AND ITS REGULATION

The structure of the NMDA receptor is reviewed in greater detail in other chapters and reviews, but a brief definition of the various modulatory sites of the NMDA receptor is perhaps pertinent in relation to the effects of ifenprodil and eliprodil and to their relations with the polyamine site.

The NMDA receptor is composed of heteromeric assemblies of NR1 and NR2 subunit families. Eight splice variants of NR1 exist and four NR2 subunits coded by different genes have been reported (Moriyoshi *et al.*, 1991; Durand *et al.*, 1992, 1993; Monyer *et al.*, 1992; Yamazaki *et al.*, 1992; Hollmann *et al.*, 1993; Karp *et al.*, 1993; Koltchine *et al.*, 1993; Kusiak and Norton, 1993; Le Bourdelles *et al.*, 1994). A futher subunit (NMDAR-L or χ-1) has recently been cloned but not yet fully characterized (Ciabarra *et al.*, 1995; Sucher *et al.*, 1995). The stoichiometry of the final assembly is not known, but native receptors composed of one NR1 and two distinct NR2 subunits have been identified (Chazot *et al.*, 1994), and similar functional recombinant triplets can be transfected (Wafford *et al.*, 1993). The receptor is unique in that it requires dual-key activation by two agonists, glutamate and glycine (Johnson and Ascher, 1987). These work at pharmacologically distinct sites each with their own particular range of agonists and competitive antagonists (Watkins, 1991; Kemp and Leeson, 1993). Receptor activation opens a channel permeable to sodium, calcium, and potassium ions (MacDermott *et al.*, 1986; Mayer and Westbrook, 1987a). The channel is gated by magnesium ions, which block it at negative potentials (below ~–40 mV). Channel block by magnesium is simply relieved by depolarization (Mayer *et al.*, 1984; Nowak *et al.*, 1984). The size of the NMDA current, as well as being related to the strength of the agonist signals, is also thus dictated by the existing depolarization state of the neuron, and in theory modulable by any other signals that either hyperpolarize or depolarize the neuron within the receptive field of the magnesium sensor. (For a review, see Mayer and Westbrook (1987b).) The NMDA channel can be blocked by compounds such as phencyclidine, 1-[2-(2-Thienyl) cyclohexyl]piperidine (TCP) or MK-801 (Wong *et al.*, 1988). The receptor is also controlled by distinct sites sensitive to zinc (Westbrook and Mayer, 1987), pH (Giffard *et al.*, 1990), redox status, and arachidonic acid (Aizenman *et al.*, 1989), and by phosphorylation (Moon *et al.*, 1994).

Polyamines (spermine and spermidine) exert four distinct effects on the NMDA receptor, two of which are stimulatory and two of which will inhibit NMDA responses (Ransom and Stec, 1988). They increase NMDA responses by increasing the affinity of the glycine site for agonists (Sacaan and Johnson, 1989). They decrease glycine site affinity for antagonists (Grimwood *et al.*, 1994; Yoneda *et al.*, 1994). They also increase NMDA responses in the presence of saturating glycine concentrations by increasing the frequency of NMDA channel opening (Rock and MacDonald, 1992a) and by lowering desensitization onset rates (Lerma, 1992) (glycine-independent potentiation). They decrease NMDA responses by lowering the affinity of the glutamate site for agonists (Pullan and Powell, 1991) (increasing affinity at this site for antagonists (Carter *et al.*, 1990)), and at high millimolar concentrations block the NMDA channel in a voltage-dependent manner (Rock and Macdonald, 1992b; Benveniste and Mayer, 1993). Certain of these effects appear to be mediated by specific subtypes, as summarized in Table 1 (Durand *et al.*, 1992, 1993; Williams, 1993b, 1995; Williams *et al.*, 1994; Zukin and Bennett, 1995). For example, only NR1 splice variants without the 5′ end insertion (NR1 a or NR1-0*xx*) show glycine-independent polyamine stimulation while both forms (NR1-a and NR1-b or NR1-0*xx* and NR1-1*xx*) show glycine-dependent polyamine stimulation and polyamine channel block. Using an NR1 receptor (NR1-a) that shows both glycine-dependent and -independent forms of polyamine stimulation as well as polyamine channel block, site-directed mutagenesis studies have shown that these effects are mediated by distinct sites on this NR1 subunit. Acidic to basic amino acid changes at E339 or E342 reduce glycine-independent stimulation by spermine without affecting glycine-dependent stimulation or channel block (Williams *et al.*, 1996). The inhibitory effect of polyamines mediated by reduction of glutamate site affinity has not been studied at all splice variants. The NR2 subunits play a major role in modifying the properties of the NR1 splice variants as all polyamine effects appear to be lost on association of NR1-a with NR2C or NR2D subunits. The NR2B association permits expression of all stimulatory and inhibitory polyamine effects, but NR2A associations only demonstrate glycine-dependent polyamine stimulation and channel block. It is important to appreciate that these molecular distinctions mean that the polyamine sites controlling their four different effects are distinctly localized at the molecular level and probably pharmacologically dissociable.

3 ACTIONS OF IFENPRODIL AND ELIPRODIL AT DIFFERENT SITES OF THE NMDA RECEPTOR

The polyamines and ifenprodil or eliprodil interact with their own particular binding sites and allosterically modify the actions of agonists and antagonists at the glutamate and glycine recognition sites. These interactions are summarized below, and their complexity illustrated in Fig. 2.

Table 1 A summary of the effects of polyamines (PA), ifenprodil and eliprodil at different NMDA receptor subunits[a]

	Glycine-independent stimulation (PA)	Glycine-dependent stimulation (PA)	Channel block (PA)	Decreased glutamate affinity (PA)	Ifenprodil sensitivity	Eliprodil sensitivity
NR1-011	+	+	+	+	+	NT
NR1-001	NT	NT	NT	NT	+	NT
NR1-010	+	+	+	NT	+	NT
NR1-000	+	+	+	NT	+	NT
NR1-111	–	+	+	NT	+	NT
NR1-101	–	+	+	NT	+	NT
NR1-110	NT	NT	NT	NT	+	NT
NR1-100	–	+	+	NT	+	NT
NR1/2A	–	+	+	–	–	–
NR1/2B	+	+	+	+	+	+
NR1/2C	–	–	–	–	–	NT
NR1/2D	–	–	–	–	–	NT

NT, not tested.

[a] For the NR1 splice variants, the effects of ifenprodil were tested on the NR1-a (NR1-0XX) homomer and in all other cases on the NR1 splice variant in association with NR2B. Data compiled from Durand *et al.* (1993), Williams (1993b, 1994, 1995), Williams *et al.* (1994), and Gallagher *et al.* (1995).

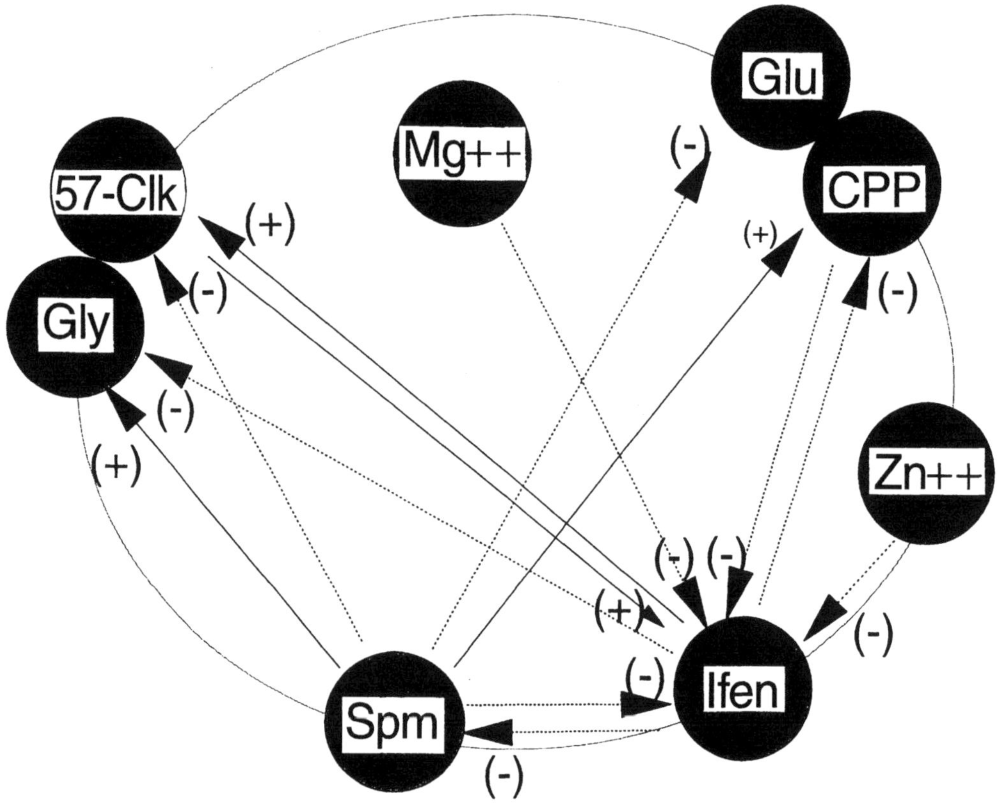

FIG. 2 A summary of the allosteric interactions between polyamines and ifenprodil and the glycine and glutamate sites of the NMDA receptor. Glutamate–glycine interactions have been omitted. Solid lines represent positive modulation (increases in binding of the target ligand) and dotted lines represent negative modulation (decreases in binding of the target ligand).

3.1 [³H]Ifenprodil and [³H]eliprodil binding

Ifenprodil displaces the σ ligand [³H]3-(3-hydroxyphenyl)-*N*-(1-propyl)piperazine ([³H](+)-3-PPP) (Contreras *et al.*, 1990; Karbon *et al.*, 1990) and at 37 °C [³H]ifenprodil binds with high affinity to a σ site (Schoemaker *et al.*, 1991, 1994) (see below). At 4 °C and in the presence of σ ligands to mask this site, high-affinity [³H]ifenprodil binding to a polyamine-sensitive site on the NMDA receptor complex can be demonstrated (Schoemaker *et al.*, 1990). [³H]Ifenprodil is totally displaced by ifenprodil, eliprodil, isoxsuprine, nylidrin, spermine and spermidine, and other compounds as listed in Table 2. NMDA channel blockers (e.g. MK-801 or phencyclidine) have no effect on [³H]ifenprodil binding although zinc and magnesium partially displace [³H]ifenprodil. Scatchard analysis has shown that spermine increases the K_d for ifenprodil without affecting B_{max}, suggesting a competitive form of interaction (Schoemaker *et al.*, 1990). Complex kinetics (biphasic association curves) have rendered more thorough analysis difficult.

The polyamine-sensitive ifenprodil site is allosterically modified by ligands acting at the glycine and glutamate sites of the NMDA receptor (Schoemaker *et al.*, 1994). Thus, competitive NMDA antagonists partially reduce [³H]ifenprodil binding in a glutamate-reversible manner. Glycine site antagonists increase [³H]ifenprodil binding in a glycine-reversible manner. Neither glutamate or glycine *per se* appear to affect [³H]ifenprodil binding (Fig. 3). Polyamine-sensitive [³H]ifenprodil binding is enriched in the synaptosomal fraction (Hashimoto *et al.*, 1992).

Polyamine-sensitive [³H]ifenprodil binding has also been demonstrated in the human frontal cortex (Schoemaker *et al.*, 1991), and shows pharmacological properties compatible with its association within the NMDA receptor complex. [³H]Ifenprodil binding to the human brain is of high affinity (K_d = 56nM; B_{max} = 40 fmol (mg tissue^{-1}), and is fully inhibited by unlabeled ifenprodil, eliprodil, spermine, and spermidine. In addition, [³H]ifenprodil binding, as in the rat cerebral cortex, is partially inhibited by the competitive NMDA antagonist, CGS19755.

The pharmacological characteristics of [³H]ifenprodil binding to membranes from the rat cerebellum were found to be markedly different from those seen in the rat or human cortex (Schoemaker *et al.*, 1992; Schoemaker and Pigasse, 1993). The affinity of [³H]ifenprodil binding was significantly lower in the cerebellum than in the rat cerebral cortex and the inhibitory effects of eliprodil (Fig. 4), polyamines, pentamidine, neomycin, Zn^{2+} and Mg^{2+} were considerably reduced. Moreover, competitive NMDA or glycine antagonists failed to affect [³H]ifenprodil binding to membranes from the adult rat cerebellum. These data clearly show a heterogeneity of polyamine-sensitive [³H]ifenprodil binding sites in the rat brain, which may reflect the different regional distribution of NMDA receptor subunits.

The pharmacological characteristics of polyamine-sensitive [³H]ifenprodil binding to membranes from the neonatal rat cerebral cerebellum, where its noncompetitive NMDA antagonist properties had previously been demonstrated (versus NMDA-evoked cGMP production) (Carter *et al.*, 1988), were found to be virtually identical to those of the adult cerebral cortex (Schoemaker *et al.*, 1992; Schoemaker and Pigasse, 1993). It would thus appear that the pharmacological properties of the polyamine-sensitive [³H]ifenprodil-binding site in the neonatal rat cerebellum resemble those of the immature or adult cerebral cortex but evolve during development. The affinity of [³H]ifenprodil binding to cerebellar membranes decreases markedly during the third postnatal week and is accompagnied by a significant increase in the maximal binding density (Fig. 5). The allosteric coupling of the [³H]ifenprodil binding site with the glutamate recognition domain of the NMDA

Table 2 The pharmacology of [³H]ifenprodil binding to a polyamine-sensitive site on the NMDA receptor

	IC_{50}(μM)	(+) Modulators	EC_{50}(μM)	E_{max} (%)
Ifenprodil	0.046	5,7-Dichlorokynurenate	0.076	155
Nylidrin	0.052	7-Chlorokynurenate	0.36	137
Eliprodil	0.13	CNQX	0.71	142
Isoxsuprine	0.14	DNQX	1.5	150
Pentamidine	3.5	HA966	>100	–
Neomycin	5.1	(–) Modulators	IC_{50}(μM)	I_{max} (%)
Spermine	8.4	CGP37849	1.7	47
Spermidine	70	D-CPP	4.75	28
Arcaine	72	CGS19755	5.42	33
Agmatine	510	D-AP5	44.7	22
Putrescine	>1000	Zn^{2+}	16	24
MK-801/TCP/PCP*	>1000	Mg^{2+}	114	16

*PCP, phencyclidine.

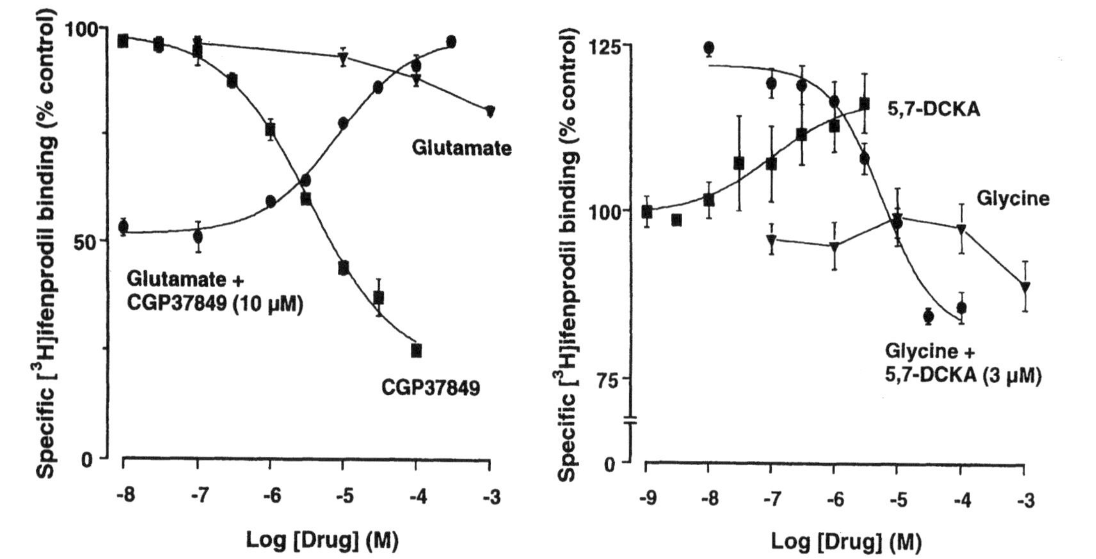

FIG. 3 Allosteric modulation of polyamine-sensitive [^{3}H]ifenprodil binding through the glutamate and glycine recognition domains of the NMDA receptor. [^{3}H]Ifenprodil binding to membranes from the rat cerebral cortex was studied at 0 °C in 5 mM Tris–HCl buffer supplemented with 3 μM of GBR12909. While glutamate fails to affect [^{3}H]ifenprodil binding, the competitive glutamate antagonist CGP37849 partially inhibits binding with an IC_{50} of 3.8 μM in a glutamate-reversible manner. Similarly, glycine fails to affect [^{3}H]ifenprodil binding under control conditions but opposes (IC_{50} = 5.2 μM) the stimulatory effect (EC_{50} = 0.15 μM) of the competitive glycine antagonist 5,7-chlorokynurenate (5,7-DCKA).

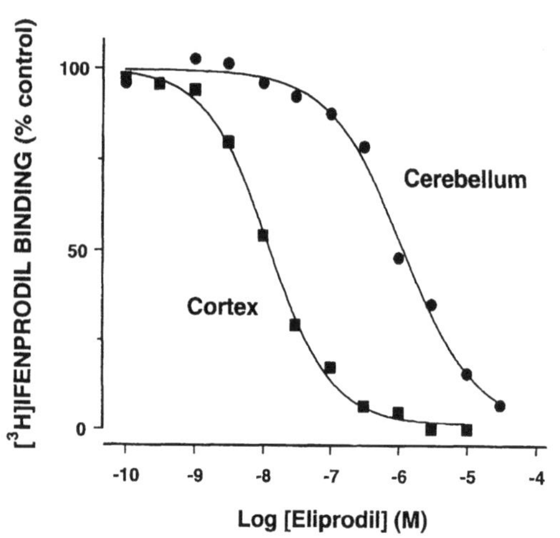

FIG. 4 Comparative effects of eliprodil against [^{3}H]ifenprodil binding in the rat cerebral cortex and cerebellum. [^{3}H]Ifenprodil binding was studied at 4 °C in 5 mM Tris–HCl and 3 μM GBR12909 using membranes from the adult rat cerebral cortex and cerebellum. Whereas eliprodil inhibits binding with an IC_{50} of 0.041 μM in the cerebral cortex, its potency is greatly reduced in the adult rat cerebellum (IC_{50} = 1.8 μM).

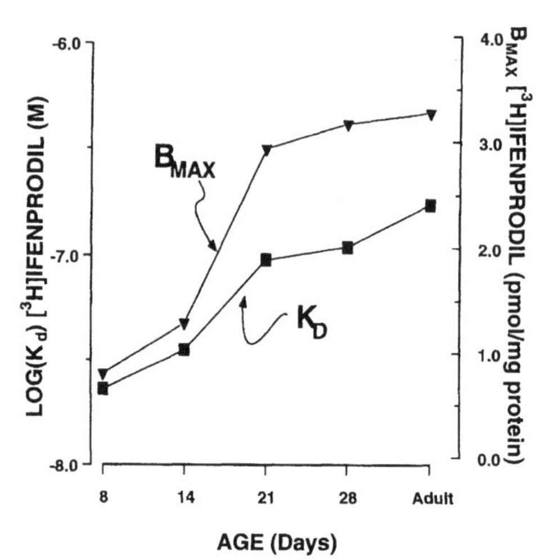

FIG. 5 Characteristics of polyamine-sensitive [^{3}H]ifenprodil binding to the rat cerebellum during development. [^{3}H]Ifenprodil binding to membranes from the cerebellum of the 8-day, 14-day, 21-day, and 28-day-old and adult (2–3-month-old) rat was studied at 0 °C in the presence of 3 μM GBR12909 in 5 mM Tris–HCl buffer (pH 7.4). Shown are the affinity (K_D) and maximal binding density (B_{max}) of specific [^{3}H]ifenprodil binding as determined by saturation analysis.

receptor, as judged from the inhibitory effects of CGP37849, is lost mainly during the fourth postnatal week. The ontogenetic development of the polyamine-sensitive [^{3}H]ifenprodil-binding site in the cerebellum, i.e. the decrease in affinity of ifenprodil, is reflected at the level of [^{3}H]MK-801 binding by the disappearance of the high-affinity phase of inhibition by ifenprodil (Schoemaker *et al.*, 1992; Schoemaker and Pigasse, 1993).

[^{3}H]Eliprodil also binds with high affinity to a polyamine-sensitive site. Briefly, [^{3}H]eliprodil binding to

Table 3 Pharmacological characteristics of polyamine-sensitive [^{3}H]eliprodil binding to the rat cerebral cortex[a]

Drug	EC_{50}(μM)	E_∞ (% control)
Eliprodil	0.063	0
Ifenprodil	0.014	0
Spermine	2.5	0
Spermidine	6.5	0
Arcaine	7.6	0
Putrescine	340	0
Neomycin	1.4	0
CGS19755	8.6	32
5,7-Dichlorokynurenate	0.26	131

[a]The pharmacological characteristics of [^{3}H]eliprodil binding were studied at 0 °C in 5 mM Tris–HCl buffer (pH 7.4) supplemented with 3 μM GBR12909. Data are shown as the EC_{50} values for the enhancement (5,7-dichlorokynurenate) or inhibition of [^{3}H]eliprodil binding, as well as [^{3}H]eliprodil binding observed in the presence of a maximally effective drug concentration in comparison to controls (E_∞).

the rat cerebral cortex is inhibited with high affinity by ifenprodil and unlabeled eliprodil as well as by different polyamines with a potency (Table 3) virtually identical to that previously demonstrated for [^{3}H]ifenprodil binding (Oblin *et al.*, 1994a). Polyamine-sensitive [^{3}H]eliprodil binding appears to be associated with the NMDA receptor, as it is stimulated by the glycine antagonist 5,7-dichlorokynurenate, and partially inhibited by the glutamate antagonist CGS19755.

3.1.1 Autoradiographic distribution of polyamine-sensitive [^{3}H]ifenprodil-binding sites

In rat brain sections, when σ sites are masked by GBR12909 and (+)-3-PPP, [^{3}H]ifenprodil binds to two sites of high and low affinity. Both are sensitive to ifenprodil, eliprodil, and to the polyamines spermine and spermidine as well as to the aminoglycoside antibiotics gentamycin, kanamycin, and neomycin. These antibiotics have been suggested as polyamine agonists (Pullan *et al.*, 1992). The high-affinity ifenprodil site is essentially localized to the forebrain, and is not observed in the cerebellum or hindbrain areas. The low-affinity site was found to be selectively sensitive to trifluoperazine, chlorpromazine, and prenylamine, and can be masked by 1 mM trifluoperazine. In the presence of trifluoperazine, the distibution of the remaining high-affinity [^{3}H]ifenprodil sites is identical to that of NR2B mRNA (Fig. 6) (Nicolas and Carter, 1994). We are not certain as to the identity of the low-affinity site which could reflect low-affinity binding of ifenprodil to other NMDA receptors, in which case one might expect trifluoperazine, chlorpromazine, and prenylamine to be selective (but weak) antagonists at certain NMDA receptor subtypes. Alternatively, it may represent a polyamine-sensitive site at another target. Trifluoperazine, chlorpromazine, and prenylamine are calmodulin antagonists, and [^{3}H]ifenprodil was found to bind with low affinity to the calmodulin protein (Nicolas and Carter, 1994), and it is possible that the low-affinity site represents a form of calmodulin.

3.2 Effects of ifenprodil on the glycine site

[^{3}H]Glycine binding in rat cortical membranes is increased by polyamines (Sacaan and Johnson, 1989) and reduced by ifenprodil (Ransom, 1991). Ifenprodil reduces the stimulatory effects of polyamines, although not in a competitive manner, while polyamines shift ifenprodil inhibition curves

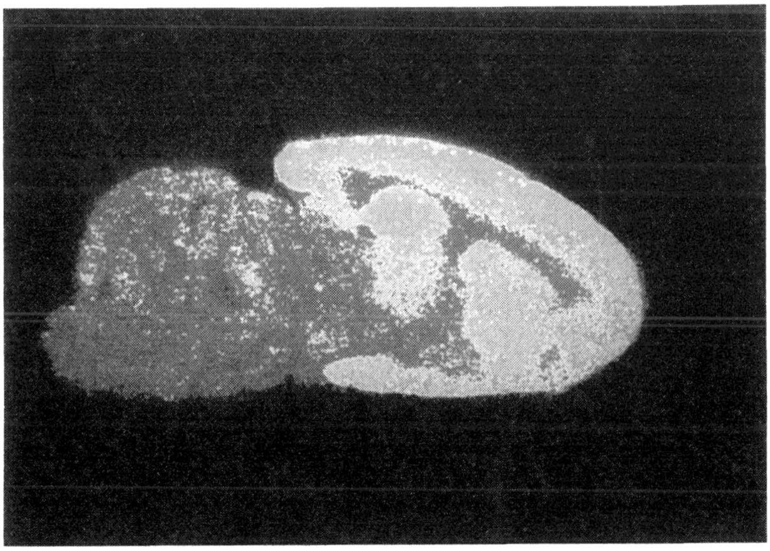

FIG. 6 The distribution of a high-affinity polyamine-sensitive [^{3}H]ifenprodil site in the rat brain. Low-affinity sites were masked with 1 mM trifluoperazine. The distribution of this site corresponds closely to that of NR2B mRNA.

to the right, and clearly reduce the affinity of ifenprodil in relation to its effects on [^{3}H]glycine binding (Ransom. 1991). These two effects, redrawn from the data from Ransom (1991), are shown in Fig. 7. It is clear from this figure that polyamines will modify the potency of ifenprodil as a modulator of glycine site affinity. This is particularly important in understanding the reasons for the different potencies of ifenprodil in functional models when this glycinergic site regulation is involved (see below and Table 4).

Ifenprodil and polyamines also have opposing effects on the binding of glycine site antagonists. Polyamines decrease and ifenprodil increases the binding of the glycine antagonist [^{3}H]5,7-dichlorokynurenate (Ogita *et al.*, 1992; (Yoneda *et al.*, 1994). Again the polyamine–ifenprodil interaction appears to be complex. Ifenprodil reduces the potency of spermidine as a displacer of [^{3}H]5,7-dichlorokynurenate, but this shift is maximal at 10 μM ifenprodil. Spermidine reduces the affinity of the [^{3}H]5,7-dichlorokynurenate site and increases the B_{max}, while ifenprodil has a minor incremental effect on B_{max}. The affinity of the [^{3}H]5,7-dichlorokynurenate site is not increased by ifenprodil. These data are summarized in Table 4.

Table 4 The effects of spermine and of ifenprodil on [^{3}H]5,7-dichlorokynurenate binding to the glycine modulatory site of the NMDA receptor

(a) The effects of ifenprodil on the inhibitory potency of spermidine as a partial displacer of [^{3}H]5,7-dichlorokynurenate

	[^{3}H]5,7-Dichlorokynurenate IC_{50} (μM)
Spermidine	500 ± 70
Spermidine + ifenprodil 10 μM	8900 ± 610*
Spermidine + ifenprodil 100 μM	6120 ± 490*

(b) The effects of spermidine or ifenprodil on the affinity and number of [^{3}H]5,7-dichlorokynurenate-binding sites

	K_d (nM)	B_{max} (pmol) (mg protein)$^{-1}$
Control	27.5 ± 2.4	2.87 ± 0.17
+ Spermidine 1 mM	199.6 ± 26.4*	7.57 ± 0.73*
+ Ifenprodil 10 μM	36.0 ± 3.6	3.83 ± 0.009*

Reproduced, with permission from Yoneda *et al.* (1994).
* $p < 0.05$ compared with spermidine alone (a) or control (b).

In our own studies (Oblin *et al.*, 1994b), spermine partially decreased [^{3}H]5,7-dichlorokynurenate binding to membranes from the rat cerebral cortex with an IC_{50} of 7.3 μM. Although ifenprodil and eliprodil, up to concentrations of 100 μM, did not affect binding under control conditions, these drugs increased binding with EC_{50} values of 0.16 and 0.45 μM, respectively, when studied in the presence of 10 μM spermine (Fig. 8). It would thus appear that a certain amount of polyamine tone has to exist for the effects of ifenprodil and eliprodil to be manifest. Ifenprodil and eliprodil have also been reported to increase the binding of the glycine antagonist [^{3}H]L-689,560, although polyamine interactions have not been reported (Grimwood *et al.*, 1992, 1994).

3.3 Effects of ifenprodil on the glutamate antagonist recognition site

Polyamines increase and ifenprodil reduces [^{3}H]3-(2-carboxypiperazine-4-yl)-propyl-1-phosphonic acid ([^{3}H]CPP) binding to the glutamate site of the NMDA receptor. The effects of spermine or spermidine are blocked by ifenprodil (Carter *et al.*, 1988). It has also been reported that polyamines reduce the affinity of agonist binding to the NMDA receptor, but the effects of ifenprodil on agonist binding have not been characterized (Pullan and Powell, 1991).

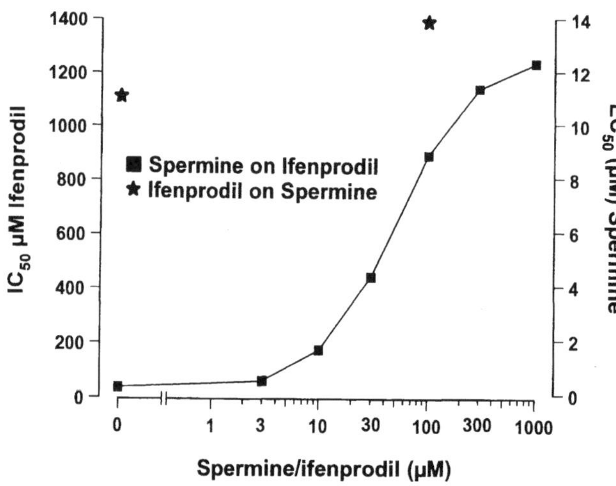

FIG. 7 The effects of spermine on the potency of ifenprodil as a partial displacer of [^{3}H]glycine binding and of ifenprodil on the potency of spermine as a stimulator of [^{3}H]glycine binding. (Redrawn, with permission, from Ransom (1991).)

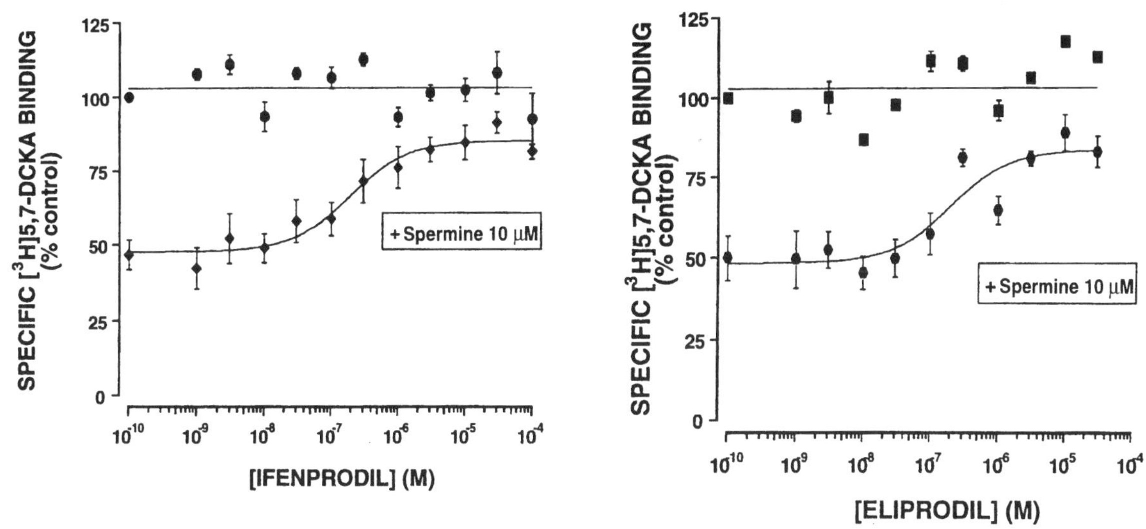

FIG. 8 Effects of ifenprodil and eliprodil on the inhibition of [^{3}H]5,7-dichlorokynurenate ([^{3}H]5,7-DCKA) binding by spermine. The effects of ifenprodil and eliprodil were studied on [^{3}H]5,7-dichlorokynurenate binding to membranes from the rat cerebral cortex under control conditions and in the presence of 10 μM spermine. Whereas both compounds fail to affect binding under control conditions, they selectively oppose the inhibitory effects of spermine, with EC_{50} values of 0.16 and 0.45 μM, respectively.

Ifenprodil does not reduce the binding of another competive NMDA antagonist, [^{3}H]CGP39653. Spermine potentiates [^{3}H]CGP39653 binding (in the presence of 7-chlorokynurenate), and this effect is ifenprodil-insensitive (Oblin and Schoemaker, 1994). Interestingly, although CGP39563 displaces [^{3}H]glutamate from transfected NR1 homomers and from NR1/2A, NR1/2B, NR1/2C, and NR1/2D dimers, [^{3}H]CGP39653 binding was only detectable at NR1/2A receptors (at which CGP39653 had the lowest K_i versus [^{3}H]glutamate (Laurie and Seeburg, 1994). Assuming that CGP39563 binds predominantly to native NMDA receptors containing the 2A subunit and ifenprodil binds to those containing the 2B subunit, a lack of effect of ifenprodil on [^{3}H]CGP39653 binding is perhaps to be expected. However, an indirect polyamine-related effect of ifenprodil on [^{3}H]CGP39653 binding can be observed via the glycine site. Glycine partially inhibits the binding of [^{3}H]CGP39653 (Sills *et al.*, 1991; Pingping *et al.*, 1993; Oblin and Schoemaker, 1994), and this effect is potentiated by spermine (presumably via its ability to increase the affinity of the glycine site for glycine). Ifenprodil and arcaine both inhibit this effect of spermine (Fig. 9) (Oblin and Schoemaker, 1994).

There are thus clear dissociations of many different

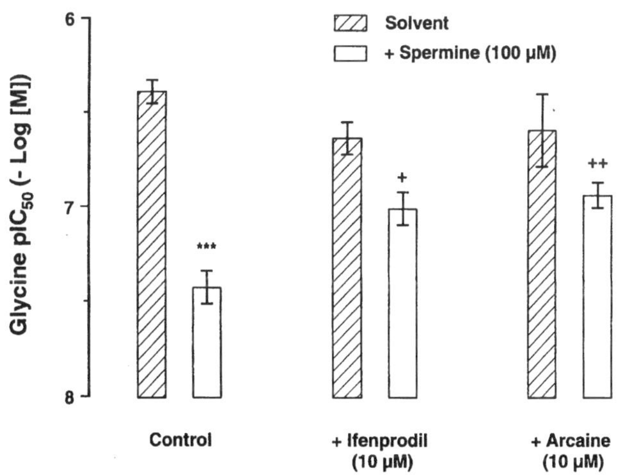

FIG. 9 Effect of putative polyamine antagonists on the inhibition of [^{3}H]CGP39653 binding by glycine. The inhibitory effects of glycine against [^{3}H]CGP39653 binding to membranes from the rat brain were studied under control conditions and in the presence of 100 μM spermine. Data are represented as mean and SEM of at least three experiments.$^{***}p < 0.001$ versus pIC_{50} glycine under control/solvent conditions;$^{+}$ $p < 0.05$ and $^{++}$ $p < 0.01$ versus pIC_{50} glycine under control/spermine conditions.

polyamine effects on NMDA antagonist site binding. Ifenprodil does not block the direct stimulatory effects of polyamines on [^{3}H]CGP39653 binding (but does block the stimulatory effects of polyamines on [^{3}H]CPP binding), and blocks the reduction in [^{3}H]CGP39653 binding mediated by polyamines via the glycine site. Given the argument that [^{3}H]CGP39653 may selectively label NR2A containing receptors, one wonders whether an indirect high-affinity effect of ifenprodil at NR2A-containing receptors could be observed via polyamine effects on the glycine site.

3.4 Effects of ifenprodil on [^{3}H]MK-801 binding

[^{3}H]MK-801 binds with high affinity to a site within the NMDA channel. In the absence of agonists that stimulate channel opening, binding to this site equilibrates extremely slowly over ~24 h. The addition of various combinations of glutamate, glycine, or polyamines opens the NMDA channel, increases access to the [^{3}H]MK-801 binding site, and thus increase its overall association rate (K_{obs}) (Kloog *et al.*, 1988; Marvizon and Baudry, 1994; Javitt and Zukin, 1995). Under subequilibrium conditions, glutamate, glycine, and the polyamines thus increase [^{3}H]MK-801 binding, which approximates to an index of NMDA channel opening. Polyamines, in addition to their acceleration of the apparent association rate, can directly increase the affinity of the MK-801-binding site, and at high concentrations will also displace [^{3}H]MK-801. Polyamine dose–response curves in such studies are thus bell shaped (Ransom and Stec, 1988). The dose–response curves generated and the effects of various antagonists can be a useful means of analyzing the interactions between the various sites on the NMDA receptor. Many of the various effects of the polyamines or ifenprodil on the NMDA receptor (shifts of glycine and glutamate site affinity and channel block) have their parallels in MK-801 binding effects. However, because of the multiple interactions between the various different sites on the NMDA receptor, and the direct effects of polyamines on the MK-801 site itself, interpretation of data collected from different laboratories each using different conditions is fraught with difficulty and has led to much contention in the field. However, in a thorough investigation on the effects of spermine and ifenprodil on the association and dissociation kinetics of [^{3}H]MK-801, Marvizon and Baudry (1994) showed that low concentrations of ifenprodil, which did not modify [^{3}H]MK-801 association kinetics (K_{obs}) in the absence of spermine, totally reversed the increase in the apparent association rate produced by spermine, thus indicating a specific and potent inhibition of at least one of the stimulatory effects of polyamines by ifenprodil. Thus, this particular action of ifenprodil is clearly polyamine-dependent in that a modulatory agonist is required before any high-affinity effects of ifenprodil can be observed.

At high concentrations, and at equilibrium, ifenprodil displaces [^{3}H]MK-801 (IC_{50} ~100 μM). This effect is unrelated to modulation of stimulatory glutamate, glycine, or polyamine sites, and probably reflects an action at the polyamine channel blocking site (Marvizon and Baudry, 1994). Equivalent channel blocking effects of high concentrations of ifenprodil (and polyamines) can be observed in electrophysiological studies (Williams, 1993b; Legendre and Westbrook, 1991).

3.5 Functional consequences of allosteric interactions between the ifenprodil, glycine, and glutamate sites

One of the major effects of ifenprodil is to reduce the affinity of the glycine site for agonists. This effect is translated in functional terms to a shift of glycine dose–response curves to the right until the maximal effects of ifenprodil (lowest possible modulable affinity of the glycine site) are reached. As modifications of glycine site affinity have no relevance if the glycine site is saturated, increasing glycine concentrations can overcome the inhibitory effects of ifenprodil when mediated by this interaction. Ifenprodil will thus not block (certain) NMDA receptors if the glycine site is saturated. The modulatory effects of ifenprodil at the glycine site appear to be present at NR1 and NR1/NR2B receptor transfects, where the inhibitory effects of ifenprodil are reduced by glycine, and can also be observed at native NMDA receptors. For example, in the immature rat spinal cord where glycine levels are saturating, ifenprodil has little effect on NMDA responses. Ifenprodil also fails to block the NMDA-evoked release of spermine and spermidine in the striatum *in vivo*. In both situations, an inhibitory effect of ifenprodil can be revealed if the effects of glycine are pharmacologically reduced by the inclusion of a glycine site antagonist in the perfusing medium (Carter *et al.*, 1992; Voltz *et al.*, 1994).

There is an evident puzzle in the glycine–ifenprodil relationships as reported in the literature. At transfected NMDA receptors (NR1/2B) the high-affinity inhibitory effect of ifenprodil is glycine-sensitive (Williams, 1993b), while in hippocampal cells in culture, where ifenprodil inhibition curves are biphasic, it is the low-affinity effect of ifenprodil that can be reversed by glycine (Legendre and Westbrook, 1991). The high-affinity effect is glycine-insensitive. It is possible that the high- or low-affinity glycine-dependent effects of ifenprodil are related to differing polyamine concentrations as would be predicted from the ability of spermine to reduce the affinity of ifenprodil as a modulator of [^{3}H]glycine binding (a lower-affinity, glycine-dependent effect of ifenprodil may simply reflect higher polyamine concentrations, see the effect of spermine on the affinity of ifenprodil as a modulator of glycine binding (Fig. 7)). It may also be related to subtype-specific effects. For example, the glycine-dependent effects of polyamines are observed at both NR2A- and

NR2B-containing receptors, for which ifenprodil has low- and high-affinity respectively. Thus, the low-affinity glycine-sensitive effects of ifenprodil in the hippocampus may reflect the presence of 2A subunits while high-affinity glycine-dependent effect of ifenprodil may reflect a higher density of 2B subunits. Nevertheless, this still leaves the high-affinity glycine-independent effects of ifenprodil, the mechanism of action and subtype definition of which remain to be characterized. Neither of the high-affinity effects of ifenprodil are related to the additional low-affinity voltage-dependent channel-blocking effects of ifenprodil observed at high (~100 μM) concentrations.

The various glycine–polyamine interactions also predict that the inhibitory effects of glycine antagonists and ifenprodil (each of which potentiate the binding of each other) should synergize, as demonstrated by *in vivo* dialysis studies (Voltz *et al.*, 1994). In contrast, certain competitive antagonists and ifenprodil reduce the binding of each other, although whether competitive antagonists can reverse the inhibitory effects of ifenprodil or vice versa has not been tested. Further work needs to be done to examine whether glutamate increases the inhibitory potency of ifenprodil (the reverse corollary of NMDA antagonists reducing the affinity of ifenprodil). In addition, it is not yet known whether ifenprodil inhibits the reduction in glutamate site affinity provoked by polyamines.

3.6 Selective antagonism or NMDA receptors containing the NR2B subunit

Ifenprodil blocks the NMDA current at the NR1 homomer with high affinity, and an ifenprodil-binding site is thus present on the NR1 subunit (Williams, 1993b; Williams *et al.*, 1993). The mutations of the NR1-a subunit that reduce glycine-independent polyamine stimulation also reduce the inhibitory effects of ifenprodil, suggesting that at least one polyamine and one ifenprodil site overlap (Williams *et al.*, 1996). Perhaps we can assume that the glycine-independent effects of ifenprodil are mediated via an NR1 subunit. Ifenprodil blocks all NR1 splice variants with high affinity (studied in association with the NR2B subunit) (Gallagher *et al.*, 1995).

At NR1A/NR2A recombinant receptors, ifenprodil blocks NMDA currents with low affinity (IC_{50} = 146 μM) in a voltage-dependent manner. Depolarization reduces the ifenprodil block (Williams, 1993a), and ifenprodil would thus appear to be a weak open-channel blocker at this subtype. This activity is reflected by the low-affinity displacement of [^{3}H]TCP or MK-801 by ifenprodil in binding studies to native membrane preparations. The effects of ifenprodil at NR1/2C or at NR1A/ε4 (ε4 being the mouse equivalent of rat NR2D) are also of low affinity (Williams *et al.*, 1993; Williams, 1994, 1995) but have not been biophysically characterized although it is supposed that they are also related to channel block.

At the NR1A/NR2B subtype, ifenprodil blocks NMDA currents with high affinity in a glycine-reversible manner (IC_{50} = 0.34 μM; Williams, 1993b). Thus the glycine-dependent high-affinity effects of ifenprodil are a property of the NR1A/NR2B combination. High-affinity effects of ifenprodil have not been observed at NR1A/2A, 2C, or 2D combinations. A high-affinity ifenprodil site on the NR2B subunit has been mapped to a position between amino acids 178 and 376 of the *N* terminus (Gallagher *et al.*, 1995). Polyamine stimulation does not map to this region of the NR2B subunit (Gallagher *et al.*, 1995). Thus, while polyamine and ifenprodil sites overlap on the NR1 subunit (Williams *et al.*, 1996), this is not the case for NR2B. There are thus two high-affinity ifenprodil sites, one on NR1 and the other on NR2B subunits. The effects of eliprodil have not been fully characterized at all NMDA receptor subtypes. However, like ifenprodil, eliprodil has only weak effects at NR1A/NR2A subtypes (IC_{50} >100 μM) and more potently blocks recombinant NR1A/NR2B receptors (IC_{50} = 0.7 μM) (Avenet *et al.*, 1996). The effects of ifenprodil and eliprodil at NR1/NR2A or NR2B receptors are shown in Fig. 10.

In HEK293 cells transfected with NR1/2A or NR1/2B subunits, ifenprodil and eliprodil displace [^{125}I]MK-801 from cells containing the NR2B subunit with low micromolar affinity. Both compounds are relatively ineffective at receptors containing the NR2A subunit (Table 5) (Besnard *et al.*, 1995).

3.7 NMDA receptor antagonism *in vitro* and *in vivo*

3.7.1 In vitro *experiments*

Ifenprodil and eliprodil block the depolarizing effects or the increased entry of calcium induced by NMDA in cell culture (native or transfected receptors) and have also been shown to block glutamate or NMDA-induced neurotoxicity. Both are ineffective in cortical wedge or hemisected spinal cord models, but this is likely due to saturating concentrations of glycine in such preparations. It is clear from the interactions discussed above that the effects of ifenprodil and eliprodil are subject to a number of conditions. Both polyamines and glycine will modulate their potency

Table 5 The relative potencies of NMDA channel blockers and of ifenprodil and eliprodil as displacers of [^{125}I]MK-801 binding to HEK293 cells transfected with NR1/NR2A and NR1/NR2B receptors

Versus [^{125}I] MK-801	NR1/NR2A K_dK_i(nM)	NR1/NR2B K_dK_i(nM)
Iodo-MK-801	3 ± 1.3	3.3 ± 1.2
MK-801	6.9 ± 1.4	7.8 ± 2.8
TCP	31.8 ± 7.9	16.7 ± 9.5
Phencyclidine	57.1 ± 27	119 ± 23.5
Ifenprodil	>30 000	1610 ± 200
Eliprodil	>10 000	3350 ± 500

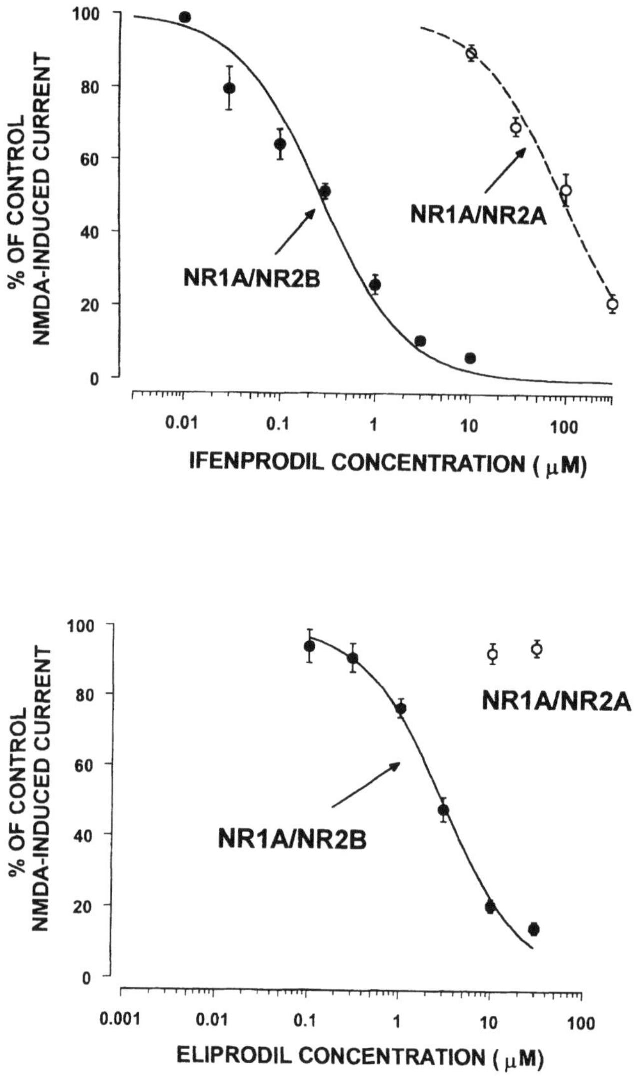

FIG. 10 Antagonism by ifenprodil and eliprodil of the NMDA current in *Xenopus* oocytes expressing NR1A/NR2A and NR1A/NR2B receptors.

or effectiveness while their potency will also depend upon the subunit composition of the particular receptor involved in the NMDA response. Not surprisingly, the *in vitro* effectiveness and potency of ifenprodil and eliprodil therefore vary considerably in different models. These differences are summarized for ifenprodil in Table 6. Both high- and low-affinity effects or partial inhibition can be observed in various preparations. The effects of glycine and the polyamines, where available, will no doubt later help to pinpoint the NMDA receptor subtype involved in these diverse responses.

3.7.2. Effects *in vivo*

In dialysis studies, ifenprodil and eliprodil block the NMDA-evoked increase in striatal dopamine release with ED_{50} values of 0.9 and 0.3 mg kg^{-1} i.p., respectively (Carter *et al.*, 1988). Ifenprodil does not *per se* block the effects of NMDA on spermine or spermidine release *in vivo* (or *in vitro*) (Nicolas *et al.*, 1994) but markedly potentiates the inhibitory effects of 7-chlorokynurenate (Fig. 11) (Voltz *et al.*, 1994). This synergism can be interpreted either as an ifenprodil-mediated increase in the affinity of the glycine site for 7-chlorokynurenate as demonstrated in binding studies (Ogita *et al.*, 1992; Yoneda *et al.*, 1994) or as a removal by 7-chlorokynurenate of the inhibitory effects of endogenous glycine on ifenprodil antagonism.

Eliprodil also antagonizes NMDA receptor-mediated functional responses *in vivo*. Thus, it blocks the excitatory effects of microiontophoretically applied NMDA on Deiters's nucleus neurons in the rat (R. Santamaria, unpublished data) and the decrease in NADH fluorescence

Table 6 The potencies of ifenprodil in different *in vitro* models related to NMDA receptor function. The effects of glycine or polyamines on those of ifenprodil, where tested, are also illustrated

Model	Region	IC_{50}	I_{max}(%)	Effect of spermine	Effect of glycine	Reference
Binding						
[^{3}H]Glycine	Cortex	0.034	44	↓ affinity	–	Ransom (1991)
[^{3}H]CPP	Cortex	0.1	40		NT	Carter *et al.* (1988)
[^{3}H]CGP39653	Cortex	>100	0			Oblin and Schoemaker (1994)
cGMP formation	Cerebellum	0.4	100	↓ affinity	NT	Carter *et al.* (1988)
Electrophysiology						
NR1	–	0.28	100	NT	NT	Williams (1993b)
NR1/NR2B	–	0.34	100	NT	Reverses	Williams (1995b)
NR1/NR2A	–	146	100	NT	NT	
NR1/NR2C	–	>10	–	NT	NT	
NR1/NR2D	–	>10	–	NT	NT	
NMDA	Cortex	1.4 (133 low)	70	NT	NT	Priestley *et al.* (1994)
NMDA	Cerebellum	1.3 (131 low)	20	NT	NT	Priestley *et al.* (1994)
NMDA	Hippocampus, high	0.75	–	NT	No effect	Legendre and Westbrook (1991)
NMDA	Hippocampus, low	161	–	NT	Glycine reverses	Legendre and Westbrook (1991)
NMDA	Spinal cord (culture)	10	100			Carter *et al.* (1988)
NMDA	Spinal cord	NT	34	NT	7-CLK reveals	Carter *et al.* (1992)
Calcium entry						
NMDA	Hippocampus	0.7	88	↑ affinity	No effect	Church *et al.* (1994)
NMDA	Cerebellum	1	50	NT	NT	Perrier and Benavides (1995)
Neurotoxicity						
Glutamate	Cortex	0.2	100	NT	NT	Graham *et al.* (1992)
NMDA	Cortex	~0.4	100	Reversed by SPM and SPD	No effect	Tamura *et al.* (1993)
NMDA	Hippocampus	0.35	100	No effect	No effect	Shalaby *et al.* (1992)
NMDA-evoked release						
Dopamine	Striatum	0.6	100	NT	NT	Nankai *et al.* (1995)
GABA	Striatum	5.6	100	NT	NT	Nankai *et al.* (1995)
Acetylcholine	Striatum	5.4	100	NT	NT	Nankai *et al.* (1995)
Spermidine	Striatum	>100		NT	7-CLK reveals	Voltz *et al.* (1994); Nicolas *et al.* (1994)
Serotonin	Cortex	0.013	50	NT	NT	Fink *et al.* (1995)
GABA	Retina	1.29	80	NT	NT	Zeevalk and Nicklas (1990)

7-CLK, 7-chlorokynurenate; GABA, γ-ammobutyric acid; SPM, spermine; SPD, spermidine; NT, not tested.

induced by intracerebroventricular injection of NMDA in the rat cerebral cortex (complete blockade at 10 mg kg^{-1} i.p.) (Nowicki *et al.*, 1989). Eliprodil also antagonizes the hyperactivity induced by bilateral intra-accumbens injection of spermine (5–20 μg) (Fig. 12). The same doses of eliprodil do not block the locomotor stimulant effects of the intra-accumbens injection of amphetamine (Sanger *et al.*, 1995).

Ifenprodil and eliprodil both show anticonvulsant effects (Table 7), and it is likely that this activity is in part a reflection of NMDA receptor antagonist activity (Perrault *et al.*, 1989, 1990; DeSarro and DeSarro, 1993). The anticonvulsant effects of these compounds are generally observed at higher doses than are necessary to provide neuroprotective activity in focal cerebral ischemia models (although ifenprodil appears to be particularly potent in the genetically epileptic rat (DeSarro and DeSarro, 1993)). These potency differences may reflect the NMDA subtype selectivity of these compounds and a differential implication of specific NMDA receptor subtypes in epilepsy and neurotoxicity. Alternatively, the lower doses needed in focal ischemia might reflect a synergy between the various properties of these compounds that might contribute to neuroprotection (NMDA and calcium antagonist activity and σ affinity). Eliprodil has also been shown to potentiate the anticonvulsant effects of the competitive NMDA

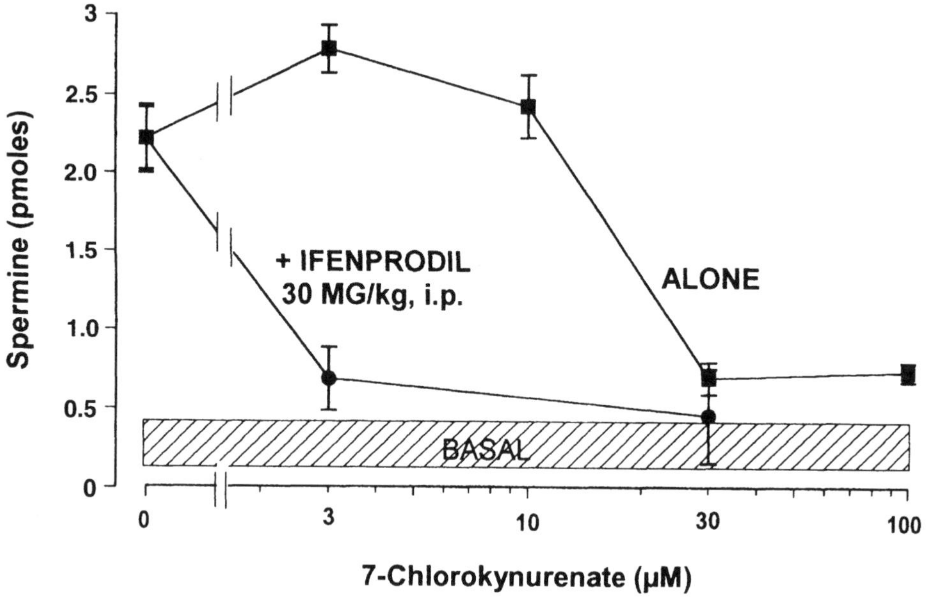

FIG. 11 The effects of 7-chlorokynurenate (3–100 µM via the dialysis fiber), ifenprodil (30 mg kg^{-1} i.p.) or their combination on the NMDA-evoked (500 µM) release of striatal spermine as measured in dialysis studies *in vivo*. Ifenprodil *per se* (30 mg kg^{-1} i.p.) has no effect on this response but markedly potentiates the effects of 7-chlorokynurenate. (Modified from data in Voltz *et al.* (1994).)

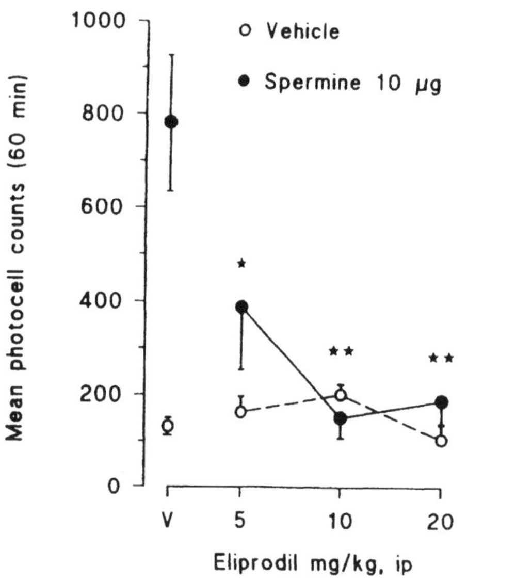

FIG. 12 The effects of eliprodil (5–20 mg kg^{-1} i.p.) on the locomotor hyperactivity induced by the intra-accumbens injection of spermine (10 µg) in rats.

antagonist CGP37849 (but not those of MK-801) and to inhibit the locomotor stimulant effects of CGP37849 (and of amphetamine) but not of MK-801 (Deren Wesolek and Maj, 1993).

Certain NMDA antagonists have been shown to display antidepressant potential in animal models (Trullas and Skolnick, 1990). Eliprodil, again at relatively high doses

Table 7 The anticonvulsant profiles of ifenprodil and eliprodil

	ED_{50} (mg kg^{-1} i.p.)	
	Ifenprodil	Eliprodil
Mouse		
Electroshock	29	15
NMDA-induced convulsions	23	26
Audiogenic seizures (DBA/2)	24	10
Intravenous NMDA (Singh *et al.*, 1991)	Tonic 5 mg kg^{-1} i.v. Clonic: >10	–
Rat		
Audiogenic seizures	0.44	–
Epilepsy prone rat (DeSarro and DeSarro, 1993)		

Data are from in-house studies or as referenced.

(20–40 mg kg^{-1}), reduces immobility in the forced swim test (but not in the tail suspension test), and produces a down-regulation of cortical β-adrenoceptors and a decrease in the affinity for agonists at the modulatory glycine site of the NMDA receptor (Layer *et al.*, 1996). These effects are common to a variety of antidepressants and to certain NMDA antagonists which also show antidepressant activity in animal models (Trullas and Skolnick, 1990; Layer *et al.*, 1996).

NMDA antagonists may be effective in animal models of hyperalgesia (Coderre and Van Empel, 1994; Millan and Seguin, 1994), and ketamine, memantine, and dextromethorphan are clinically effective in certain types of chronic pain (Eide *et al.*, 1994; Price *et al.*, 1994;

Eisenberg *et al.*, 1995). Ifenprodil does not produce analgesia in a rat model of chronic pain (forepaw formalin injection) (Coderre and Van Empel, 1994).

Recently, systemically administered ifenprodil or intrastriatally administered (+)-HA966 have been shown to reverse the parkinsonian symptoms in marmosets produced by the dopaminoselective neurotoxin 6-hydroxydopamine (Mitchell *et al.*, 1995). A pilot study with ifenprodil in parkinsonian patients did not, however, reveal symptomatic improvement (Montastruc *et al.*, 1992). The potential benefits of NMDA receptor blockade in Parkinson's disease nevertheless merit further investigation.

Given that ifenprodil and eliprodil are selective antagonists for receptors containing the NR2B subunit, their relative potencies in different behavioral models presumably provide some indication of the role of these subtypes in the particular response involved. Thus, as these NR2B receptor antagonists do not produce amnesia, analgesia, psychostimulation, or hypertension, presumably NR2B-containing receptors are less implicated in these phenomena. This analysis is of course complicated by the other pharmacological effects of these agents, but both compounds could be very useful for tracking the identities of different receptors involved in the diverse beneficial and undesirable effects of NMDA antagonists.

Table 8 Pharmacological characteristics of [^{3}H]Ifenprodil and [^{3}H]3-PPP binding to σ sites in the rat cerebral cortex[a]

Drug	[^{3}H]Ifenprodil IC_{50} (μM)	[^{3}H]3-PPP IC_{50}(μM)
Ifenprodil	0.009	0.009
Eliprodil	0.022	0.003
Haloperidol	0.024	0.002
Nylidrin	0.049	0.028
DTG	0.071	0.021
Pentazocine	0.15	0.038
BMY14802	0.15	0.12
(+)-3-PPP	0.23	0.022
WB4101	0.24	0.061
Phencyclidine	1.1	0.73
TCP	1.2	0.9
(–)-3-PPP	1.4	0.16
(–)SKF10047	2.5	0.88
(+)SKF10047	23	0.29
MK-801	>10	>100
Spermine	>1000	>1000
Spermidine	>1000	>1000

[a][^{3}H]Ifenprodil binding (0.5–1 nM) to the σ site in rat cortical membranes was studied at 37 °C.

4 OTHER SITES OF ACTION OF IFENPRODIL AND ELIPRODIL

4.1 Ifenprodil and eliprodil as σ ligands

At 37 °C, [^{3}H]ifenprodil labels a σ site with high affinity. This site has the characteristics of the σ_2 site (Hashimoto and London, 1993, 1995; Hashimoto *et al.*, 1994; Schoemaker *et al.*, 1994) and is displaced with high affinity by ifenprodil, eliprodil, haloperidol, and 1,3-di(2-tolyl)guanidine (DTG) (Table 8). Polyamines do not displace [^{3}H]ifenprodil from this site. [^{3}H]Ifenprodil σ sites are mainly localized in the microsomal fraction, while polyamine-sensitive NMDA receptor-linked sites are concentrated in synaptic membranes, and the two sites are clearly distinct. Certain σ ligands have nevertheless been shown to modulate hippocampal CA_3 NMDA responses *in vivo* (Monnet *et al.*, 1992, 1994; Debonnel, 1993; Bergeron *et al.*, 1995) with a distinctive agonist and antagonist σ pharmacology. Very low intravenous doses of σ ligands (micrograms per kilogram) potentiate NMDA responses with bell-shaped dose response curves. DTG, (+)-pentazocine, BD-737, JO-1784, and L-687,384 potentiate the effects of NMDA in this model, and haloperidol, BMY-14802, and (+)-3-PPP act as antagonists. These effects have not been reported or characterized *in vitro*, and ifenprodil or eliprodil have not been tested in this *in vivo* model. Other studies have implicated σ sites in the control of motor behavior (Walker *et al.*, 1990). For example, the injection of DTG or of other σ ligands into the red nucleus induces dystonia, and their unilateral injection into the substantia nigra produces contralateral turning behavior (Walker *et al.*, 1988; Matsumoto *et al.*, 1990). The turning response to intranigral DTG, produced via dopaminergic activation, is antagonized in a dose-dependent manner by ifenprodil and eliprodil (Bastianetto *et al.*, 1995a,b).

A σ binding site (perhaps one of a large family) has recently been identified and cloned from liver tissue. It is a 27 kDa phenylalkylamine-binding polypeptide with a high affinity for emopamil and other related calcium antagonists and for ifenprodil, haloperidol, DTG, and pentazocine. It is a protein with four transmembrane segments, and is concentrated in the endoplasmic reticulum. It has no known channel or receptor sequence homology, but bears some ressemblance to bacterial or eukaryotic drug transporters and carries a sodium-binding site as observed in other sodium-dependent transporters. Its function remains a mystery, although it has been stressed that many of the drugs that bind to this protein possess anti-ischemic properties (Moebius *et al.*, 1993, 1994; Hanner *et al.*, 1995). The protein itself is obviously not an integral part of the NMDA receptor, and it is not clear how or why drugs with affinity for such a protein could modulate NMDA responses *in vivo*. This site, or σ sites in general, can have little relevance to the mechanism of action of ifenprodil as a polyamine site NMDA antagonist as described above, although its involvement in the neuroprotective effect of ifenprodil cannot be excluded. *In vivo* autoradiographic studies have shown that intravenously injected [^{3}H]ifenprodil labels a site pharmacologically akin to σ_2 sites and that [^{3}H]ifenprodil is fully displaced by the systemic injection of haloperidol, ifenprodil, eliprodil, BMY14802, and DTG, but not by +SKF10047, (–)-3-PPP, or (–)-butaclamol (Benavides *et al.*, 1992). The relevance of this site to the neuroprotective effects of ifenprodil is not clear as haloperidol, BMY14802 or DTG do not provide neuroprotection in a mouse focal cerebral ischemia model (Gotti *et al.*, 1990; Poignet *et al.*, 1992).

4.2 Calcium channel antagonism

Ifenprodil and eliprodil are also relatively potent calcium channel antagonists. Calcium antagonist effects of ifenprodil were initially reported in the perfused mesenteric artery and in the isolated rat anococcygeus muscle, where ifenprodil, in the low micromolar range, blocks constriction or contraction induced by potassium chloride or calcium (Adeagbo, 1984; Adeagbo and Magbagbeola, 1985; Honda and Sakai, 1987; Honda *et al.*, 1988, 1989a). Ifenprodil (10 μM) totally blocks KCl induced $^{45}Ca^{2+}$ uptake into rat brain synaptosomes (Honda *et al.*, 1989b).

In cultured cortical neurons, depolarization evokes a calcium current (studied with barium as the permeant cation) which is partially blocked by the L-type calcium channel blocker nimodipine and partially by the N-type calcium antagonist ω-contoxin-GVIA. L- and N-type calcium channels together carry ~70% of this current which is totally blocked by cadmium. Eliprodil partially blocks the depolarization-induced current (I_{max} = 87%) with an IC_{50} of 1.48 μM. Thus, eliprodil blocks both L- and N- type neuronal calcium channels and also blocks a proportion of the calcium current involving another type(s) of calcium channel (Biton *et al.*, 1994) (Fig. 13). In the same neurons, eliprodil blocked NMDA responses (in the presence of 10 μM D-serine) with an IC_{50} of 0.67 μM (I_{max} = 80%).

In dissociated cerebellar purkinje cells, a large proportion of the depolarization-evoked calcium current (82%) is carried by ω-agatoxin-IVA-sensitive P-type calcium channels, the remainder being blocked by the L- and and N-type antagonists nimodipine and ω-conotoxin-GVIA. Eliprodil totally blocks this current, with an IC_{50} of 2.2 μM (Biton *et al.*, 1996) (Fig. 14), and thus blocks P-, L-, and N-type neuronal calcium channels in this preparation. In hippocampal pyramidal neurons, ifenprodil blocks high-voltage activated calcium channels (predominantly L and N) with an IC_{50} of 17 μM. In the same neurons, NMDA responses are blocked by ifenprodil with an IC_{50} of 0.7 μM (Church *et al.*, 1994). Ifenprodil and eliprodil block human N channels with IC_{50} values of 50 and 10 μM, respectively, and human P channels with IC_{50} values of 60 and 9 μM, respectively (Bath *et al.*, 1995). Overall these data show that ifenprodil is a more potent NMDA antagonist than eliprodil and more active as an NMDA than as a calcium antagonist. However, eliprodil is a more potent calcium antagonist than ifenprodil and its potencies as an NMDA and calcium antagonist are similar.

Although post occlusion treatment with calcium channel blockers (nimodipine, diltiazem, flunarizine, verapamil) in a mouse model of focal ischemia is not particularly efficacious (Gotti *et al.*, 1990), a number of calcium channel antagonists with less hypotensive potential have been shown to be effective in focal cerebral ischemia models. These include nivaldipine, emopamil, and lifarizine (Nakayama *et al.*, 1988; Kucharczyk *et al.*, 1991; Li *et al.*, 1994). It is likely that neuronal calcium channel blockade plays an important part in the neuroprotective effects of eliprodil. It might also be considered that the additive or synergistic effects of NMDA receptor antagonism and calcium channel block could contribute to the fact that at

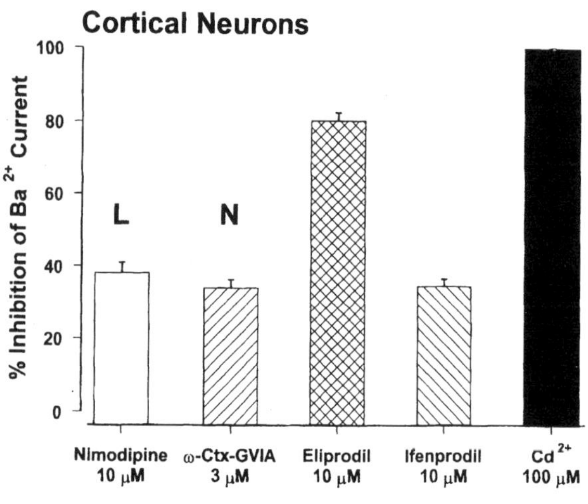

FIG. 13 The effects of nimodipine (10 μM), ω-conotoxin-GVIA (3 μM), eliprodil (10 μM), ifenprodil (10 μM), and cadmium chloride (100 μM) on calcium channels expressed in cultured cortical neurons. See Biton *et al.* (1994) for details.

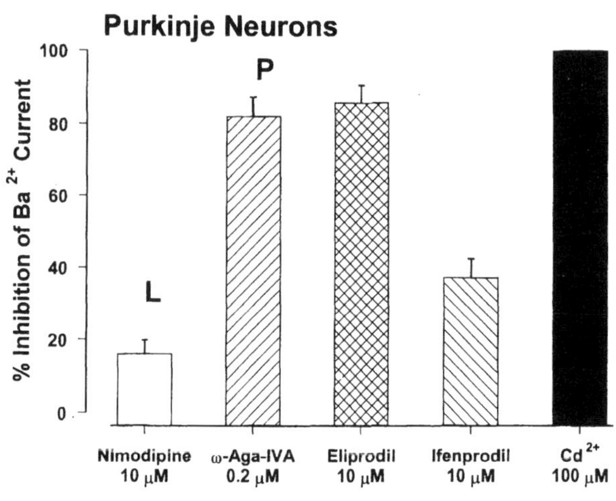

FIG. 14 The effects of nimodipine (10 μM), ω-agatoxin-IVA (0.2 μM), eliprodil (10 μM), ifenprodil (10 μM), and cadmium chloride (100 μM) on P-type calcium channels in dissociated cerebellar Purkinje cells. See Biton *et al.* (1996) for further details.

neuroprotective doses few NMDA (or calcium) antagonist linked side-effects are observed.

4.3 Other receptors

The NMDA receptor and calcium channel antagonist effects (and possibly σ properties) of ifenprodil and eliprodil are those most likely to contribute to the neuroprotective effects of these drugs. Effects at other receptors have also been reported, and these drugs clearly have multiple properties. These include (for ifenprodil) a relatively high affinity for α_1 adrenoceptors (Carron *et al.*, 1971; Chenard *et al.*, 1991) and a moderate affinity for α_2, $5HT1_A$, 5HT2 (Chenard *et al.*, 1991), $5HT_3$ (McCool and Lovinger, 1995), and histamine (H_1) receptors (Chenard *et al.*, 1991).

5 NEUROPROTECTIVE EFFECTS

5.1 Neuroprotective effects *in vitro*

Ifenprodil and eliprodil block the neurotoxic effects of L-glutamate and NMDA in hippocampal or cortical neurons in culture with IC_{50} values in the low micromolar range (Graham *et al.*, 1992; Shalaby *et al.*, 1992), and eliprodil (1–10 μM) antagonizes the diminution of the excitatory postsynaptic potential recorded from the CA1 area of rat hippocampal slices subjected to hypoxia (R. Santamaria, unpublished data). In cortical neurons the neuroprotective effects of ifenprodil on NMDA receptor mediated toxicity were reduced by spermidine, but not by glycine (Tamura *et al.*, 1993). The lack of effect of glycine may be important in assigning the neuroprotective effects of ifenprodil to its glycine-independent block of NMDA receptors, but more detailed studies are necessary to characterize this possible distinction. Ifenprodil does not block the neurotoxic effects of kainate in the hippocampus (Tamura *et al.*, 1993) or of kainate or quisqualate in cortical culture (Graham *et al.*, 1992). Ifenprodil and other NMDA antagonists also block the neurotoxic effects of the anti-HIV drug D-aspartate-β hydroxamate in glial/neuronal cortical coculture (Lockhart *et al.*, 1993). Ifenprodil (IC_{50} = 0.28 μM) and other NMDA antagonists reduce (by about 80%) the neurotoxic effects of hypoglycemia/anoxia in neuronal culture. Other σ ligands with or without NMDA antagonist properties were effective versus *in vitro* ischemia-induced neuronal death in this model, but σ ligands without affinity for the NMDA receptor did not block NMDA-related neurotoxicity (Lockhart *et al.*, 1995). This distinction is important as it shows that the neuroprotective effects of ifenprodil versus NMDA are mediated by its NMDA antagonist properties, but also that its σ effects could contribute to its neuroprotective effects in cerebral ischemia.

Such neuroprotective effects are to be expected given the NMDA antagonist potential of ifenprodil and eliprodil. However, given their subtype selectivity, it is important to note that the neuroprotective effects of ifenprodil or eliprodil versus NMDA are observed in the low micromolar range. Thus the NMDA receptor subtypes that ifenprodil and eliprodil block with high affinity are involved in NMDA receptor-mediated neurotoxicity, at least in cultured cortical or hippocampal neurons.

In the embryonic chick retina there is evidence for receptor heterogeneity in relation to NMDA-mediated toxicity and the effects of ifenprodil. NMDA provokes the death of retinal amacrine cells, all of which are protected by MK-801 or CGS19755. Ifenprodil protects most amacrine neurons (IC_{50} = 1.26 μM) but leaves a subset of amacrine neurons unprotected. Furthermore, the proportion of ifenprodil-insensitive amacrine cells increases with age, suggesting the development of an ifenprodil-resistant NMDA receptor population. Interestingly, ifenprodil inhibited kainic acid-induced neurotoxicity in this model, although at much higher concentrations (10–500 μM) (Zeevalk and Nicklas, 1990, 1992).

5.2 Neuroprotective effects *in vivo*

5.2.1 *Focal cerebral ischemia*

Eliprodil protects against neuronal loss in mouse, rat, and cat models of permanent focal ischemia (middle cerebral artery (MCA) occlusion), providing 40–60% protection in each model (Fig. 15) (Gotti *et al.*, 1988, 1990; Scatton *et al.*, 1994a). Ifenprodil also produces extensive protection in the mouse and cat MCA models. Eliprodil is effective via the intravenous (rat, cat) and intraperitoneal routes of administration (mouse) and is also orally active (rat). In the mouse and rat model of focal ischemia, eliprodil is still effective even if its first administration is delayed by up to 3 and 1h, respectively (Fig. 16) (Gotti *et al.*, 1990; D. Duverger, J-P. Nowicki, J. Benavides, B. Scatton, unpublished data. Eliprodil is also very effective at reducing cortical infarction volume (–57%) in a rat model of transient (2 h) focal cerebral ischemia (D. Duverger, J-P. Nowicki, J. Benavides, B. Scatton unpublished data). As with most other types of treatment, the volume of striatal infarction is not affected (Figs 16 and 17).

5.2.2 *Cerebral trauma*

Cerebral trauma can be modeled in anesthetised rats by the brief and timed application of a pressure pulse (1.5 atm for 10 ms) via a fluid-filled tube apposed to the parietal cortical dura mater (fluid percussion injury). This procedure produces a reproducible cortical lesion which has been extensively characterized by Toulmond *et al.* (1993a,b). Eliprodil, administered 5 min after the percussion impact, provides dramatic protection in this model (60% reduction of the cortical lesion). The window of therapeutic opportunity in this model for eliprodil appears to be considerably longer than in focal ischemia, and significant protection is still observed even when the first administration of eliprodil is delayed by 18 h (Toulmond *et al.*, 1993b) (see Fig. 16).

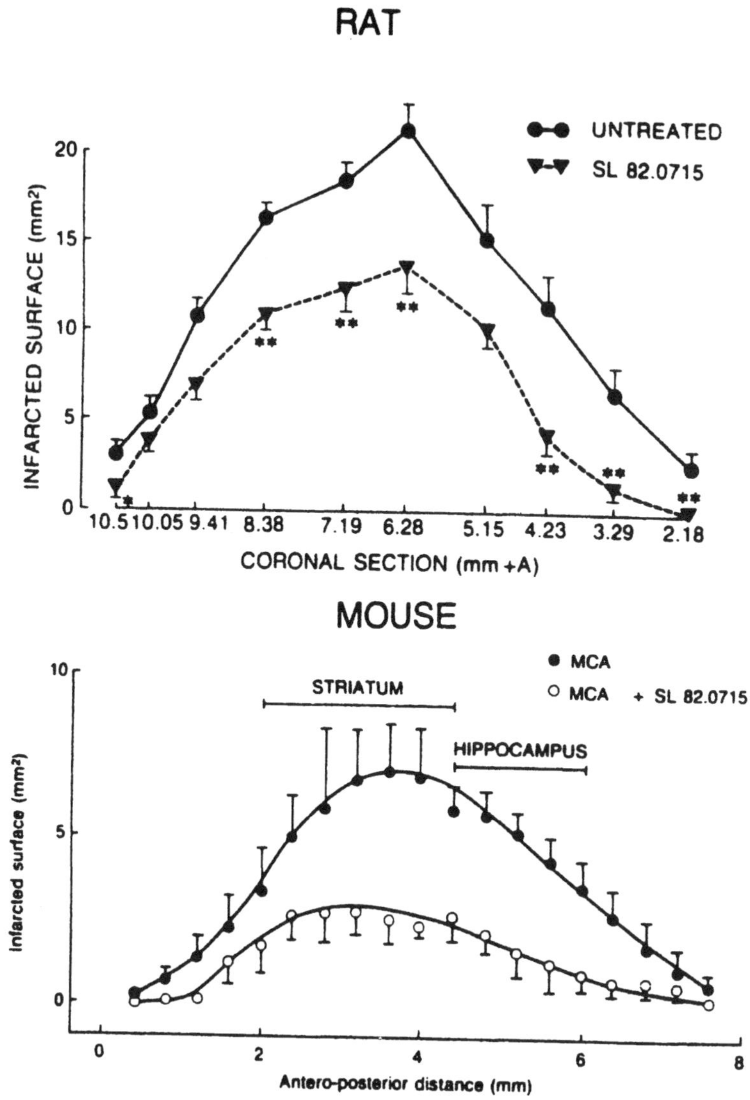

FIG. 15 The neuroprotective effects of eliprodil (SL82.0715) in mouse, rat, and cat (page 75) models of focal ischemia (MCA occlusion). The protective effects of ifenprodil in the cat model are also shown. The curves represent the cross-sectional area of the infarct at different anteroposterior planes read from transverse brain sections.

5.2.3 *Methamphetamine-induced neurotoxicity*

The systemic injection of methamphetamine in mice produces an irreversible destruction of nigrostriatal dopaminergic neurons. Methamphetamine toxicity is blocked by a number of NMDA antagonists including the channel blockers MK-801, phencyclidine, and ketamine, and the competitive NMDA antagonists NPC12626 and CGS19755. Ifenprodil and eliprodil are also effective in this model, with respective ED_{50} values of 0.074 and 0.081 mmol kg^{-1} (~30 mg kg^{-1} i.p.) (Sonsalla *et al.*, 1991). These doses are considerably higher than those needed to provide neuroprotection in the focal ischemia model in mice. Given the wide-ranging potencies of ifenprodil and eliprodil at different NMDA receptor subtypes, one wonders whether the NMDA (or other) receptors involved in ischemia and methamphetamine-related neurotoxicity are the same. The reason for the implication of NMDA receptors in methamphetamine neurotoxicity is not clear but the ability of NMDA antagonists to prevent such toxicity illustrates that their clinical uses may not be limited to stroke or cerebral trauma. The use of such drugs as preventive or decelerative therapy in Parkinson's disease might well be considered (Sonsalla, 1995).

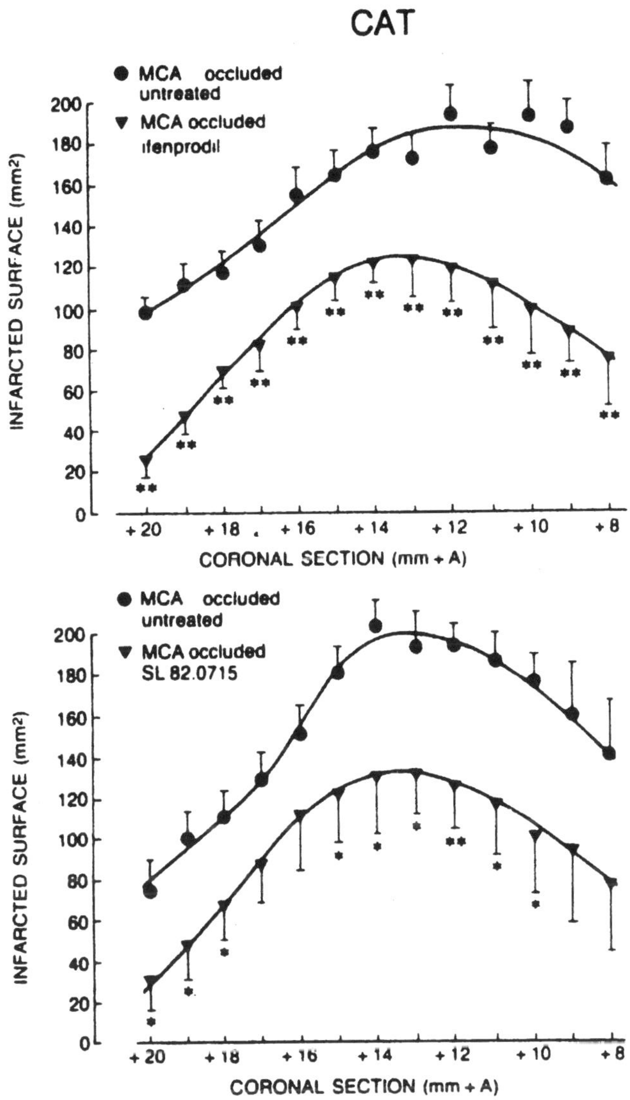

FIG. 15 (*contd.* from page 74)

6 BEHAVIORAL PHARMACOLOGY AND SIDE-EFFECT PROFILE

While NMDA antagonists acting at various different sites on the NMDA receptor are all neuroprotective in animal models of stroke, many suffer from a side-effect profile that may preclude their clinical development. There is a strong argument that certain side-effects are a small price to pay for the salvage of a large area of brain tissue in stroke, but a compound with fewer side-effects is obviously a better alternative.

NMDA receptor activation has been shown to be involved in the genesis of long-term potentiation in the hippocampus (Collingridge and Lester, 1989), a phenomenon implicated in memory mechanisms (Izquierdo, 1993). Many NMDA antagonists produce amnesic effects (Nabeshima *et al.*, 1986; Benvenga and Spaulding, 1988; Morris, 1989; Sanger and Joly, 1991; Sierocinska *et al.*, 1991). NMDA receptors in brainstem cardiovascular control centers control blood pressure and heart rate (Fix *et al.*, 1993), which can be markedly increased by certain NMDA antagonists (Lewis *et al.*, 1989). Such effects may limit their development in stroke. Drugs such as phencyclidine also produce confusion, disorientation, and acute psychotic episodes in the clinic, reflected in animal models by the induction of hyperactivity and stereotyped behavior. Noncompetitive NMDA antagonists such as phencyclidine and ketamine have been subject to considerable abuse, presumably related to their psychotomimetic actions (Gorelick and Balster, 1995). Certain NMDA antagonists

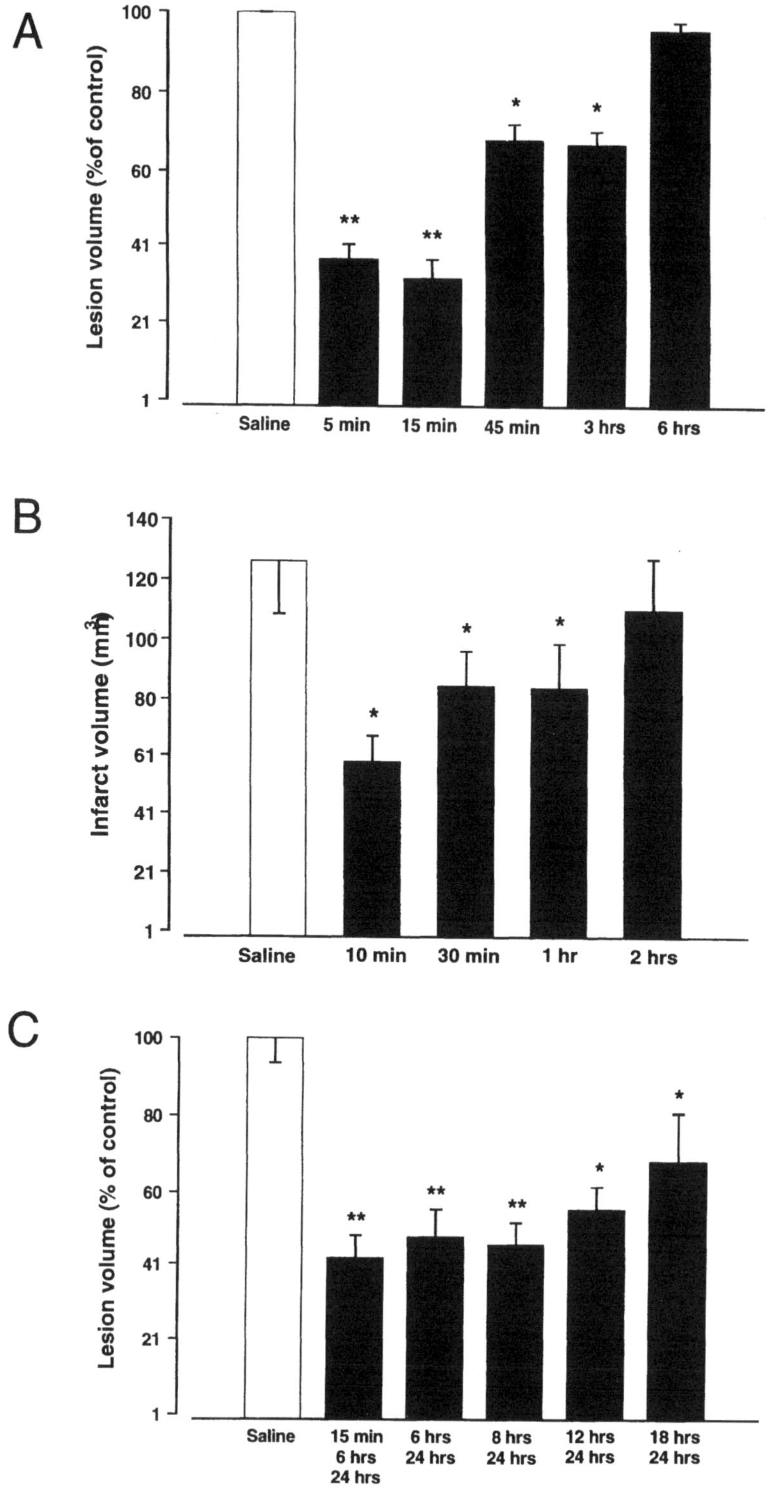

FIG. 16 Windows of therapeutic opportunity for eliprodil in permanent (mouse) and transient (rat) focal cerebral ischemia models and in a rat model of cerebral trauma. (A) Mouse permanent focal ischemia. Administration of eliprodil (10 mg kg^{-1} i.p.) was delayed for various times after MCA occlusion as indicated by the *x* axis. (B) Rat transient focal ischemia. The first drug administration (1 mg kg^{-1} i.v.) was delayed for 10 min, 30 min, 1 h, or 2 h after occlusion. *p <0.05 compared to nontreated animals. (C) Fluid percussion induced cerebral trauma in the rat. Eliprodil (10 mg kg^{-1} i.p.) treatment times after the impact are shown on the *x* axis. *p <0.05 and **p <0.01 compared to nontreated animals.

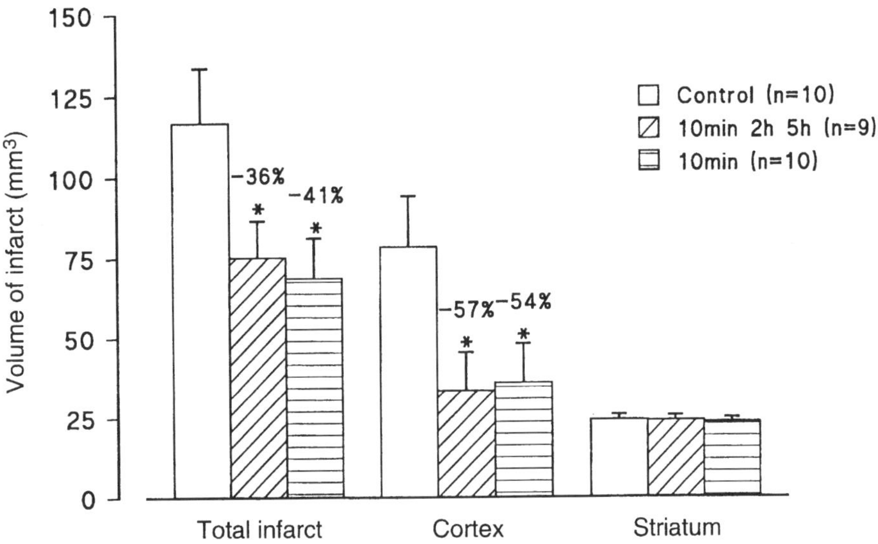

FIG. 17 The neuroprotective effects of eliprodil (1 mg kg^{-1} i.v.) in a transient model of focal ischemia in the rat. The MCA was occluded with a nylon filament for 2 h. Eliprodil was administered either as a single dose 10 min after occlusion, or thrice, 10 min, 2 h, and 5 h after occlusion. *p <0.05 compared to nontreated animals.

provoke a marked increase in cerebral glucose consumption and cerebral blood flow in limbic regions (McCulloch and Iversen, 1991, 1995; Hargreaves *et al.*, 1993b) that may be related to such psychostimulant effects. In the clinic, ketamine triggers psychosis in schizophrenic patients and produces a similar activation of limbic cerebral blood flow (Lahti *et al.*, 1995). In rats, it has also been noted that in areas in which the stimulation of glucose use is greatest (for example the cingulate cortex), neuronal vacuolization can occur (Olney *et al.*, 1989; Olney, 1994). There is some debate as to the relevance of this phenomenon to neuronal health, but it is a worrying sign of the potentially neurotoxic effects of some NMDA antagonists particularly if used chronically. A number of NMDA antagonists also provoke motor incoordination and ataxia likely related to antagonism of cerebellar NMDA receptors (Carter, 1994).

Ifenprodil and particularly eliprodil have been extensively studied in relation to their side-effect potential (Fig. 18). Neither compound produces locomotor stimulation, and indeed both tend to reduce locomotor activity at high doses in mice. They do not impair motor coordination as measured in the Rotarod test except at very high doses (Perrault *et al.*, 1989; Carter *et al.*, 1991; Ginski and Witkin, 1994). Neither drug interferes with the learning of a passive avoidance response in mice (Sanger and Joly, 1991) at doses 30 times those needed to provide maximal protection in the mouse focal ischemia model. Moreover, in contrast to MK-801, eliprodil (10 μM) does not block long-term potentiation in rat hippocampal slices (R. Santamaria, unpublished data). Ifenprodil and eliprodil do not appear to have abuse potential as neither substitute for the phencyclidine cue in drug discrimination studies (Fig. 19). Eliprodil itself has no reinforcing effects in rhesus monkeys (Jackson and Sanger, 1988; Sanger and Jackson, 1989; Balster *et al.*, 1994).

Eliprodil has no effect on heart rate or blood pressure in the conscious normotensive rat (Nowicki *et al.*, 1989). Eliprodil has no stimulatory effect on cerebral glucose metabolism or cerebral blood flow in rats and does not produce the type of neuronal vacuolization observed with MK-801 or phencyclidine (Duval *et al.*, 1992; Cudennec *et al.*, 1994). Indeed, it blocks the increased cerebral expression of heat shock protein induced by MK-801 (Wang *et al.*, 1995a). σ ligands have been shown to block MK-801-mediated neurotoxicity *in vivo* (Olney, 1994), and the σ ligand haloperidol also blocks the NMDA antagonist induced expression of heat shock protein (Sharp *et al.*, 1992). L-type calcium channel antagonists also block the phencyclidine-induced expression of heat shock protein (Sharp *et al.*, 1994). The ability of eliprodil to block the neurotoxic effects of MK-801 may thus be related to its high affinity at σ sites and/or to its calcium channel-blocking properties.

Thus, while eliprodil is neuroprotective *in vivo*, it does not produce the battery of side-effects (at neuroprotective doses) that are associated with a wide range of other NMDA antagonists. This profile is likely a reflection of multiple properties including the selective antagonism of NMDA receptors containing NR2B subunits and the additional possession of calcium channel antagonist effects. The σ properties of these compounds might also contribute to their neuroprotective effects and to their low side-effect profile, but until the function of the σ site(s) is adequately defined this can remain but a speculative possibility.

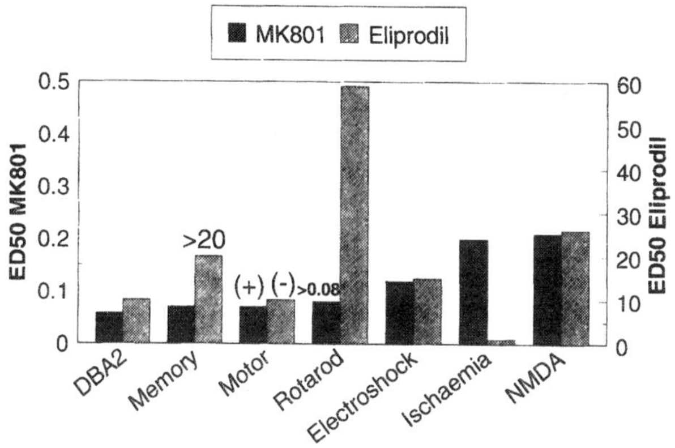

FIG. 18 The relative behavioral profiles of eliprodil and MK-801 in a series of tests in mice. ED_{50} values (mg kg^{-1}) for the various tests are shown on the *y* axes. DBA2, audiogenic seizures in DBA/2 mice; memory, passive avoidance test; Rotarod, motor coordination test; motor, locomotor activity as measured in a photocell chamber; electroshock, electrically induced seizures; ischemia, MCA occlusion; NMDA, NMDA-induced seizures. (+), stimulation; (–), inhibition. * doses of greater than 0.08 mg kg^{-1} were not achievable because of locomotor stimulation. >20, no effects on memory were observed up to a dose of 20 mg kg^{-1}. The sedative effects of higher doses interfere with the lever-pressing necessitated by the passive avoidance test.

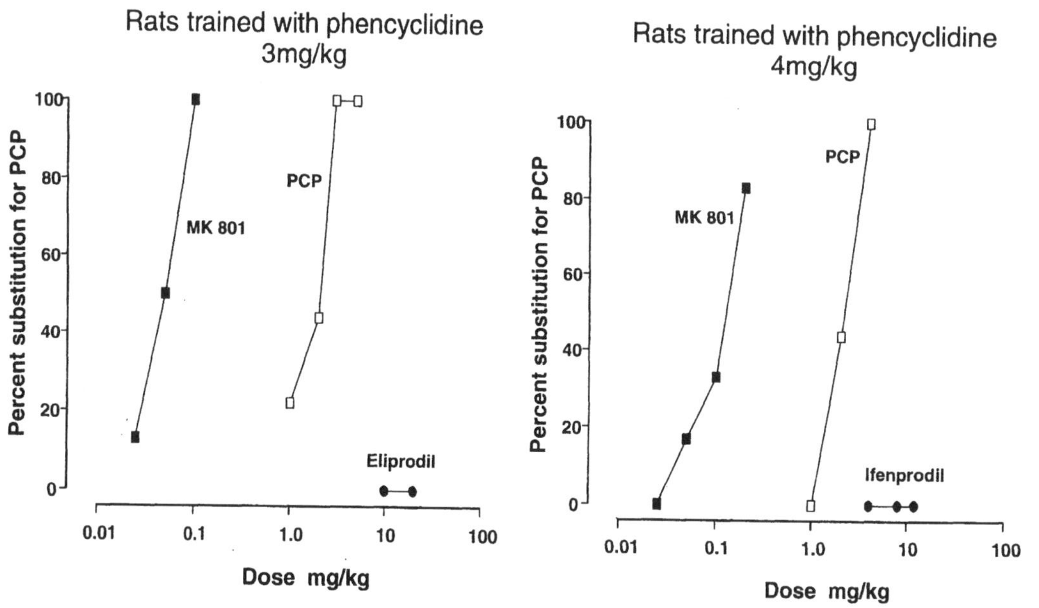

FIG. 19 A comparison of the abilities of MK-801, phencyclidine (PCP), eliprodil, or ifenprodil to substitute for phencyclidine in drug discrimination studies in the rat.

All drugs have side-effects. While eliprodil possesses none of the worrisome effects associated with nonselective NMDA receptor antagonism, it is no exception to this rule. In cats and in man, high intravenous or arterial doses of eliprodil induce a reversible prolongation of the cardiac QT_c interval. In the clinic, this factor sets the highest dose that can be administered intravenously. However, this effect occurs at doses higher than the expected neuroprotective doses calculated from plasma 'area under the curve' (AUC) values found in rat at neuroprotective doses. The molecular origin of this effect is unknown.

7 PHARMACOKINETICS

The high lipid solubility of eliprodil provides the drug with a marked pharmacokinetic advantage in terms of cerebral penetration. Indeed, animal pharmacokinetic studies with [^{14}C] eliprodil have shown that after intravenous or oral administration in the rat, this compound has a rapid brain penetration (brain to plasma ratio ~20 after intravenous injection) and is eliminated slowly from this organ. The pharmacokinetics of eliprodil are therefore well suited for achieving maximal concentrations in the lesioned tissue within a minimal delay after drug administration. The plasma half-life of eliprodil is 2.2 and 7–9 h after intravenous administration in the rat and monkey, respectively. During pharmacokinetic studies in young healthy subjects, AUC and C_{max} values increased linearly with the dose. The absolute bioavailability following oral administration (10 mg) is 47% of the administered dose. In single-dose studies (intravenous or oral) in humans, the mean apparent elimination half-life was relatively long ($t_{1/2\beta}$ = 16–20 h) and dose-independent. Following repeated intravenous or oral administration, a terminal half-life of approximately 50 h was measured. The principal metabolic pathway of eliprodil in man is glucuroconjugation (Scatton *et al.*, 1994a,b).

8 CLINICAL TRIALS

8.1 Phase I

The clinical safety profile of eliprodil has been evaluated in 189 young, middle-aged or elderly healthy human volunteers following single and repeated administration (Giroux *et al.*, 1994; Patat *et al.*, 1994; Scatton *et al.*, 1994a,b). During these studies, subjects received eliprodil at various doses (0.2–60 mg), routes of administration (intravenous or oral), and drug regimens (single doses and up to 7 days repeated doses b.i.d.). Safety was assessed by clinical examination and reporting of adverse events, laboratory evaluations (hematology, chemistry, urine analysis) and cardiovascular evaluations. Overall, few adverse effects were reported. The most common adverse event was headache, and, at doses higher than or equal to 40 mg p.o., dizziness and somnolence. Specifically, no episodes of psychosis were observed. There were no clinically significant alterations of blood pressure or heart rate. However, at the highest doses investigated, asymptomatic dose-related increases were noted in the duration of the QT_c interval (the mean prolongation of QT_c did not exceed 50 ms after single intravenous doses of up to 6 mg). Moreover, eliprodil (30 mg single oral dose) did not produce either sedation or stimulation or any dysphoric or euphoric effects. In addition, eliprodil did not impair short-term memory (Sternberg memory scanning test) or long-term memory (delayed free recall and recognition of pictures). These clinical data confirm the predictions from animal studies in relation to lack of NMDA receptor antagonist-related side-effects. Animal studies appear to be reliably predictive of subsequent clinical problems in man, as those compounds which have so far ventured into the clinic with a known side-effect potential have unfortunately confirmed their therapeutic problems (Sveinbjornsdottir *et al.*, 1993; Grotta *et al.*, 1995; Muir and Lees, 1995).

8.2 Phase II safety studies

The clinical safety of eliprodil has also been assessed in 114 acute stroke patients suffering mainly from an infarction located within the MCA territory (70%) without coma or stupor. This multicenter, randomized, double-blind, placebo-controlled trial was performed with two dose regimens of eliprodil: 3 mg b.i.d. for 3 days (i.v.), followed by 10 mg b.i.d. for 11 days (p.o.) and 1.5 mg bid for 3 days (i.v.), then 5 mg bid for 11 days (p.o.) with a 3 month follow-up. Clinical examination, cardiovascular survey and reporting of adverse effects were performed during the treatment and the follow-up periods. Eliprodil was well tolerated at both dose regimens, and no unexpected events were noted. The percentage of patients presenting at least one emergent event or serious clinical adverse events was comparable in all three groups either during the treatment period or within 8 days after discontinuation of the study drug. No psychotomimetic events or torsades de pointes were reported nor clinically significant changes in laboratory parameters. No statistically significant effect of eliprodil on blood pressure, heart rate, and QT_c interval were observed.

Eliprodil is currently in phase II/III clinical trials in stroke and head trauma in Europe and in the USA. At the time of writing, these studies are still blinded, and we are unable to comment on clinical data for neuroprotection.

While eliprodil is in clinical development, ifenprodil has already been marketed (Vadilex). It is used to improve the peripheral circulation, and a recent clinical trial has reported efficacy in peripheral arterial occlusive disease (where clinical tolerance over a 6 month period (60 mg, dermal patch) was reported as excellent) (Branchereau and Rouffy, 1995). Thus, a certain amount of clinical experience is available that at least gives an idea of the potential side-effect profile of this compound.

In a double-blind trial of ifenprodil (20 mg orally three times a day; n = 42) versus placebo (n = 45) in patients with chronic cerebrovascular disorders, no relevant side-effects were observed (Sinforiani *et al.*, 1988). In a Japanese study, transient decreases in blood pressure were observed following the administration of a high intravenous dose of 10 mg (probably a reflection of α_1 receptor antagonism) (Sawada and Kawai, 1975). The compound thus does not evoke the hypertensive effects associated with certain NMDA antagonists in man.

9 CONCLUSIONS

Ifenprodil and eliprodil are noncompetitive NMDA receptor antagonists whose actions are mediated via a polyamine-sensitive site on the NMDA receptor. There is still much to be learnt concerning the exact nature of their interactions with polyamines, but their close link with this site is indisputable. They are in addition selective antagonists of NMDA receptors containing NR2B subunits. This subtype selectivity explains why these compounds do not block NMDA responses in all models and could also partly explain the low side-effect profile of this type of compound. Both compounds should be useful tools to investigate the role of NR2B-containing NMDA receptor subtypes. Their NMDA antagonist properties must contribute to their neuroprotective potential in animal models of cerebral ischemia and trauma, but it is also likely that their other properties contribute to their neuroprotective effects. They are both potent σ ligands, although until the role of these sites is more firmly established it is impossible to comment on this property in relation to its contribution to neuroprotective potential. Both ifenprodil and eliprodil are also relatively potent voltage-operated calcium channel antagonists, a property that very likely contributes to their neuroprotective effects.

In clinical terms, debate about which mechanism contributes most to neuroprotective potential, while an intellectual challenge, is probably an irrelevance. The bottom line from the animal studies is that these compounds are neuroprotective without producing unacceptable side-effects.

If the neuroprotective potential of eliprodil or other NMDA antagonists in man lives up to the expectations raised by animal studies, the treatment of neurological diseases will experience a major revolution. Before a clear clinical indication of efficacy, other potential uses of this compound are speculative, but any demonstrable efficacy of this or other compounds in stroke would no doubt precipitate studies into possible uses to prevent or diminish the progression of other neurodegenerative disorders including Alzheimer's, Huntington's, and Parkinson's diseases.

ACKNOWLEDGMENTS

We would like to thank Keith Williams for the many preprints, and the many other colleagues who have contributed to work on these compounds. This chapter is dedicated to Sebastian Pigasse.

7 (3*S*,4a*R*,6*R*,8a*R*)-6-[2-(1(2)*H*-Tetrazole-5-yl) ethyl]decahydroisoquinoline-3-carboxylic Acid (LY293558) and its Racemate (LY215490): A Selective and Competitive AMPA/Kainate Receptor Antagonist

DAVID LODGE[1] AND DARRYLE D. SCHOEPP[2]

[1]*Lilly Research Centre Ltd, Eli Lilly and Company, Erl Wood Manor, Windlesham, Surrey GU20 6PH, UK*
[2]*Lilly Research Laboratories, Eli Lilly and Company, Indianapolis, IN 46285, USA*

1 SUMMARY

During the structure–activity development of series of decahydroisoquinoline-based *N*-methyl-D-aspartate (NMDA) antagonists, some compounds in the series showed activity at α-amino-3-hydroxy-5-methyl-4-isoxazolepropionic acid (AMPA) receptors. Of these, (3*S*,4a*R*,6*R*,8a*R*)-6-[2-(1(2)*H*-tetrazole-5-yl)ethyl]decahydroisoquinoline-3-carboxylic acid (LY293558) (Fig. 1) was one of the most potent and selective for AMPA receptors *in vitro* and *in vivo*. LY215490 is the racemic mixture. LY293558 is centrally active following parenteral administration in animals, with no NMDA receptor antagonist activity at *in vivo* doses which block AMPA receptors, and a pharmacology consistent with effects of other known AMPA antagonists. LY293558 possesses neuroprotectant activity against AMPA- and ischemia-induced neuronal injury in multiple animal models including focal ischemia in the rat and cat, and spinal ischemia in the rabbit. Thus, LY293558 may have clinical utility as a neuroprotectant in patients subjected to an ischemic neuronal event that involves glutamate excitotoxicity.

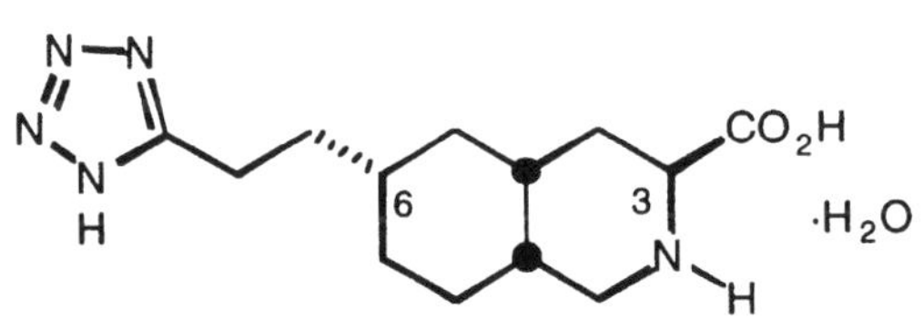

FIG. 1 Structure of LY293558 monohydrate.

2 LY293558 AS A NOVEL NEUROPROTECTANT COMPOUND

There are no drugs currently available to treat neurological damage resulting from cerebral ischemia. One therapeutic strategy for treating cerebral ischemia involves drugs that prevent the excessive stimulation of excitatory amino acid (glutamate) receptors in the central nervous system which leads to neuronal degeneration via a process termed 'excitotoxicity'. Considerable scientific literature has implicated glutamate excitotoxicity in the pathophysiology of

EXCITATORY AMINO ACIDS – CLINICAL RESULTS WITH ANTAGONISTS
ISBN 0-12-546820-2

ischemic neuronal injury (Choi, 1988b). The rationale is that ischemia leads to loss of ATP and hence depolarization of cells. This in turn results in glutamate release by reversal of glutamate uptake transporters and by calcium-mediated exocytosis. The high concentrations of glutamate activate neuronal glutamate receptors allowing further calcium accumulation intracellularly with potentially lethal effects on neuronal cells.

The receptors activated by the released glutamate include ionotropic and metabotropic subtypes. Of the ionotropic receptors, both the AMPA and NMDA subtypes are known to be permeable to calcium and hence would be likely to contribute to neurotoxicity. Indeed, glutamate receptor antagonists acting at either the NMDA or AMPA-preferring subtypes of glutamate receptors have been demonstrated to be neuroprotectant in animal models of ischemic injury (Benveniste, 1991; Gill, 1994). However, NMDA antagonists may have a side-effect profile that is not ideal (Fix, 1994), and AMPA antagonists offer a newer glutamate antagonist approach to treat ischemic neuronal injury in humans.

3 *IN VITRO* GLUTAMATE RECEPTOR PROFILE

The relative affinities of LY293558 for subtypes of glutamate receptors were determined by measuring displacement of ^{3}H-labelled receptor ligands that selectively bind to glutamate receptor subtypes (Schoepp *et al.*, 1991). LY293558 potently displaces specific binding of the AMPA receptor ligands [^{3}H]AMPA and 6-cyano,7-nitro-quinoxaline-2,3-dione ([^{3}H]-CNQX) to rat brain membranes with IC_{50} values of 1.4 and 1.0 μM, respectively. LY293558 shows relative selectivity for AMPA receptors, since displacement of ligands for NMDA ([^{3}H]CGS19755) and kainate ([^{3}H]kainate) receptors requires relatively higher concentrations of LY293558 (IC_{50} >10 μM). LY293558 does not have appreciable affinity at metabotropic (G-protein coupled) glutamate receptors as seen by lack of displacement of 1S,3R-aminocyclopentanedicarboxylate (ACPD)-sensitive [^{3}H]-glutamate binding to rat brain membranes (Wright *et al.*, 1994). (Table 1 and Fig. 2). A receptor binding profile for affinity at other major neurotransmitter receptors demonstrated that LY293558 has no appreciable affinity (tested at 10 μM) for adrenergic, dopaminergic, serotonergic, cholinergic, GABAergic, or histaminergic receptors.

Table 1 *In vitro* affinities of LY293558 for excitatory amino acid receptor subtypes in rat brain membranes

Ligand	IC_{50}(μM)	*n*
AMPA receptor		
[^{3}H]AMPA (5 nM)	1.4 ± 0.1	3
[^{3}H]CNQX (2 nM)	1.0 ± 0.3	3
NMDA receptor		
[^{3}H]CGS 19755 (10 nM)	12.1 ± 2.0	3
Kainate receptor		
[^{3}H]Kainate (5 nM)	28.1 ± 1.7	3
Metabotropic receptor		
[^{3}H]Glutamate (ACPD-sensitive) (10 nM)	>100	3

Data adapted from Schoepp *et al.*, (1995).

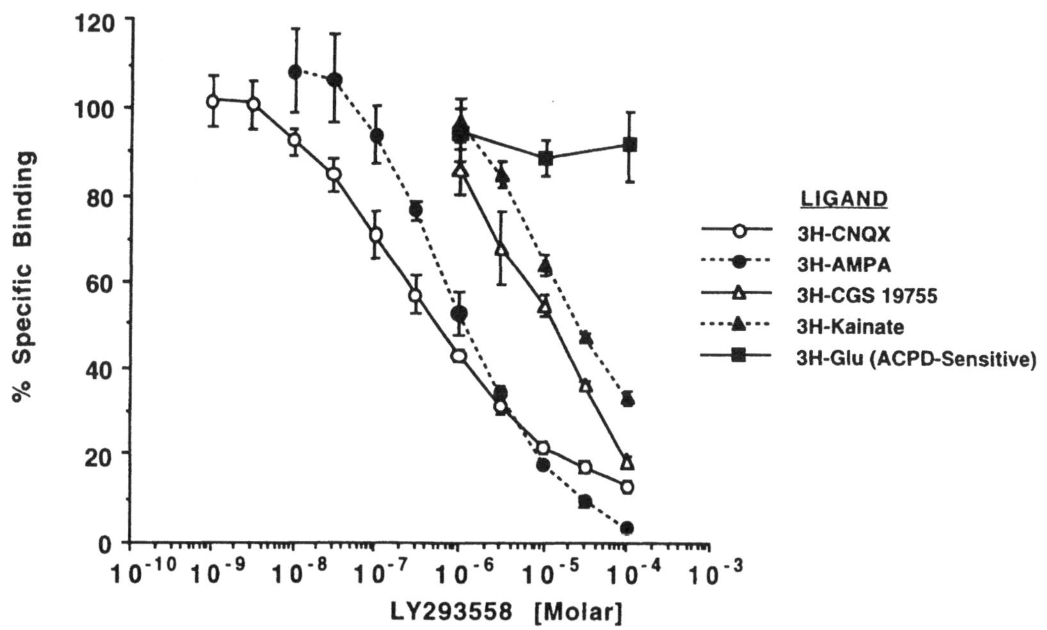

FIG. 2 Displacement of excitatory amino acid receptor ligands by LY293558 in rat brain membranes. (Data adapted from Ornstein *et al.* (1993) and Schoepp *et al.* (1995).)

Functional antagonism of AMPA receptors in rat brain slices by LY293558 was demonstrated by electrophysiological and biochemical measures (Harrison and Simmonds, 1985; Vezzani *et al.*, 1987). In rat cerebral cortical slices, LY293558 fully blocks AMPA (40 μM)-induced depolarization with an IC_{50} of 2.3 ± 0.8 μM (*n* = 4). LY293558 fully antagonized AMPA (100 μM)-induced release of [^{3}H]-norepinephrine from superfused slices of the rat hippocampus with an IC_{50} of 4.7 ± 0.6 μM (Desai *et al.*, 1995). In both assays, LY293558 produced near-parallel shifts of the AMPA concentration–response curves and Schild plots close to unity. In neither assay did LY293558 have any significant intrinsic activity.

These binding and functional *in vitro* studies indicate that LY293558 is a selective and competitive AMPA receptor antagonist, with no agonist activity.

4 *IN VIVO* AMPA RECEPTOR ANTAGONISM

The *in vivo* efficacy and selectivity of LY293558 as an AMPA receptor antagonist in the rat central nervous system (CNS) was assessed electrophysiologically. The action potential discharge rate of spinal neurons in anesthetized rats was recorded from the center barrel of seven barrelled glass microelectrodes, and was increased by ionotophoretic application of AMPA and NMDA (Anis *et al.*, 1983). LY215490 (racemic LY293558) administered iontophoretically selectively blocked the AMPA- and kainate-induced, but not the NMDA-induced, excitations (Fig. 3). In subsequent experiments, LY293558, 2, 5, and 10 mg kg^{-1}, was administered to the rat by intravenous bolus injection, and produced dose-related reductions in responses to AMPA (see Table 2). The onset of this effect was gradual, and reached a peak at approximately 30 min after administration. Significant but incomplete recovery from these effects of LY293558 was observed within 2 h of administration. In contrast, even the highest dose of LY293558 (10 mg kg^{-1} i.v.), which nearly abolished the responses to AMPA, had no effect on NMDA-induced excitation. These data demonstrate that parenteral administration of LY293558 blocks central AMPA receptors over the dose range of 2–10 mg kg^{-1} i.v. LY293558 had no significant NMDA receptor antagonist activity even at doses which completely suppressed AMPA receptor responses.

5 EFFECTS OF LY293558 ON CNS EXCITABILITY

The ability of parenterally administered LY293558 to prevent maximal electroshock seizures (MES) in mice was determined as an index of protection from excessive central glutamatergic neuronal activation (Leander *et al.*, 1992). At 5 min postadministration, LY293558 protects mice from MESs with an ED_{50} of 2.9 mg kg^{-1} i.v. (95%

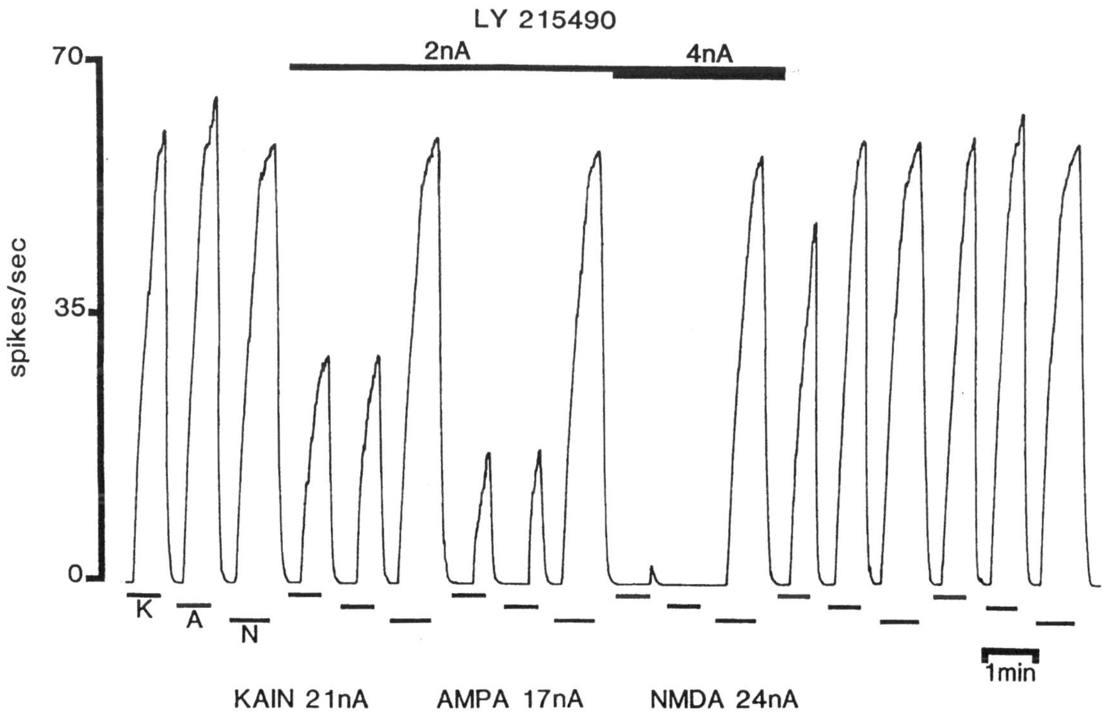

FIG. 3 Selective reduction of responses of a spinal neuron to AMPA and kainate by the iontophoretic ejection of LY293558 (as the racemate LY215490). On spinal neurons, AMPA and kainate increase the firing rate by an action on AMPA receptors.

Table 2 *In vivo* antagonism of AMPA receptor-mediated depolarizations in rat spinal neurons by LY293558.[a]

LY293558 ($mg\ kg^{-1}$ i.v.)	Reduction of response (mean ± SE) (%) AMPA	NMDA	*n*
2	18 ± 2	4 ± 1	6
5	49 ± 4	8 ± 1	6
10	86 ± 7	8 ± 4	5–6

[a]Control responses to AMPA and NMDA were made nearly equal and submaximal by adjusting the ejecting currents of these two agonists. Percentage reductions were calculated by comparing the mean of at least two responses at the peak effect of LY293558 with those in the immediate preinjection period.

confidence interval (CI) 1.8–6.0 $mg\ kg^{-1}$ i.v.) and an ED_{50} of 6.0 $mg\ kg^{-1}$ i.p. (95% CI 2.6–10.8 $mg\ kg^{-1}$ i.p.). Somewhat higher doses of LY293558 produced motor impairment in mice, as indexed by impairment in the horizontal screen test. The ED_{50} for LY293558 in the horizontal screen test was 10.0 $mg\ kg^{-1}$ i.p. (95% CI 9.3–10.7 $mg\ kg^{-1}$ i.p.). This effect of LY293558 has a rapid onset (5 min) and a duration of action of 1.5–3 h (Ornstein *et al.*, 1993; Schoepp *et al.*, 1995).

The anticonvulsant effects of LY293558 against MES in mice indicates that parenterally administered LY293558 rapidly enters the CNS to block excessive neuronal excitability. At about twofold higher doses, LY293558 produces CNS depression as indicated by impairment of motor abilities.

In a study by Browne and McCulloch (1994), LY293558 produced dose-related decreases in cerebral glucose utilization in multiple brain regions of the rat. This occurred over doses of 10, 30, and 100 $mg\ kg^{-1}$ i.v., and the pattern of these effects were like the other structurally different AMPA receptor antagonist compound 2,3-dihydroxy-6-nitro-7-sulfamoylbenzoyl[*f*]quinoxaline (NBQX), and were unlike effects observed with NMDA receptor antagonists which increase glucose utilization in certain limbic regions. This study confirmed the CNS depressant effects of these doses and indicates that LY293558 may not have neuropathological liabilities associated with NMDA receptor antagonist compounds. Consistent with this, LY293558 at doses of 10, 25, 50, and 75 $mg\ kg^{-1}$ i.v. produced no vacuolization of the rat cingulate region under conditions where prominent vacuolization can be observed with MK-801 (A.S. Fix, P.L. Ornstein and D.D. Schoepp, unpublished data).

6 *IN VIVO* NEUROPROTECTION AGAINST AMPA-INDUCED EXCITOTOXICITY

The *in vivo* neuroprotectant action of LY293558 against AMPA-induced neuronal degeneration was assessed in rats (Schoepp *et al.*, 1989). Unilateral stereotaxic infusion of AMPA (50 nmol) into the striatum of anesthetized rats results in a delayed loss of cholinergic and GABAergic neurons, which can be quantified 6–7 days post-AMPA infusion by decreases in choline acetyltransferase (ChAT) and glutamic acid decarboxylase (GAD) enzyme activities, respectively. To examine potential neuroprotection by LY293558 from this delayed brain injury, six repeated doses of LY293558 (25 or 50 $mg\ kg^{-1}$ i.p.) were administered at 2 h intervals (0, 2, 4, 6, 8, and 10 h post-AMPA) to provide central AMPA receptor antagonism for about a 12 h period of time. This LY293558 treatment produced highly significant protection (p <0.05) from AMPA-induced neuronal degeneration (see Fig. 4) (Schoepp *et al.*, 1996, in press).

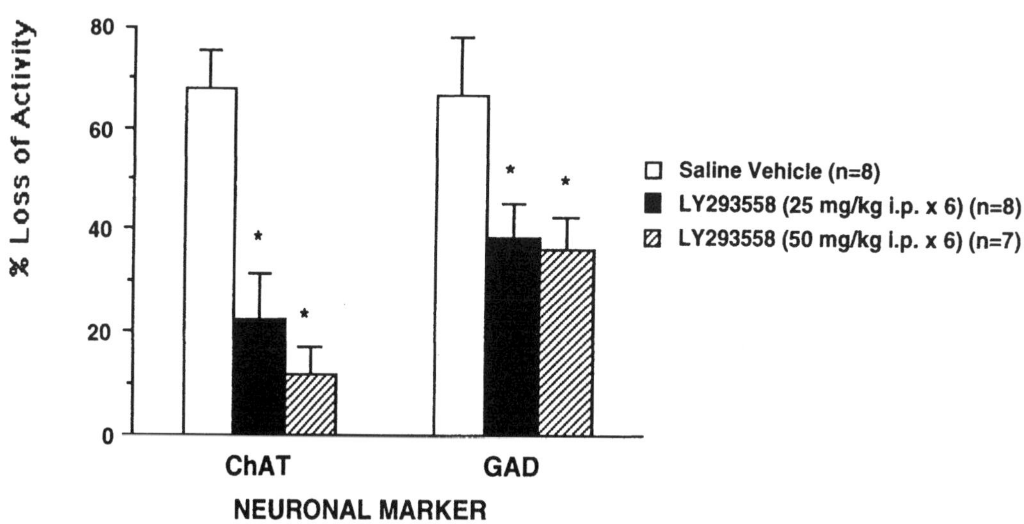

FIG. 4 *In vivo* LY293558 neuroprotection against AMPA-induced striatal neuronal degeneration in the rat (*p <0.05). (Data adapted from D. D. Schoepp *et al.* (unpublished).)

7 NEUROPROTECTANT ACTIVITY IN ANIMAL MODELS OF CEREBRAL ISCHEMIA

Treatment with the AMPA antagonist LY293558 was examined for neuroprotection against irreversible ischemic brain damage in a model of focal cerebral ischemia in the anesthetized cat (Park *et al.*, 1988). LY293558 (15 mg kg^{-1} i.v.) or saline vehicle was given in a bolus at 30 min prior to permanent occlusion of one middle cerebral artery (MCA) (Bullock *et al.*, 1994). A continuous infusion of LY293558 (7 mg kg^{-1} h^{-1}) or saline vehicle was initiated at 25 min before MCA occlusion and maintained for 6 h after MCA occlusion. Treatment with LY293558 produced a significant reduction in the volume of ischemic damage (20% protection when compared to vehicle-treated cats, $p < 0.02$) in the cerebral hemisphere and cerebral cortex (see Table 3). The volume of ischemic damage in the caudate nucleus, which has no collateral blood flow, was not affected by LY293558. The reduction in ischemic damage by LY293558 could not be attributed to alteration in physiological variables such as temperature and blood pressure, as these were assessed and not altered by LY293558 treatment during the postischemia period.

Table 3 Reduction of ischemic injury by LY293558 following MCA occlusion in the anesthetized cat

	Volume of ischemic damage (mm^3)	
Brain region	Vehicle control	LY293558
Cerebral hemisphere	3423 ± 80	2822 ± 215*
Cerebral cortex	2825 ± 102	2285 ± 208*
Caudate nucleus	263 ± 7	250 ± 11

*$p < 0.02$.
Data adapted from Bullock *et al.* (1994).

In a rat focal ischemia model (Gill *et al.*, 1991), LY293558 (given as racemic LY215490) was administered in two repeated (30 mg kg^{-1} i.v.) doses at 30 min pre- and 30 min post-permanent oclusion of the left MCA. This treatment resulted in a significant reduction (about 30%; $p < 0.0001$, ANOVA) in the volume of ischemic tissue in the hemisphere and cortical areas, with no protection of the caudate (Fig. 5) (Gill and Lodge, 1994). This experiment resulted in the death of two control-treated and two drug-treated animals. Neuroprotection appeared to be accompanied by CNS depressant activity, since the animals administered 2 × 30 mg kg^{-1} of LY215490 had some respiratory problems when extubated. There were no statistically significant differences between control- and drug-treated groups for body temperature, blood pressure, P_aCO_2, or P_aO_2 measured postocclusion.

LY293558 was also examined in a global spinal ischemia model in rabbits (Kocklar *et al.*, 1988). The duration of spinal ischemia required to produce permanent paraplegia in rabbits was assessed in saline vehicle-injected rabbits and in animals that received a bolus dose of 16 mg kg^{-1} i.v. of LY293558 5 min postocclusion, then 2.2 mg kg^{-1} i.v. for 1 h (Bowes *et al.*, 1994). When assessed at 18 h postischemia, the ET_{50} (time of ischemia producing 50% paraplegia) was 30.4 ± 5.0 min in the

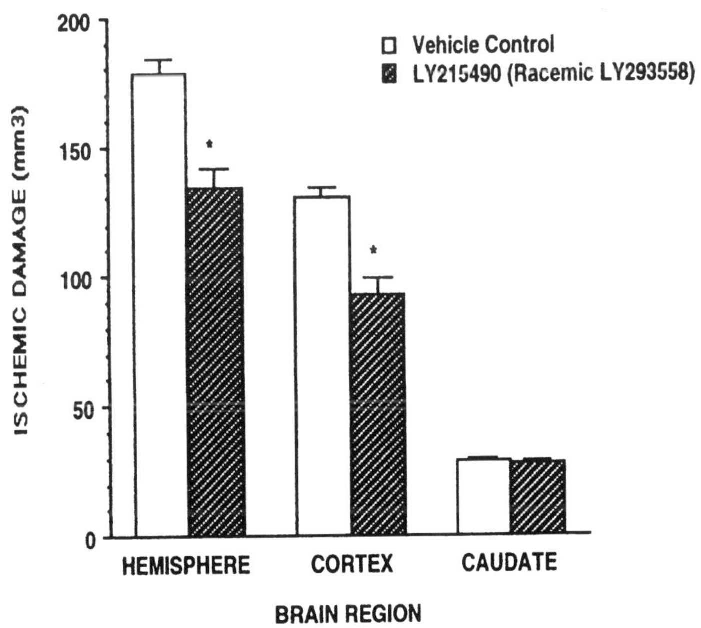

FIG. 5 Reduction of ischemic injury by LY293558 (as the racemate LY215490) following permanent MCA occlusion in the anesthetized rat (*$p < 0.0001$). (Data adapted from Gill and Lodge (1994).)

control group and 50.1 ± 3.2 min in the LY293558 group (p <0.05). Animals treated with LY293558 displayed mild sedation and ataxia for several hours after drug administration, but all animals appeared to have recovered by the 18 h neurological evaluation period. Thus, LY293558 treatment produced a positive effect on neurological outcome following ischemia in the rabbit species.

Temporary bilateral carotid artery occlusion in the gerbil was used to assess the ability of LY293558 to protect neurons from global ischemic damage in the hippocampus (Gill *et al.*, 1987). Male gerbils were divided into three groups: a sham group (n = 5) and two ischemic groups (n = 10), one of which received LY293558 and the other remained untreated. The sham animals were treated identically other than they had no carotid artery occlusion. Five days later the animals were deeply anesthetised with halothane, and the brains perfused with 10% formal saline and prepared for histology. The number of viable cells in the CA1 hippocampal pyramidal layer was counted at three levels equivalent to the frontal planes 1.5, 1.7 and 1.9 mm rostral to the interaural line. Values given in Table 4 are the number of cells surviving per millimeter in the CA1 field at the 1.7 mm level; the other levels gave almost identical results. Differences between untreated and LY293558-treated groups were compared statistically using a Student's t-test. Five dose regimes were used, the initial dose of LY293558 being given 30 min before, immediately after, 30 min after, 60 min after, or 90 min after the carotid occlusion. In each group, the initial dose of 20 mg kg^{-1} was administered intraperitoneally followed by a further four intraperitoneal injections at 3-hourly intervals of 10 mg kg^{-1}. This dosing paradigm was associated with CNS depressant activities including ataxia, twitching, and general and respiratory depression.

As can be seen in Table 4, the results showed that this dose regime offered significant neuroprotection in the first three treatment groups. When the initial dose was delayed to 60 min after the occlusion, although there was a greater number of neurons surviving than in the control group, the large variation prevented this reaching significance at the 5% level (A. Bond and D. Lodge, unpublished data).

8 OVERVIEW AND DISCUSSION

As described above, the following pharmacological features of LY293558 have been established:

(1) LY293558 is selective and relatively potent as a competitive antagonist at the AMPA subtype of glutamate receptor *in vitro*.
(2) LY293558 is a highly effective AMPA receptor antagonist *in vivo* over the dose range of 2–10 mg kg^{-1} i.v. *In vivo* doses of LY293558 (10 mg kg^{-1} i.v.) which produce near-maximal block of AMPA receptor responses produce no significant NMDA receptor antagonist activity.
(3) Following parenteral administration in mice, LY293558 potently (ED_{50} 2.9 mg kg^{-1} i.v.) blocks excessive neuronal excitation induced by maximal electroshock. About threefold higher doses of LY293558 produced motor impairment as indexed by the horizontal screen test in mice. This central effect demonstrated that LY293558 has a rapid onset and a 1–4 h duration of action.
(4) LY293558 was clearly neuroprotective in a rat model of AMPA-induced excitoxic damage, as well as multiple cerebral ischemia animal models. This includes focal ischemia in the cat (15 mg kg^{-1} i.v. bolus, then 7 mg kg^{-1} h^{-1} i.v. for 6 h) and rat (2 × 30 mg kg^{-1} i.v., as the racemic compound), global spinal ischemia in the rabbit (16 mg kg^{-1} i.v. bolus, then 2.2 mg kg^{-1} i.v. for 1 h), and global ischemia in the gerbil (20 mg kg^{-1} i.p. then 4 × 10 mg kg^{-1}).
(5) The above studies demonstrate that LY293558 was clearly neuroprotectant in animal models of both focal (rat and cat) and global (gerbil, rabbit) ischemia. However, some degree of CNS depression, at least mild to moderate ataxia and sedation, would be expected to be associated with neuroprotectant doses of LY293558.

These data indicate that LY293558 has a preclinical pharmacological profile generally consistent with the known effects of other AMPA receptor antagonists such as NBQX, including anticonvulsant and, at higher doses, CNS depressant activity (Gill 1994). LY293558 was also

Table 4 Effect of LY293558 on hippocampal neuronal damage following 5 min of bilateral carotid artery occlusion in the gerbil[a]

Time of initial dose	Unoccluded sham	Untreated	LY293558
30 min before	255 ± 4 (5)	10 ± 1 (10)	132 ± 29 (9)**
Immediately after	186 ± 11 (5)	9 ± 1 (10)	71 ± 30 (8)*
30 min after	270 ± 10 (5)	9 ± 2 (10)	36 ± 11 (8)*
60 min after	310 ± 8 (5)	12 ± 2 (10)	85 ± 39 (9)
90 min after	270 ± 10 (5)	9 ± 2 (10)	14 ± 6 (9)

*p <0.05, **p <0.01.

[a]LY293558 was dosed at 20 mg kg^{-1} i.p. then at 10 mg kg^{-1} i.p. a further four times at 3 h intervals. The initial dose was given at the time indicated. Viable CA1 neurons were counted per millimeter in hematoxylin and eosin-stained sections of the brains removed when the gerbils were killed 5 days later.

demonstrated to have neuroprotectant properties in multiple animal models of ischemia (Table 5). LY293558 also has other desirable qualities, including a high aqueous solubility allowing for parenteral administration, a rapid onset of action, and a reasonable (1–4 h) duration of effect. This compound offers a way to test the role of non-NMDA receptors in ischemic neuronal injury in both animals and humans.

Table 5 Summary of animal ischemia studies

Animal model	Dose administered	Effect observed	Reference
Rat MCA occlusion (permanent)	2×30 mg kg^{-1} i.v. as (±) racemic compound at 30 min pre- and 30 min postocclusion	Significant (p <0.0001) decrease infarct volume (31% in cerebral cortex)	Gill and Lodge (1994)
Cat MCA occlusion (permanent)	15 mg kg^{-1} i.v. 30 min preocclusion then 7 mg kg^{-1} h^{-1} for 6 h	Significant (p < 0.02) decrease infarct volume (19% in cerebral cortex)	Bullock *et al.* (1994)
Rabbit spinal ischemia	16 mg kg^{-1} i.v. 5 min after recirculation (55 min postocclusion on average) then 2.2 mg kg^{-1} for 60 min	Significant (p < 0.01) increase in ischemic time to produce permanent paraplegia	Bowes *et al.* (1994)
Gerbil bilateral carotid artery occlusion (5 min)	20 mg kg^{-1} i.p., then 10 mg kg^{-1} every 3 h × 4	Significant protection of CA1 hippocampal neurons when treatment was initiated up to 30 min postocclusion (p <0.05)	A. Bond and D. Lodge (unpublished data)

8 The NBQX Story

LARS NORDHOLM, MALCOLM SHEARDOWN AND TAGE HONORÉ

Novo Nordisk, Novo Nordisk Park, DK-2760 Måløv, Denmark

1 SUMMARY

2,3-Dihydroxy-6-nitro-7-sulfamoylbenzo[*f*]quinoxaline (NBQX) was the first selective α-amino-3-hydroxy-5-methyl-4-isoxazolepropionic acid (AMPA) antagonist to demonstrate neuroprotective properties in animal models of focal and global cerebral ischemia. Furthermore, the compound has shown activity in additional models where excitatory amino acid antagonists may have beneficial effects (epilepsy, emesis, anesthesia, and Parkinson's disease).

In high doses (>12.5 mg kg^{-1} i.v.), NBQX acts as a neurodepressant with anesthetic and, to some extent, respiratory depressant effects. Potentiation of anesthetics has been demonstrated. NBQX does not possess abuse liability of the phencyclidine or opiate type.

The compound does not have a major impact on cardiac functions.

Continuous infusion for up to 4 weeks with 72 mg kg^{-1} day^{-1} (rats) or 24 mg kg^{-1} day^{-1} (dogs) does not induce signs of toxicity. Higher doses induce pharmacological effects as mentioned above, and, furthermore, lead to reversible nephrotoxicity due to up-concentration and precipitation of NBQX crystals in the kidneys.

No treatment-related clinical signs were observed when NBQX was administered to man in doses up to 0.03 mg kg^{-1} i.v. over 30 min. Due to the unfavorable combination of physical chemical properties and pharmacokinetics of the compound, no further escalation of the dose has been tried.

2 PHARMACOLOGY

2.1 *In vitro* studies and structure–activity relationship

The first quinoxalinedione to show any affinity at the high-affinity AMPA receptor site was compound (**a**) (Fig. 1). This unsubstituted compound displaced AMPA binding with an IC_{50} of 33 μM. The introduction of a nitro group in position 5 (compound (**b**) increased the potency at the AMPA receptor 30-fold, giving an IC_{50} value of 1.2 μM. The substitution of another electron-donating moiety such as the sulfonamido group (compound (**c**)) also improved potency compared to the unsubstituted compound (IC_{50} = 7.9 μM). Compounds (**b**) and (**c**) gave IC_{50} values of 9.2

EXCITATORY AMINO ACIDS – CLINICAL RESULTS WITH ANTAGONISTS
ISBN 0-12-546820-2

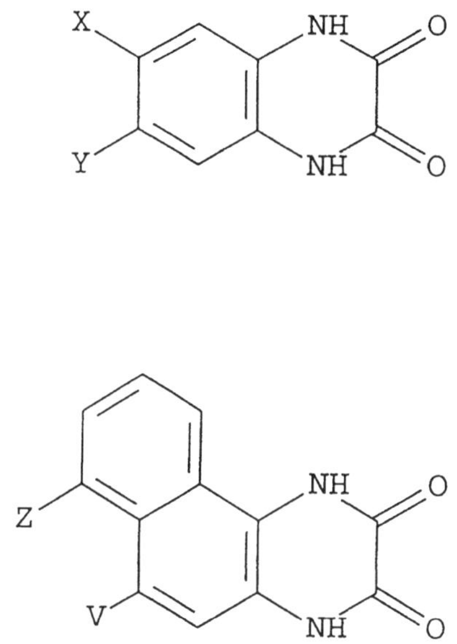

a: X=H, Y=H

b: X=NO_2, Y=H

c: X=SO_2NH_2, Y=H

d: X=NO_2, Y=NO_2

e: X=CN, Y=NO_2

f: Z=SO_2NH_2, V=NO_2

g: Z=SO_2NH_2, V=H

h: Z=SO_2NH_2, V=NH_2

i: Z=SO_2NH_2, V=Br

FIG. 1 Structures of quinoxalinediones closely related to NBQX.

and 32 μM, respectively, at the high-affinity kainate receptor site. A double substitution at positions 6 and 7 (compound (**d**), 6,7-dinitroquinoxaline-2,3-dione (DNQX)) further increased AMPA receptor affinity 60-fold in comparison to the unsubstituted compound, giving an IC_{50} at the AMPA receptor of 0.5 μM. Compound (**e**) with a cyano group at position 6 and a nitro group at position 7 (6-cyano-7-nitroquinoxaline-2,3-dione (CNQX) produced an IC_{50} value of 0.3 μM at the AMPA site. DNQX and CNQX, however, both showed considerable potency for the high-affinity kainate site with IC_{50} values of 2 and 1.5 μM, respectively. These compounds although being potent AMPA/kainate antagonists were also relatively potent *N*-methyl-D-aspartate (NMDA) antagonists. This activity was found to be related to their affinity for the NMDA-associated glycine site, and the IC_{50}-values were 9.5 μM (DNQX) and 14 μM (CNQX), respectively. The ratios, therefore, of glycine/AMPA affinity were 46 for CNQX and 19 for DNQX. These binding data were supported in a functional pharmacology assay using quisqualate-, kainate-, and NMDA-stimulated spreading depression in the chicken retina (Sheardown, 1989, 1993). In this assay CNQX gave IC_{50} values of 0.9, 1.3, and 17.2 μM against quisqualate, kainate, and NMDA, and for DNQX the IC_{50} values (μM) were 1.64, 1.44, and 4.2, respectively.

The incorporation of an extra benzene ring in the basic structure with a sulfonamide at position 7 and a nitro group at position 6 (Fig. 1, compound (**f**), NBQX) produced a compound of even higher affinity for the AMPA receptor (IC_{50} = 0.15 μM) but with a lower affinity at the kainate site (IC_{50} = 4.8 μM), and no affinity at the glycine site (IC_{50} > 90 μM). In the chick retina spreading depression NBQX was very potent against quisqualate (IC_{50} = 0.48 μM) and less potent against kainate (IC_{50} = 12 μM), thus being 24 times more potent against quisqualate than kainate responses. As predicted by the binding studies, NBQX did not inhibit NMDA-evoked responses in the chick retina.

Substituting the sulfonamido group in NBQX with H (compound (**g**) reduced affinity at the AMPA receptor (IC_{50} = 3.7 μM) but increased affinity at the kainate receptor (IC_{50} = 2.8 μM) and at the 3-(2-carboxypiperazine-4-y) propyl-1-phosphonic acid (CPP) site of the NMDA receptor (IC_{50} = 6.27 μM).

Substitution in the same position with either an amino group (compound (**h**) or a bromo group (compound (**i**) reduced AMPA receptor affinity approximately 40 times (IC_{50} = 6.7 and 6.1 μM, respectively). However, these two compounds were devoid of any ability to displace CPP binding.

NBQX is thus far the best compound with respect to potency and selectivity. The compound was tested in a wide range of other receptor assays without showing any inhibitory activity.

2.2 *In vivo* studies

The discovery of the quinoxalinedione family as competitive antagonists of AMPA-mediated responses gave the possibility of testing the importance of the AMPA system in different disease models. In particular, NBQX (Fig. 2), being a potent and almost specific AMPA antagonist, has proved a valuable tool in predicting the future for the use of AMPA antagonists as drugs.

2.2.1 *Epilepsy*

Excitatory amino acid transmitters are involved in the initiation of seizures and their propagation. Recent

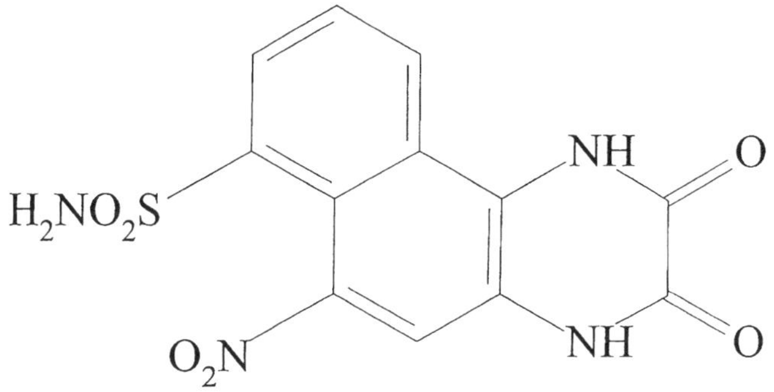

FIG. 2 2,3-dihydroxy-6-nitro-7-sulfamoylbenzo[*f*]quinoxaline (NBQX).

Molecular weight:	336.3 g mol^{-1}
Dissociation constants:	$pK_{aI} = 6.4$, $pK_{aII} = 10.5$
Solubility in aqueous solution (approx.):	0.1 mg ml^{-1} (pH 7), 1 mg ml^{-1} (pH 8), 30 mg ml^{-1} (pH 9)
Partition coefficient (log *p*):	–0.08 (pH 3.0), –1.18 (pH 7.6), –2.25 (pH 2.25)
Shelf-life:	at least 3 months at 4 °C in the dark
Source:	Novo Nordisk internal reports

evidence indicates potential roles for the AMPA receptors in experiments on amygdala-kindled rats, i.e. a model of partial epilepsy. NBQX potently increased the focal seizure threshold and inhibited seizure spread from the focus (Loescher *et al.*, 1993) NBQX protected against seizure induced by maximal electroshock (MES) and pentylenetetrazol. However, NBQX produced motor impairment at doses similar to those that were protective in the MES test (Yamaguchi *et al.*, 1993).

NBQX administered intraperitoneally 30 min. before sound stimulation protected against convulsion, with an ED_{50} of 3.5 mg kg^{-1} in mice sensitive to sound-induced seizures (DBA/2 mice) (Swedberg *et al.*, 1990). Furthermore, NBQX showed anticonvulsant activity against sound-induced seizures in genetically epilepsy-prone rats (GEPR-9), with an ED_{50} of 5.7 mg kg^{-1} i.p. (30 min) and against photically induced myoclonus in *Papio papio*, with an ED_{50} of 3.2 mg kg^{-1} i.p. (1 h). NBQX was without adverse behavioral effects at these doses (B. S. Meldrum, unpublished data).

2.2.2 *Spasticity*

NBQX, in a dose-dependent manner, reduced electromyographic activity in genetically spastic rats with a minimal effective dose of 1.7 mg kg^{-1} i.p. In doses up to 30 mg kg^{-1} i.p., NBQX did not affect the exploratory motility of normal Wistar rats (Turski *et al.*, 1992).

2.2.3 *Emesis*

NBQX dose dependently suppressed or prevented cisplatin-induced emesis in ferrets. A combination of a submaximal dose of NBQX and a submaximal dose of the 5-HT_3 antagonist zacopride completely abolished the emetic response (Fink-Jensen *et al.*, 1992).

2.2.4 *Anesthesia*

NBQX dose dependently prolonged the sleeping time of hexobarbital (Dall *et al.*, 1993). No indications of interactions on hexobarbital elimination of either isomer was found. It is therefore likely that NBQX acts synergistically with hexobarbital to depress the central nervous activity.

NBQX influences halothane minimum alveolar concentration (MAC) in the rat (McFarlane *et al.*, 1992). A maximal 58% reduction of halothane MAC was achieved with an NBQX loading dose of 42 mg kg^{-1} followed by a continuous infusion rate of 36 mg kg^{-1} h^{-1} (control = 1.02 ± 0.07%; NBQX = 0.43 ± 0.12%; $p < 0.01$). Mean arterial pressure was not affected by any dose of NBQX. The righting reflex was impaired in the high-dose NBQX group only.

2.2.5 *Amyotrophic lateral sclerosis*

Amyotrophic lateral sclerosis (ALS) is a neurodegenerative disorder affecting motor neurons. Cerebrospinal fluid (CSF) from patients with ALS was significantly neurotoxic when added to a rat neuronal culture, with a neuronal survival rate of only 47% compared with 80% or so for control CSF (Couratier *et al.*, 1993a). This neurotoxicity was blocked by CNQX. No neuroprotective effects were obtained with NMDA antagonists. ALS CSF contains a specific neurotoxic factor which is sensitive to AMPA antagonism. The possible neuroprotective effect of CNQX is suggested to be mediated by blockage of non-NMDA receptor activation of interneurons in the dorsal horn (Paleckova *et al.*, 1992).

2.2.6 *Parkinson's disease*

The excitation mediated by L-glutamate in the subthalamic nucleus, internal pallidal segment, and reticular part of substantia nigra is thought to play a role in the development of akinesia and muscular rigidity in Parkinsons's disease (Mitchell *et al.*, 1986; Miller and Delong *et al.*, 1987; Carpenter, 1981; Graybiel, 1990; Klockgether and Turski, 1989).

However, recent experiments have not been able to confirm the potential of NBQX in Parkinson's disease: intranigra infusion of CNQX at a dose of 1.0 nmol neither affected the contralateral rotation following systemic apomorphine administration into unilaterally dopamine-denervated rats nor the spontaneous rotation (Burns *et al.*, 1993).

NBQX treatment did not reverse the parkinsonism nor changed the responses to the selective D_2 dopamine receptor agonist (+)-PHNO or the partial dopamine D_1 receptor agonist CY208–243 in two parkinsonian monkeys (Luquin *et al.*, 1993).

2.2.7 *Arthritis*

The experimental arthritis in rats of the knee joint which leads to joint swelling, increased joint temperature, limping, guarding, and a decrease in paw withdrawal latency (PWL) to radiant heat (hyperalgesia) was significantly reduced after administration of CNQX in the spinal cord. Four hours after initiation of the arthritis, the degree of joint inflammation and PWL times returned to baseline levels (Sluka *et al.*, 1994).

2.2.8 *Cerebral ischemia*

Because of the excitotoxic properties of glutamate and in light of the findings of Benveniste *et al.* (1984) that glutamate is released during periods of cerebral ischaemia it has been suggested that excitatory amino acid receptors play a role in the loss of neuronal cells following cerebral ischemia. AMPA receptors occur in high density in both the hippocampus and the cerebral cortex (Honore *et al.*, 1988), and it was therefore a powerful working hypothesis that the AMPA receptor may be involved in the excitotoxic effects of glutamate, and that AMPA receptor antagonists may have neuroprotectant properties in models of cerebral ischemia.

Global ischemia NBQX was shown to be a neuroprotectant by Sheardown *et al.* (1990). In this study the compound was administered 1 or 2 h following a 5 min period of global cerebral ischemia in Mongolian gerbils. The compound produced a highly significant reduction of neuronal loss in the CA1 region of the hippocampus in both cases. These findings were extended to 6 and 24 h delay following the period of ischemia, and a highly significant reduction in neuronal loss following NBQX administration at both time points was observed (Sheardown *et al.*, 1993). In both of these studies the compound was not very potent, requiring three injections of a 30 mg kg^{-1} dose (administered intraperitoneally) (Fig. 3). It has been shown by

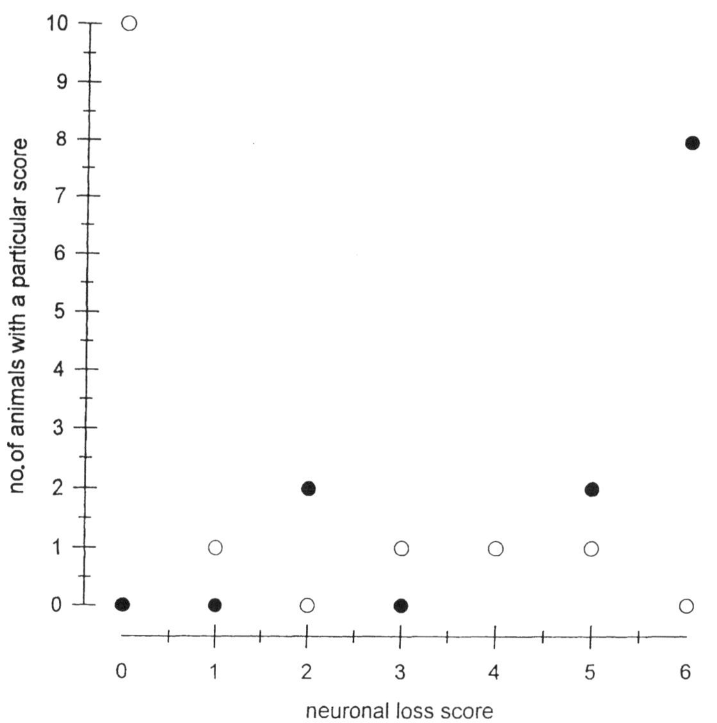

FIG. 3 Protective effect of NBQX (3 × 30 mg kg^{-1} i.p.) in the gerbil global ischemia model. The neuronal loss in hippocampus CA1 is scored on a scale from 0 to 6, where 6 is the worst outcome. ●, 5 min ischemia; ○, 5 min ischemia + NBQX starting 24 h after ischemia.

Judge et al (1991) that NBQX can prevent the increase in locomotor activity observed in mongolian gerbils following a period of global ischaemia.

The neuroprotective effect of NBQX in the CA1 region of the hippocampus following global ischemia has since been demonstrated in the rat four-vessel occlusion (Sheardown *et al.*, 1990; Buchan *et al.*, 1991a; Diemer *et al.*, 1992; Le Peillet *et al.*, 1992; Li and Buchan, 1993), the rat two-vessel occlusion (Nellgård and Wieloch, 1992; Lippert *et al.*, 1994), and the rat neck cuff model (Kaiser *et al.*, 1992; Diemer *et al.*, 1992). Li and Buchan (1993) demonstrated that NBQX was effective when administered up to 12 h following ischemia. Some of the above studies also showed protection in other brain regions such as the striatum and the CA3 region of the hippocampus (Le Peillet *et al.*, 1992), the neocortex (Nellgård and Wieloch, 1992; Le Peillet *et al.*, 1992), and the cerebellum (Diemer *et al.*, 1992).

Neuroprotection has also been demonstrated in a non-rodent species. Redmond *et al.* (1995) showed that NBQX given intravenously following hypothermic circulatory arrest in the dog resulted in both a neuroprotective effect and a reduction in neurological deficit.

Global ischemia is a model of the type of ischemia observed clinically following cardiac arrest. The data reviewed thus far may suggest, therefore, that AMPA antagonists could be of clinical benefit in surviving cardiac arrest patients or in cardiac surgery.

An overview of global ischemia experiments carried out with NBQX is outlined in Table 1.

Focal ischemia In focal ischemia (stroke) models, NBQX is also an effective neuroprotectant. Protection in the rat temporary middle cerebral artery (MCA) occlusion was first shown by Buchan *et al.* (1991b) in spontaneous hypertensive rats, where they obtained ~32% reduction in infarct volume. In the same study, MK-801 (dizocilpine) was ineffective. Gill *et al.* (1992) and Smith and Meldrum (1993) have both shown a reduction in infarct volume using permanent occlusion of the rat MCA. Smith and Meldrum demonstrated a time window after onset of ischemia of between 1 and 2 h. Gill *et al.* (1992) and Xue *et al.* (1994) have shown that NBQX can reduce the infarct volume following MCA occlusion when administered intravenously.

NBQX also reduces infarct volume following occlusion of the mouse MCA. A. J. Hansen (unpublished data) found a 43% reduction in infarct volume in mice following permanent MCA occlusion using three doses of 30 mg kg^{-1} i.v.

The data from these MCA studies indicate that AMPA antagonists may be of clinical benefit in the treatment of ischemic stroke.

An overview of focal ischemia experiments with NBQX is outlined in Table 2.

Spinal and cerebral trauma Wrathall *et al.* (1992, 1994) have shown that NBQX in a dose-dependent manner reduces both tissue loss and functional impairment following spinal cord trauma, and von Euler *et al.* (1994) have demonstrated that NBQX could also reduce the degeneration caused by focal spinal cord ischemia. Such a compound could prove to be of benefit in the treatment of spinal injury.

NBQX is also effective in protecting against damage caused by cerebral trauma. Bernert *et al.* (1992) have demonstrated that NBQX can reduce the hippocampal CA1 neuronal loss observed following severe cerebral trauma, suggesting a further potential clinical use for AMPA antagonists in head injury.

Brain edema Cerebral ischemia and trauma is often complicated by the apperance of brain edema. In a model of brain edema it was shown that blockade of AMPA receptors by NBQX could reduce brain edema in two experimental models of edema following opening of the blood–brain barrier (Westergren and Johansson, 1993).

Conclusion In conclusion, NBQX can reduce the neuronal loss and neurological deficits observed following a wide range of ischemic and traumatic insults. This suggests that AMPA antagonists may be of benefit in the treatment of cardiac arrest, complications following cardiac surgery, stroke, head injury, and spinal injury.

2.3 Toxicity and side-effect profile

Most of the information in this section is unpublished data from Novo Nordisk internal reports. However, a few relevant published papers are included as well.

2.3.1 *Acute toxicity*

In mice (five males and five females) the maximum non-lethal dosage by single intravenous bolus administration over 30 s was higher than 200 mg kg^{-1}, with a no-effect level of 6.25 mg kg^{-1}. Observed effects were first ataxia and recumbancy (12.5 mg kg^{-1}), with increasing dose, decreased motor activity, loss of righting reflex, decreased respiration (100 mg kg^{-1}), ptosis, and prostration were seen. Other effects were sedation, lacrimation, mydriasis, and decreased body temperature. All the observed effects disappeared within 8 h even in the highest dose studied (200 mg kg^{-1}). Autopsy was performed 14 days after dosing. No abnormalities were observed.

A similar study was performed in rats with bolus administration over 60 s. Observations were almost identical to observations in mice with the same tolerability. NBQX showed pronounced anesthetic properties. Also in this study, autopsies were performed 14 days following treatment without observation of treatment-related

Table 1 Neuroprotective effects of NBQX in global ischaemia models – type of protection indicated without quantitative data

Species	Model	Dosage regimen	Protection	Reference
Gerbil	2-VO (5 min)	3 × 30 mg kg^{-1} i.p. pre/postischemia	CA1	Sheardown *et al.* (1990)
	2-VO (5 and 10 min)	3 × 30 mg kg^{-1} i.p. (60, 70, 85 min after ischemia)	CA1 Postischemic hyperactivity	Judge *et al.* (1991)
	2-VO (5 min)	3 × 30 mg kg^{-1} i.p.		Sheardown *et al.* (1993)
		6 h after ischemia	CA1	
		12 h after ischemia	CA1	
Rat	4-VO (10 min)	3 × 30 mg kg^{-1} i.p. (0, 15, 30 min after ischemia)	CA1	Buchan *et al.* (1991a)
	4-VO (20 min)	10 mg kg^{-1} i.v. + 10 mg kg^{-1} h^{-1} (3 h)	Striatum Cortex CA3	Le Peillet *et al.* (1992)
	Neck cuff + hypotension (10 min)	3 × 30 mg kg^{-1} i.p. (0, 30, 60 min after ischemia)	CA1	Kaiser *et al.* (1992)
		15 min postischemia 15 mg kg^{-1} over ½ h + 30 mg/kg over ½ h + 2 mg kg^{-1} h^{-1} for 23 h[a]	CA1	F. Kaiser (unpublished data)
		72 μg day^{-1} for 7 days using a micro-osmotic pump with delivery to a brain ventricle	CA1	A. Fink-Jensen and J. B. Hansen (unpublished data)
	4-VO + hypotension (10 min)	3 × 30 mg kg^{-1} i.p. pre/postischemia	CA1	Diemer *et al.* (1992)
		(0, 10, 25 min after ischemia)	CA1	
		(60, 70, 85 min after ischemia)	CA1	
	2-VO + hypotension (10 min)	30 mg kg^{-1} i.p. 0 min after ischemia + 4.5 mg kg^{-1} h^{-1} i.v. (6 h)	CA1 Neocortex	Nellgård and Wieloch (1992)
	4-VO (10 min)	90 mg kg^{-1} i.v.	No protection	Li and Buchan (1993)
		3 × 30 mg kg^{-1} i.p. (0, 15, 30 min)	CA1	
		6 h delay	CA1	
		12 h delay	CA1	
		24 h delay	No protection	
	2-VO + hypotension (10 min)	3 × 30 mg kg^{-1} i.p. (0, 10, 25 min)	CA1	Lippert *et al.* (1994)
		Combined with dizocilpine	CA1	
Piglet	Hypothermic circulatory arrest	25 mg kg^{-1} i.v. before cardiopulmonary bypass	No protection	Aoki *et al.* (1994)
Dog	Hypothermic circulatory arrest	3 mg kg^{-1} i.v. for 3 h + 1.5 mg kg^{-1} for 2 h Start 2 h after cardiopulmonary bypass	Neuronal Neurological	Redmond *et al.* (1995)
	Elevated intracranial pressure	Infusion 1 mg kg^{-1} over 3 min + 3 mg kg^{-1} h^{-1} for 4 h Start 5 min postischemia	No protection	Lanier *et al.* (1993)

CA1/3, CA1 or CA3 – area of the hippocampus; 2-VO, two-vessel occlusion; 4-VO, four-vessel occlusion.
[a]The dosage regimen gives the same plasma concentration profile as 3 × 30 mg kg^{-1} i.p.

abnormalities. Microscopically, examinations of the kidneys, jejunum, and ileum did not reveal any abnormalities even following doses of 200 mg kg^{-1}.

Acute studies in dogs were performed with short-term infusion (<24 h). At doses of 1.3 mg kg^{-1} h^{-1} for 6 h or more, dose-related subdued behavior or agitation, salivation, frequent micturations, loss of balance, or slight ataxia were observed. These symptoms started at NBQX plasma concentrations of 700 ng ml^{-1}.

2.3.2 *Subacute toxicity studies*

Continuous infusion of 72 mg kg^{-1} day^{-1} for 4 weeks was clinically well tolerated by rats. No clinically or

Table 2 Neuroprotective effects of NBQX in focal ischemia models – protection refers to the percentage reduction in infarct volume unless stated otherwise

Species	Model	Dosage regimen	Protection (%)	Reference
SHR	MCAO (2 h)	3 × 30 mg kg^{-1} i.p.	32	Buchan *et al.* (1991b)
		(90, 120, 150 min after onset of ischemia)	24	Xue *et al.* (1994)
		Same regimen + MK-801	NS	Xue *et al.* (1994)
		7.5 mg kg^{-1} h^{-1} i.v. for 4 h Start 1 h after onset	35	
	MCAO (1 h)	3 × 30 mg kg^{-1} i.p. (30, 60, 90 min after onset of ischemia)	NS	DeGraba *et al.* (1994)
		5 mg kg^{-1} h^{-1} i.v. for 6 h Start 0 h after ischemia	NS	
	MCAO permanent	Same regimen	NS	
Rat	MCAO permanent	2 × 3 mg kg^{-1} i.v. (0, 1 h)	NS	Gill *et al.* (1992)
		2 × 10 mg kg^{-1} i.v. (0, 1 h)	NS	
		2 × 30 mg kg^{-1} i.v. (0, 1 h)	27	
		30 mg kg^{-1} i.v. (0 h) + 10 mg kg^{-1} h^{-1} for 4 h	35	
		3 × 10 mg kg^{-1} i.p. (0, 30, 60 min)	NS	Smith and Meldrum (1993)
		3 × 20 mg kg^{-1} i.p. (0, 30, 60 min)	70	
		3 × 30 mg kg^{-1} i.p. (0, 30, 60 min)	75 + neurological	
		(60, 90, 120 min)	58% + neurological	
		(120, 150, 180 min)	neurological	
	Embolic stroke model	3 × 30 mg kg^{-1} i.p. (90, 120, 150 min)	77	Meden *et al.* (1993) Overgaard *et al.* (1993)
		+ TPA treatment (120 min) Same regimen	Neurological	Overgaard *et al.* (1993)
Rat pups	Hypoxic–ischemic brain damage	2 × 15 mg kg^{-1} i.p. (0, 1 h after ischemia)	NS	Hagberg *et al.* (1994)
		2 × 20 mg kg^{-1} i.p. (0, 1 h after ischemia)	28	
Mouse	MCAO	2 × 50 mg kg^{-1} i.p. (15 min before and 15 min after ischemia)	16	Lippert *et al.* (1994)
		Same regimen + MK-801 1 h preischemia	32	
		3 × 30 mg kg^{-1} i.v. (0, 30, 60 min)	43	Hansen (unpublished data)
		3 × 30 mg kg^{-1} i.v. (60, 90, 120 min)	NS	

MCAO, middle cerebral artery occlusion; SHR, spontaneous hypertensive rat; TPA, recombinant tissue plasminogen activator; NS, no significant protection (p >0.05).

histopathologically treatment-related abnormalities were registered at this dose. Steady state plasma concentrations were of the order of 1000 ng ml^{-1}. Higher doses (166 or 500 mg kg^{-1} day^{-1} for 14 days) led to renal lesions (tubulo-interstitial nephritis). The lesion were directly correlated with precipitation of NBQX crystals in the tubuli.

The crystal-induced tubular nephritis was a result of an active tubular secretion of the acidic NBQX, leading to concentrations which exceeded the solubility of the compound. Attempts to prevent tubular secretion with probenecid were not successful. Increasing the diuresis simultaneously with NBQX treatment did not improve the histopathological outcome to any significant extent. The nephrotoxicity was shown to be reversible (B. Guldhammer, unpublished data).

NBQX administered intraveneously by continuous infusion at dose levels up to 24 mg kg^{-1} day^{-1} to beagle dogs for 4 weeks did not induce any signs of toxicity. A dose level of 48 mg kg^{-1} day^{-1} was terminated after 17 h due to severe clinical signs (recumbancy, loss of balance, tremor, subdued behavior, and vomiting). Plasma concentrations of NBQX were in the range 1200–2100 ng ml^{-1} after

48 mg kg^{-1} day^{-1} and 800–1100 ng ml^{-1} after 24 mg kg^{-1} day^{-1}.

2.3.3 *Safety pharmacology*

NBQX showed low toxicity to the isolated rabbit heart. A drug concentration of 40 µg ml^{-1} did not induce effects on cardiac contractility or heart rate, whereas coronary flow decreased by 30%. The electrocardiogram (ECG) was unchanged (Nielsen-Kudsk *et al.*, 1991).

Intravenous administration of 16 mg kg^{-1} over 30 min to pentobarbital anesthetized rats caused a minor decrease in heart rate and blood pressure. Plasma concentrations were at the 10 µg ml^{-1} level.

Cardiovascular and respiratory effects in anesthetized cats revealed a pronounced interaction between pentobarbital and NBQX, leading to respiratory arrest at plasma concentrations of 10 µg ml^{-1} or higher (0.1–0.3 mg kg^{-1} min^{-1} for 35 min). Also, interaction with NMDA antagonists resulting in respiratory depression has been described (Foutz *et al.*, 1994). Cardiovascular parameters such as blood pressure, heart rate, response to carotid occlusion and ECG were only affected as a consequence of respiratory arrest.

Potentiation of anesthetic effects has been demonstrated when NBQX is combined with halothane (McFarlane *et al.*, 1992) or hexobarbital (Dall *et al.*, 1993).

NBQX does not possess abuse liability of the phencyclidine or opiate type demonstrated in a rat discrimination test (Swedberg *et al.*, 1991).

2.4 Pharmacokinetics

NBQX pharmacokinetics has been studied in a number of species, including man. A summary of pharmacokinetic parameters following intraveneous administration is outlined in Table 3.

One of the major problems with NBQX is, as already mentioned, the low solubility and thus the risk of precipitation of the compound in the kidneys. Despite the fact that NBQX has a low solubility in aqueous media, it is a very polar compound. This results in a poor blood–brain barrier permeability. When measuring CSF concentrations in gerbils and rats following administration of NBQX, less than 1% of the corresponding plasma concentrations can be recorded (Sheardown *et al.*, 1993; L. Nordholm, unpublished data). This means that NBQX has to be administered in relatively high doses in order to get an anti-ischemic effect. It has been suggested that the effect is caused by other factors such as metabolites formed following intraperitoneal administration or even unknown processes resulting from induction of a peritonitis. Metabolite profiling did not support the theory of an active metabolite, as the metabolites isolated were much less potent on the AMPA receptor than NBQX (L. Nordholm, unpublished data). Furthermore, a direct administration of NBQX into rat brain ventricles did demonstrate neuroprotection in a global model of cerebral ischemia (A. Fink-Jensen and A. J. Hansen, unpublished data). A more reliable theory is that NBQX is precipitating in peritoneum following intraperitoneal administration, making a sustained-release type administration form. This theory is supported by the fact that intravenous infusion of the compound by use of a dosage regimen resulting in the same plasma concentration profile as intraperitoneal administration demonstrates protection in global cerebral ischemia (F. Kaiser, unpublished data).

3 HUMAN PHARMACOLOGY

A single intravenous dose tolerance study in healthy normal volunteers with preliminary pharmacokinetic assessment has been performed (Ingwersen *et al.*, 1994).

The study was carried out as a double-blind, placebo-controlled investigation with six male volunteers on the active drug and two on a placebo at each dose level (age 18–37 years). NBQX was administered by an infusion pump over 30 min. Two dose levels were investigated, 0.0075 and 0.03 mg kg^{-1}. In order to facilitate diuresis and thereby prevent any theoretical possibility of precipitation of the substance in the kidneys, the volunteers received 1 l of 5% glucose infusion solution during the last hour before administration of NBQX (30 ml/75 kg).

The volunteers underwent physical examination, measurements of blood pressure, pulse rate, body temperature, ECG, and assessment of adverse events. Laboratory examinations included hematology, biochemistry, urinalysis (including sediment analysis), and analysis of NBQX in plasma and urine.

Table 3 Key pharmacokinetic parameters for NBQX following intraveneous administration

Parameter	Mice	Rats	Dogs	Man
R_{in} (mg kg^{-1} h^{-1})	3, 10, 30[a]	0.5–10	0.2–5	0.015–0.06
$t_{1/2}$ (hours)	1–4	0.8	1–3	0.75
CL (l kg^{-1} h^{-1})	0.6–1.0	3.2	1.5	0.22
CL_R (l kg^{-1} h^{-1})		0.9		0.13
V_Z (l kg^{-1})	1–4	4	1–3	0.24

Data compiled from Dalgaard *et al.* (1994) and Ingwersen *et al.* (1994).
[a] Bolus administration.

No drug-related effects were encountered from any of the parameters investigated. Examination of urine for parameters such as microalbumin, creatinine, α_1 microglobulin, and *N*-acetylglucosamide as well as microscopic assessment did not reveal any indication of kidney damage. The highest NBQX concentration measured in urine was 3.28 μg ml^{-1}. NBQX was well tolerated at the dose levels investigated with concomitant forced diuretic conditions.

The maximal plasma concentration measured just before termination of infusion was linearly correlated to dose (41.2 and 177 ng ml^{-1}, respectively), which was also the case for the area under the curve (AUC) (35.9 and 134 h × ng ml^{-1}, respectively). Other pharmacokinetic parameters (see Table 3) were independent of the dose level.

A major part of the NBQX dose was excreted unchanged in urine (38.5–76.8%), and there were no signs of any known metabolites in chromatograms from plasma or urine samples. The study was carried out with non-labeled substance, and the presence of metabolites can therefore not be totally excluded. It is, however, very probable that the remaining part of the dose was excreted in faeces, as has been observed in studies with ^{14}C-labeled NBQX in rats (L. Nordholm, unpublished data).

The maximal plasma concentration was one-quarter of the plasma concentration, leading to significant clinical signs in the 4 week dog toxicity study, and one-fifth less than the plasma concentration measured at the no-effect level in the 4 week rat toxicity study. Anti-ischemic effects of NBQX have, on the other hand, only been demonstrated in considerably higher doses. Plasma concentrations in successfully performed ischemic studies in rats are estimated to be at least 5 μg ml^{-1}. Studies with an AMPA antagonist in stroke patients should therefore be performed with as high a dose as possible, and minor adverse events such as sedation are acceptable. It is, however, not feasible to treat this population with doses of NBQX which potentially are nephrotoxic. As the solubility of NBQX is very pH-dependent, with a very low solubility in acidic solution, it was decided to measure the solubility of NBQX in a large population (approximately 100 subjects) of human urine samples (S. H. Ingwersen, unpublished data). It was concluded that the solubility in a number of cases was so low that this could give problems with precipitation in the kidneys of patients in the target population. Doses demonstrating clinical signs in volunteers (e.g. sedation) could, furthermore, be on the edge of the urinary concentration, causing precipitation of crystals in the tubuli.

Further clinical development of NBQX was therefore stopped.

4 DISCUSSION

A wide range of animal models has demonstrated a potentially beneficial effect of NBQX in a number of indications. Being an excitatory amino acid antagonist with neurodepressant activitities, AMPA antagonists will probably best find their position as drugs for the treatment of ischemic or traumatic injuries or as anesthetic tools.

NBQX is as efficacious as the NMDA antagonists in most models of focal ischemia and does not seem to have the psychotomimetic side-effect problem.

In models of global ischemia, NBQX shows protection even when treatment is started several hours after the ischemic insult. Experiments carried out in nonrodent models of hypothermic circulatory arrest indicate, in fact, that treatment will only be beneficial if postponed 2 h relative to cardiopulmonary bypass (J. M. Redmond, personal communication).

NBQX is, however, not an ideal drug candidate due to the unfavorable physical chemical properties. AMPA antagonists with completely different structures have fortunately also demonstrated neuroprotective effect in models of cerebral ischemia (Le Peillet *et al.*, 1992; Madden *et al.*, 1992). It is, therefore, our hope that a suitable AMPA antagonist will soon find its way to clinical trials and demonstrate a neuroprotective effect in patients suffering from stroke or other ischemic brain disorders.

9 Riluzole in Amyotrophic Lateral Sclerosis

ERIK LOUVEL

Rhône-Poulenc Rorer Japan, Research and Development Division, 13–1, Kachidoki 1-chome, Chuo-Ku, Tokyo 104, Japan

1 SUMMARY

Riluzole is the first drug that has been demonstrated to increase the survival of patients suffering from amyotrophic lateral sclerosis (ALS), a rapidly fatal neurodegenerative disease affecting motoneurons (MNs), in two double-blind placebo-controlled studies since the disease was described by Charcot in 1874. In animal pharmacological studies, riluzole has shown potent neuroprotective properties in models of global and focal ischemia. It has also been shown to increase survival of transgenic mice expressing the human mutated SOD-1 (superoxide dismutase 1) considered as the first accurate model of familial ALS. In terms of mechanism(s) of action, riluzole may involve a direct, but noncompetitive, blockade of excitatory amino acid (EAA) receptors, an inhibition of glutamic acid release, an activation of a G-dependent signal transduction pathway, and a blockade of voltage-dependent Na channels in the inactivated state, or other ones still unknown. The synergy of these, neither exclusive nor necessarily independent, mechanisms would endow riluzole with its unique pharmacological, clinical, and safety profile.

2 INTRODUCTION

ALS is a rapidly progressive and fatal neurodegenerative disease which affects the central and peripheral MNs. The

EXCITATORY AMINO ACIDS – CLINICAL RESULTS WITH ANTAGONISTS
ISBN 0-12-546820-2

princeps clinical description of the disease was given by Charcot more than one century ago (Charcot, 1874). The disease is characterized by a rapidly evolutive amyotrophy, fasciculations, and pyramidal signs. The involvement of respiratory muscles leads to respiratory failure and death. The annual incidence is 2.4 cases/100 000. The median survival is currently 3 years after the onset of the first clinical symptoms. So far, the etiology of the disease is unknown. The age of onset and the site of onset of the first symptoms (either limb onset or bulbar onset) are prognostic factors. From 5 to 10% of ALS are autosomal dominant forms (familial ALS (FALS)). Recently, mutations affecting the SOD-1 gene have been reported (Rosen *et al.*, 1993) in about 20% of FALS.

3 CHEMICAL STRUCTURE

Riluzole, 2-amino-6-trifluoromethoxybenzothiazole (Fig. 1), has been discovered and developed by Rhone-Poulenc Rorer. It belongs to the benzothiazole chemical family.

Riluzole is rapidly and extensively absorbed when administered orally. It easily crosses the blood–brain barrier.

4 NEUROPROTECTIVE PROPERTIES

4.1 Neuroprotective effects *in vitro*

Riluzole has demonstrated neuroprotective properties in several *in vitro* models of neurodegeneration. It prevented anoxic injury in cultured rat cerebellar neurons and inhibits cell death, with an EC_{50} of 30 μM (Dessi *et al.*, 1993). In a model of EAA-related neurotoxicity on MN-enriched rat cultures, riluzole showed protective effects against glutamic acid with a minimal effective concentration of 0.1 μM (Estevez *et al.*, 1995).

Rothstein and co-workers have proposed a model of chronic neurotoxicity on rat organotypic spinal cord cultures in which the toxicity is induced by a long-term inhibition of glutamate uptake by transport inhibitors such as *threo*-hydroxyaspartic acid or pyrrolidine dicarboxylic acid (Rothstein *et al.*, 1993). In this model, riluzole (0.1 mM) was shown to be neuroprotective in increasing MN survival (Rothstein and Kluncl, 1995). Another model has been developed by Couratier and co-workers (1993b)

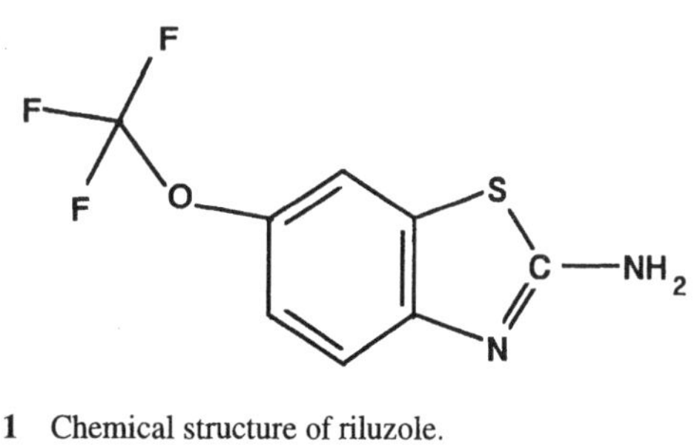

FIG. 1 Chemical structure of riluzole.

using the cerebrospinal fluid (CSF) of patients suffering from ALS. ALS CSF has been shown to be neurotoxic for cortical neurons in rat primary cultures. The toxic factor has not been identified yet. However, its toxicity is mediated via α-amino-3-hydroxy-5-methyl-4-isoxazolepropionic acid (AMPA) kainate receptors because it is reversed by 6-cyano-7-nitroquinoxaline-2,3-dione (CNQX) and not by MK-801. In this model, riluzole (0.5 μM) attenuated neuronal death and increased the cell survival rate from 45 to 61% (Couratier *et al.*, 1994).

4.2 Neuroprotective effects *in vivo*

Riluzole is active in various models of global cerebral ischemia. In a model of hypobaric hypoxia model in mice, riluzole increased dose dependently the survival time. This neuroprotection was statistically significant from a dose of 4 mg kg^{-1} i.p. (unpublished data). In a model of global ischemia by transient bilateral clamping of both carotid arteries in Mongolian gerbils, riluzole (4 mg kg^{-1} i.p., 0.5, 4.5, 24.5, and 28.5 h after clamping) prevented hippocampal neuronal damage in the ischemic gerbils in comparison with controls (Malgouris *et al.*, 1989; Pratt *et al.*, 1992). The effect of the product against hypoxia was also assessed on the animal electrocorticogram (ECoG). Riluzole, at doses of 2, 4, and 8 mg kg^{-1} i.p. administered after clamping, provided a significant dose-dependent protection in reducing the ischemia-related slow waves and in increasing the proportion of high-frequency activities (Pratt *et al.*, 1992). Using the model of transient global ischemia by the four-vessel occlusion in rats, it is possible to measure the release of glutamic acid by microdialysis. Riluzole at a dose of 8 mg kg^{-1} i.p. administered 20 min before ischemia initiation, completely abolished the release of glutamic acid. In addition, the product significantly reduced the lesion size in the treated ischemic animals in comparison to the nontreated ischemic animals (M. Plotkine, O. Ghribi, A. Doble, R. G. Boulu, unpublished data). In a model of definitive focal ischemia provoked by occlusion of the middle cerebral artery (MCAO), riluzole (8 mg kg^{-1} i.p.) administered 0.5 and 24.5 h after occlusion significantly decreased the volume of the cortical infarct from 99.2 ± 8.1 to 71.3 ± 8.6 mm^3 (Pratt *et al.*, 1992). Wahl and co-workers (1993) have shown that single doses of riluzole (4 and 8 mg kg^{-1} i.p. administered 30 min after the cauterization) decreased the size of the cortical lesion on a dose-dependent manner: 94 ± 12 and 73 ± 15 mm^3 respectively in treated animals versus 139 ± 8 mm^3 in nontreated ischemic animals. A histotoxic anoxia can be induced by potassium cyanide (KCN). KCN, an irreversible ligand of the mitochondrial cytochrome *c*-oxidase complex 4, blocks the mitochondrial respiratory chains. Pretreatment by riluzole protected mice from the effects of histotoxic anoxia induced by KCN, with an ED_{50} of 4.6 mg kg^{-1} i.p. (unpublished data).

Until recently, there was no accurate model of ALS. In 1993, Rosen and co-workers showed that some forms of autosomal dominant FALS are linked to mutations of the gene coding for the Cu/Zn-binding superoxide dismutase (SOD-1). This finding has led to the development of a transgenic mice strain overexpressing the human mutated gene and has opened new avenues to dissect the chain of pathological events leading to MN degeneration. Transgenic mice expressing high levels of human SOD-1 containing a $Gly^{93} \rightarrow$ Ala substitution develop an MN disease (Gurney *et al.*, 1994). The clinical features are accompanied by a severe loss of choline acetyltransferase (ChAT)-containing neurons mainly in the ventral spinal cord. Riluzole (100 $\mu g\ ml^{-1}$ in drinking water) increases the survival of mice overexpressing the mutated human SOD-1 (Gurney *et al.*, 1996).

5 MECHANISM(S) OF ACTION

An overview of riluzole mechanism(s) of action was recently made by Doble (1996). Mizoule and colleagues have been the first to report that riluzole may block the glutamatergic transmission (Mizoule *et al.*, 1985). The original work was based upon *in vivo* studies reporting that riluzole prevented L-glutamate and kainate-induced seizures. Since riluzole was shown not to bind directly to any of the known binding sites on EAA receptors (Benavides *et al.*, 1985; Debono *et al.*, 1993), the question of the intimate mechanism(s) of action of the product was raised. Electrophysiological studies on EAA receptors expressed in *Xenopus* ooytes have shown that riluzole interacts directly, but noncompetitively, with the NMDA receptor (IC_{50} = 18 μM), and to a lesser extent with the kainate receptor (IC_{50} = 167 μM) (Debono *et al.*, 1993). *In vitro*, Hubert and co-workers (1994) reported that riluzole blocks glutamic acid and NMDA-induced calcium mobilization in cultured rat cerebellar granule cells. This effect is abolished by pertussis toxin, and this suggests that riluzole may activate a G-dependent transductionnal process.

Riluzole is a potent and highly specific blocker of inactivated Na channels since it preferentially blocks inactivated rather than resting or open Na channels (efficacy ratio >300) in frog (*Rana esculenta*) sciatic nerves (Benoit and Escande, 1991). The same profile is observed in recombinant receptors expressed in *Xenopus* ooytes (K_i = 0.2 μM) (Hebert *et al.*, 1994). Riluzole binds to the batratoxin site on the voltage-dependent Na channel with an IC_{50} of 10 μM. Through this mechanism of action, riluzole may prevent neuron overactivation. However, some riluzole effects, such as inhibition on glutamic acid release or antagonism of an increase in intracellular calcium concentration, are observed in the presence of tetrodotoxin, and thus in the absence of Na channel activity. So, it is unlikely that this molecular mechanism underlies all of the neuroprotective properties of riluzole (Doble, 1996).

Riluzole inhibits the evoked release of glutamic acid, both *in vitro* and *in vivo* (Hubert and Doble, 1989; Cheramy *et al.*, 1992; Martin *et al.*, 1993). On cultured rat cerebellar granule cells this effect remains in the presence of tetrodotoxin (TTX) (Martin *et al.*, 1993), and is blocked by pertussis toxin (Doble *et al.*, 1992), which suggests that at least two different molecular mechanisms are involved. This inhibition of glutamic acid release is not sufficient *per se* to explain some riluzole postsynaptic properties (i.e. blocking properties on the effects of iontophoretically applied EAAs on MN firing in the rat facial nucleus *in vivo*, or antagonistic properties *vis-à-vis* glutamic acid-induced calcium mobilization *in vitro*), and cannot explain all of the neuroprotective properties of riluzole.

Despite the exact molecular mechanisms of riluzole not being fully known, riluzole appears to be associated (Doble, 1996) with a direct, but noncompetitive, blockade of EAA receptors, an inhibition of glutamic acid release, an activation of a G-dependent signal transduction pathway, and a blockade of voltage-dependent Na channels in the inactivated state. These various mechanisms are not necessarily independent and exclusive. Their relative weights in the neuroprotective properties of riluzole remain to be determined. However, and as suggested by Doble (1996), their synergy would endow riluzole with its unique pharmacological and safety profile.

6 RATIONALE OF CLINICAL TESTING

At the time of the clinical program initiation, the excitotoxic hypothesis was one hypothesis among several, and was considered as highly speculative. Since then, and despite the fact that ALS etiology is still unknown and that the excitotoxic hypothesis still remains speculative, there is now further evidence suggesting that EAAs might play a role in ALS.

6.1 Excitotoxic hypothesis

The possible role of glutamate as a trigger of the pathophysiological cascade leading to neurodegeneration was suggested by Lucas and Newhouse (1957) and Olney (1969), and is the origin of the 'excitotocixity' concept (Olney *et al.*, 1971). This glutamate neurotoxicity has been demonstrated to be calcium-dependent (Choi, 1985). An activation of glutamate receptors induces an increase in intracellular calcium concentrations. More recently, Lipton and Rosenberg (1994) have proposed that an overstimulation of glutamate receptors may play a role, as a final common pathway, in diverse neurodegenerative processes. There is currently no direct evidence of a trigger role for glutamate in ALS. In addition, and until the recent clinical results obtained with riluzole, all the clinical trials performed with products interfering with the glutamatergic transmission (branched amino acids, dextrometorphan, lamotrigine) have failed to demonstrate an efficacy in ALS. Moreover, the recent findings of the mutated SOD-1 gene in patients with FALS does not support *a priori* this hypothesis. However, the growing understanding that interactions between EAAs and free radicals (i.e. EAA receptor activation can generate free radical formation; Lafon-Casal *et al.*, 1993) or vice versa, allows a comprehensive approach of MN degeneration in which EAAs may play a role (Louvel *et al.*, 1996). Whatever the exact role of EAAs (trigger, secondary event, or toxic factor on selectively vulnerable MNs), there is now some evidence for a role of EAAs in the pathological process leading to the clinical features.

Three rare disorders in which MNs are affected, Guam disease, lathyrism, and domoic acid intoxication, provide significant insights into the role of excitotoxic agents in MN degeneration.

Gam disease is a disorder associating ALS, parkinsonism, and dementia, affecting the Chamorro population of the Guam and Rota islands, and is related to the consumption of *Cycad circinalis* L., a traditional source of food (Spencer *et al.*, 1987). The cycad flour contains an EAA, β-*N*-methylamino-L-alanine (BMAA). BMAA activates NMDA receptors at high concentrations, and non-NMDA and metabotropic receptors at low concentrations (Spencer *et al.*, 1994). Repeated administration of BMAA to monkeys induced upper and lower MN deficit with parkinsonian features and behavioral disorders associated with degenerative changes of motor neurons in the cerebral cortex and in spinal cord (Spencer *et al.*, 1987).

Lathyrism is a form of upper MN disease induced by consumption of the chick-pea *Lathyrus sativus* (Spencer *et al.*, 1986). This vegetable contains a potent EAA, β-*N*-oxalylamino-L-alanine (L-BOAA). L-BOAA is a glutamate agonist with a preferential affinity for the AMPA receptor (Spencer *et al.*, 1994). Monkeys fed with L-BOAA develop signs of corticospinal deficit (Spencer *et al.*, 1986). Excitotoxicity by excessive activation of AMPA receptors is clearly in this disorder the trigger of a pathophysiological cascade of events leading to MN degeneration. Domoic acid (DOA) is responsible for cognitive disorders and a lower MN deficit in consumers of mussels contaminated by the toxin (Teitelbaum *et al.*, 1990). DOA is a potent kainate agonist. When administered to rats, kainic acid induced MN degeneration in the spinal cord, and hindlimb paraplegia (Hugon *et al.*, 1989). Despite the fact that the mechanism(s) underlying the selective distribution of neuronal lesions is (are) unknown in Guam disease, lathyrism, and DOA intoxication, these disorders illustrate that MNs present a selective vulnerability to exogenous excitotoxic agents.

An increase of glutamate levels in the serum of ALS patients was reported by Plaitakis and Caroscio (1987). These authors suggested that a defect in the glutamate metabolism may underlie ALS. An increase of glutamate and aspartate levels in the CSF was reported by Rothstein and colleagues (Rothstein *et al.*, 1990) but not confirmed by others (Perry *et al.*, 1987). Recently, the excitotoxicity hypothesis has been strengthened by new findings. Rothstein and colleagues (1992) described a defect in synaptosomal high-affinity glutamate uptake in the brain (motor cortex) and the spinal cord of patients who died from ALS in comparison with controls. Failure of this uptake system would result in abnormal concentrations of glutamate in the extracellular compartment, leading to neuronal degeneration. Moreover, a chronic inhibition of glutamate release in spinal cord cultures induces MN degeneration. The MN degeneration was reversed by non-NMDA antagonists (GYKI-52466 and CNQX) and not by an NMDA antagonist (MK-801) (Rothstein *et al.*, 1993). Recently, it has been shown that the glial glutamate transporter (GLT-1) concentration was severely decreased (90% decrease of the GLT-1 protein in comparison with controls) in a quarter of ALS motor cortex specimens (Rothstein *et al.*, 1995).

The toxicity of ALS CSF toward rat cortical neurons and the protecting effects of AMPA/kainate antagonists against this toxicity (Couratier *et al.*, 1993b) further support the excitotoxic hypothesis for ALS.

6.2 Survival as the clinical end-point

The progressive and selective degeneration of upper and lower MNs is the pathological hallmark of ALS (Oppenheim, 1984). The degeneration of MNs leads to the unique clinical phenotype of the disease. The course of ALS is much shorter than that of the more common neurodegenerative diseases. The specificity of the lesion, the linearity of the process, and the rapidity of the evolution make ALS a model of neurodegenerative disease.

The rapidly fatal outcome of the disease favors a survival analysis. This criteria is fully reliable in contrast with the scores to rating scales which are subject to the rater's subjectivity. Raters' subjectivity increases the variability of the measures thereby decreasing both the statistical power of clinical trials and the probability to show a positive effect. The possibility of using a hard point – survival – as a clinical end-point, in contrast to soft measures, makes the efficacy evaluation of a potential treatment in ALS much easier. This also makes a trial in the disease technically feasible.

7 CLINICAL RESULTS

7.1 Preliminary determination of a neuroprotective dose

Before initiating a long-term double-blind placebo-controlled study of riluzole in ALS it was necessary to choose an effective neuroprotective dose using a route of administration compatible with a chronic treatment. It was not possible to test several doses *a priori* on a long-term basis without any preliminary proof of efficacy. The centrally active dose has been determined in a model of hypobaric hypoxia in healthy volunteers by the neuroprotective effect of three doses of riluzole (50, 100, and 200 mg day^{-1})

in 12 healthy male volunteers subjected to hypobaric hypoxia. The conditions of hypobaric hypoxia were created in a low-pressure chamber.

This study was a double-blind, placebo-controlled, cross-over study. Each 4.5 day treatment period was followed by an 8 day wash-out period. The volunteers were exposed to four hypobaric hypoxic periods.

Hypobaric hypoxia was obtained by decreasing the air pressure to simulate an altitude of 5000 m. An intermediate level was planned at 3000 m to allow for the volunteers' adjustment and to avoid the adverse effects associated with too fast a rise. The plateau at 5000 m lasted 60 min. Hypobaric hypoxic conditions result in significant quantitative electroencephalogram (qEEG) abnormalities such as a significant increase of slow activities and α activity (Kraaier *et al.*, 1988), a significant decrease in total power, and a significant shift of the maximal peak position (dominant frequency) toward the slow frequencies. Hypobaric hypoxic conditions also decrease the performance in psychometric tests.

Riluzole was administered during 4 days prior to the day of the experimental situation, with the last dose administered just before the initiation of the hypobaric hypoxia. The qEEG was recorded in normoxic conditions (before rising) and in hypoxic conditions 10 and 45 min after the arrival at 5000 m. The hemoglobin oxygen saturation was significantly decreased in hypoxic conditions (by 19–27%).

In hypoxic conditions, significant qEEG abnormalities were observed in the placebo group as anticipated. Riluzole showed a dose-related neuroprotective effect on the hypobaric hypoxic qEEG modifications by comparison with placebo (unpublished data). This effect was statistically significant from 100 mg day^{-1}. All doses of riluzole administered appeared to be well tolerated during the 4.5 day treatment period. The lowest dose showing a statistically significant neuroprotective effect (100 mg day^{-1}) was therefore selected as the daily dosage regimen to perform the first trial in ALS.

7.2 First pivotal clinical study in ALS

7.2.1 Aim

The aim of the study was to compare the efficacy and safety of riluzole (50 mg orally b.i.d.) to placebo in patients with either limb onset or bulbar onset ALS.

7.2.2 Methods

A double blind, parallel group, placebo-controlled study was initiated. Based on the survival time of ALS patients, a follow-up period of at least 12 months was chosen. The patients were stratified *a priori* based upon the location of the first symptoms of the disease (site of onset, i.e. bulbar onset or limb onset form).

The inclusion criteria allowed the enrollment of outpatients with typical ALS. Patients between 20 and 75 years of age, with a disease duration of less than 5 years, and 60% or more remaining vital capacity were eligible. Patients with a tracheostomy or pending tracheostomy, associated dementia or major psychological disorders, monoclonal gammopathy, multiple neuroconduction block, or other life-threatening disease were excluded. Patients with significant renal or hepatic impairment were also excluded.

The primary efficacy outcomes were the survival and the rate of deterioration in functional status. Survival was defined as death or tracheostomy. The functional status, measured by the modified bulbar and limb Norris scales (Norris *et al.*, 1974; Lacomblez *et al.*, 1989), were evaluated at entry and then carried out at regular intervals (every 2 months). The secondary end-points included the rate of change in functional status measured by the Medical Research Council manual-testing scale (MRC scale) (Briggs *et al.*, 1991), the evaluation of the respiratory function (Forced vital capacity, FVC), the clinical global impression (CGI), and assessment of fasciculations, cramps, stiffness, and tiredness by visual analog scales (VASs).

7.2.3 Demographic data

A total of 155 patients were included in the study (123 with the limb onset form (79.4%) and 32 with the bulbar onset form (20.6%). Seventy-seven patients were randomly assigned to the riluzole group (62 patients with the limb onset form and 15 with the bulbar onset form), and 78 to the placebo group (61 and 17, respectively). The patients characteristics are reported in Table 1. No statistically significant difference was observed in patients' characteristics between the treatment groups.

7.2.4 Efficacy on survival

The Kaplan–Meier survival curves are shown in Fig. 2. A statistically significant increase in survival was demonstrated for patients treated with riluzole at 12 months (p = 0.010; stratified Wilcoxon test) and at study completion (p = 0.030; stratified Wilcoxon test). The Kaplan–Meier estimates of the probability for surviving at

Table 1 First pivotal study in ALS – patients' characteristics

	Placebo (n = 78)	Riluzole (n = 77)
Sex (M/F)	46/32	45/32
Onset form (limb onset/bulbar onset)	62/15	61/17
Age (years)	58.1 ± 11	56.8 ± 11
Weight (kg)	65.1 ± 12	56.81 ± 12
Duration of disease (years)	2.3 ± 1.8	2.2 ± 1.7
Limb function score (/63)	40.8 ± 16	41.7 ± 16
Bulbar function score (/39)	30.1 ± 11	30.7 ± 10
Vital capacity ratio (%)	86 ± 18	92 ± 17
Muscle testing score (/110)	79.1 ± 19	82.7 ± 17

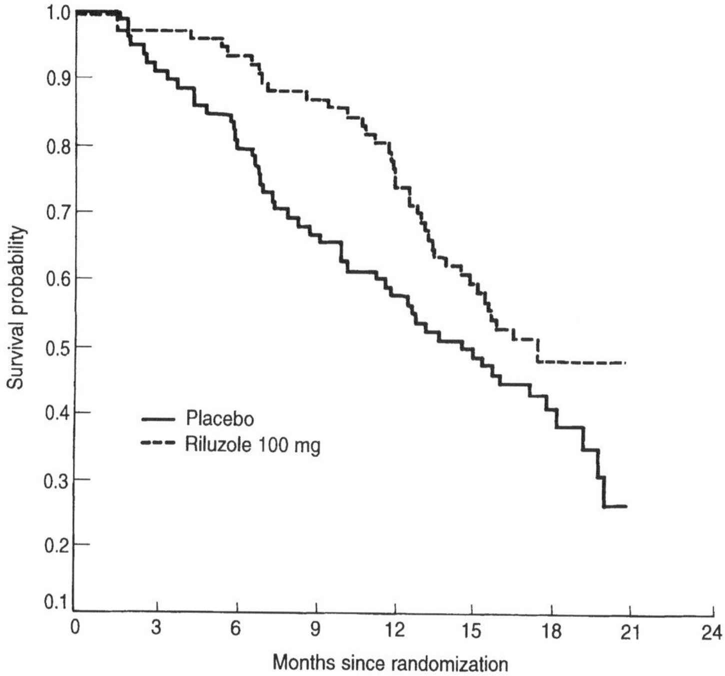

FIG. 2 First pivotal study in ALS – Kaplan–Meier survival curves.

1 year are 58 and 74% in the placebo group and the riluzole group, respectively. They are 37.2 and 49.4% in the placebo group and the riluzole group, respectively, at study completion (Bensimon *et al.*, 1994).

Therefore, survival was significantly longer in ALS patients treated with riluzole compared to those treated with placebo. This increase was mostly due to significantly greater survival in patients with bulbar onset disease (p = 0.014 at 12 months; p = 0.013 at study completion). The median survival in the placebo group was 239 days whereas the median survival was not reached in the riluzole group after 476 days (more than half of the riluzole patients were still alive at 12 months). In the limb onset patients, there was only a trend toward an increased survival (p = 0.17 at 12 months and p = 0.355 at study completion).

7.2.5 *Efficacy on functional status* (Fig. 3)

A significantly slower rate of deterioration in muscle strength was shown in the riluzole group, regardless of the site of onset, in comparison with the placebo group (p = 0.028, two-sided test). The difference in scores on the MRC scale between the two groups was 11.5 ± 5.2 (33.4%). No statistically significant difference was detected for the Bulbar Norris Scale or the Limb Norris Scale.

7.2.6 *Safety*

Adverse events were recorded for the duration of the study. Additional safety parameters were investigated such as blood pressure, ECG, hematological, and biochemical parameters. The dose-related adverse effects recorded in this study are reported along with those of the second pivotal study.

7.3 Second pivotal clinical study in ALS

7.3.1 *Aim*

The aims of the study were to confirm the efficacy results obtained in the first study, to determine the optimal daily dose in ALS, and to assess the long-term safety of a higher dose (200 mg day^{-1}).

7.3.2 *Methods*

This study was a double-blind, placebo-controlled, four parallel group study. Doses of 50 mg day^{-1} (25 mg b.i.d.), 100 mg day^{-1} (50 mg b.i.d.), and 200 mg day^{-1} (100 mg b.i.d.) were chosen to reproduce the results previously obtained and to bracket the dose shown to be effective in the first study. The follow-up duration period was 18 months to assess the effect of riluzole over time. As in the first study, the patients were stratified by the onset form of disease (bulbar or limb onset).

The inclusion criteria were similar to those used in the first study. Diagnosis was established according to the criteria of the World Federation of Neurology defined at the El-Escorial Consensus Conference (Rowland and McLeod, 1988). The patients meeting the criteria of 'probable' or 'definite' ALS were eligible.

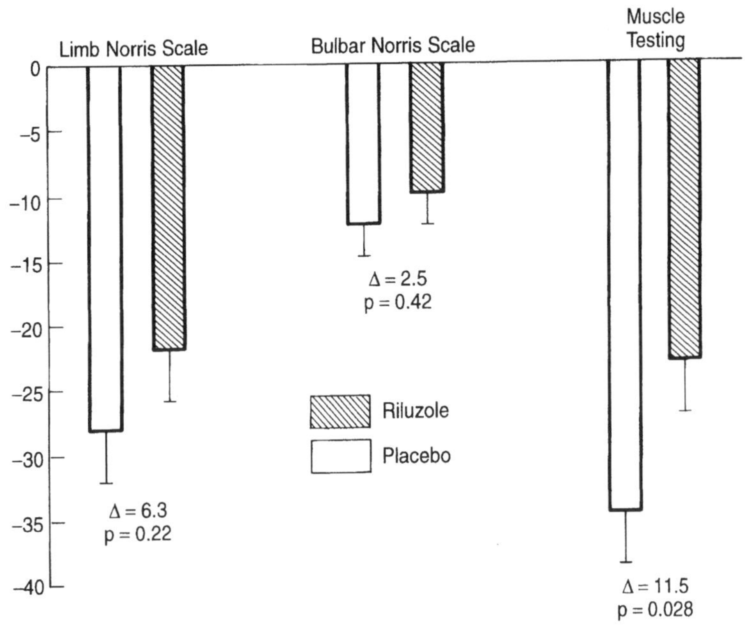

FIG. 3 First pivotal study in ALS – rate of deterioration of functional scores.

As in the first study, the primary efficacy end-point was survival, defined in this study as death, tracheostomy, or intubation with artificial ventilation. The secondary efficacy outcomes were the MRC scale, the Bulbar and Limb Norris Scales, the FVC, the VAS (fasciculations, cramps, stiffness, and tiredness), and the CGI.

7.3.3 *Demographic data*

A total of 959 patients were included in the study (664 with the limb onset form (69.2%) and 295 with the bulbar onset form (30.8%)). Only 40 patients (4.2%) were diagnosed to have FALS. The number of patients per treatment group and the patients' characteristics are shown in Table 2. There were no statistically significant difference between the four treatment groups in terms of disease onset form, demographic data, or functional status at entry.

7.3.4 *Efficacy on survival*

The Kaplan–Meier survival curves are shown in Fig. 4. A statistically significant increase in survival was demonstrated for patients treated with riluzole 100 and 200 mg day^{-1} at 12 months (p = 0.016 and 0.027, respectively; stratified Wilcoxon test) and at study completion (p = 0.050 and 0.061, respectively) (Lacomblez *et al.*, 1996). Despite the fact that the effect at 50 mg day^{-1} did not reach statistical significance (p = 0.23), the survival for patients on riluzole (whatever the daily dose) was greater than that for the placebo group. No additional significant benefit was

Table 2 Second pivotal study in ALS – patients' characteristics

	Placebo (n = 242)	Riluzole 50 mg day^{-1} (n = 237)	Riluzole 100 mg day^{-1} (n = 236)	Riluzole 200 mg day^{-1} (n = 244)
Sex (M/F)	152/90	144/93	143/93	136/108
Onset form (limb onset/bulbar onset)	168/74	166/71	165/71	165/79
Familial ALS (n)	11	11	10	8
Age (years)	56.0 ± 11.5	57.1 ± 10.7	56.9 ± 10.9	56.8 ± 10.8
Weight (kg)	68.1 ± 13.1	67.6 ± 13.0	68.1 ± 13.4	67.1 ± 11.5
Duration of disease (years)	1.8 ±1.4	1.9 ± 1.2	1.7 ± 1.2	1.8 ± 1.2
Limb function score (/63)	44.1 ± 15.2	44.4 ± 15.0	44.7 ± 14.6	45.8 ± 14.0
Bulbar function score (/39)	30.3 ± 9.6	31.6 ± 9.3	31.1 ± 9.4	31.0 ± 9.5
Vital capacity ratio (%)	87.6 ± 18.2	88.6 ± 18.9	88.4 ± 19.1	88.2 ± 19.4
Muscle testing score (/110)	87.9 ± 16.8	88.6 ± 15.7	89.1 ± 14.8	88.7 ± 14.6

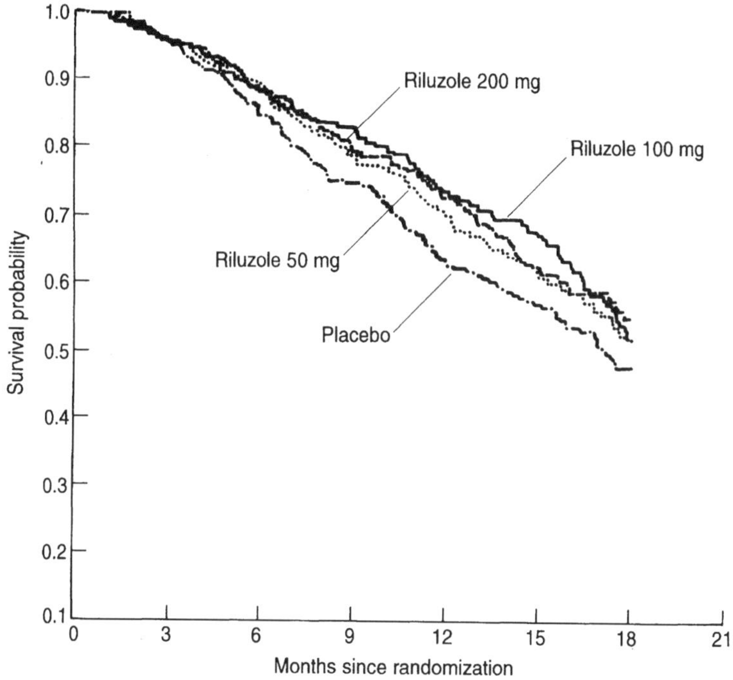

FIG. 4 Second pivotal study in ALS – Kaplan–Meier survival curves.

seen with 200 mg day^{-1} in comparison with 100 mg day^{-1}. The dose–effect relationship ($p = 0.04$, trend test) at study completion is shown in Fig. 5.

Since prognostic factors have been identified in ALS (i.e. onset form, sex, age, disease duration, and weight), a Cox model was used to adjust for important prognostic variables. Statistically significant increases in survival at study completion were suggested by this analysis for all patients treated with riluzole 50, 100, and 200 mg day^{-1} ($p = 0.044$, 0.002, and 0.0004, respectively).

The survival rates for each treatment group at 12 months and at study completion are plotted in Fig. 6.

At 12 months, the survival rate in the riluzole 100 mg day^{-1} group was 73.7%, compared to 62.8% in the placebo group. At study completion the rates were 56.8 and 50.4%, respectively.

7.3.5 *Efficacy on functional status*

The results of the first study suggested that riluzole provided a benefit regarding muscle deterioration. These previous findings were not observed in the second study. One possible explanation to explain this discrepancy between the results is the poor intercenter and intercountry reliability of functional scales. In the first study, seven centers and two countries were involved, while the second study

Table 3 Adverse effects (percentage of patients) reported in the two pivotal trials which appeared to demonstrate a dose–response relationship

		Riluzole		
	Placebo ($n = 320$)	50 mg/day ($n = 237$)	100 mg/day ($n = 313$)	200 mg/day ($n = 244$)
Asthenia	12	15	19	20
Nausea	11	12	16	20
ALT elevations (>3 times the ULN[a])	2	6	10	15
Vomiting	2	4	4	5
Vertigo	1	3	2	5
Somnolence	1	1	2	4
Circumoral paresthesia	0	1	2	3

ULN, upper limit of the normal range.

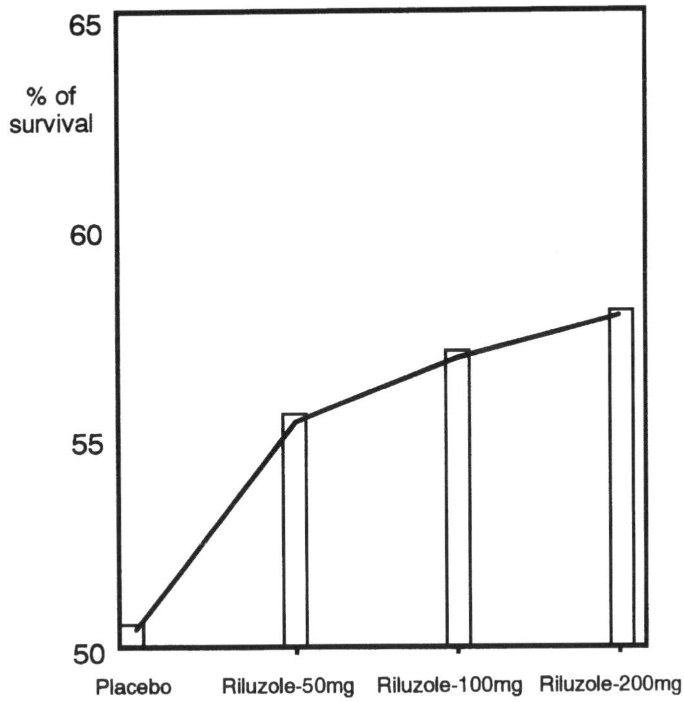

FIG. 5 Second pivotal study in ALS – dose–effect relationship on survival rate at 18 months.

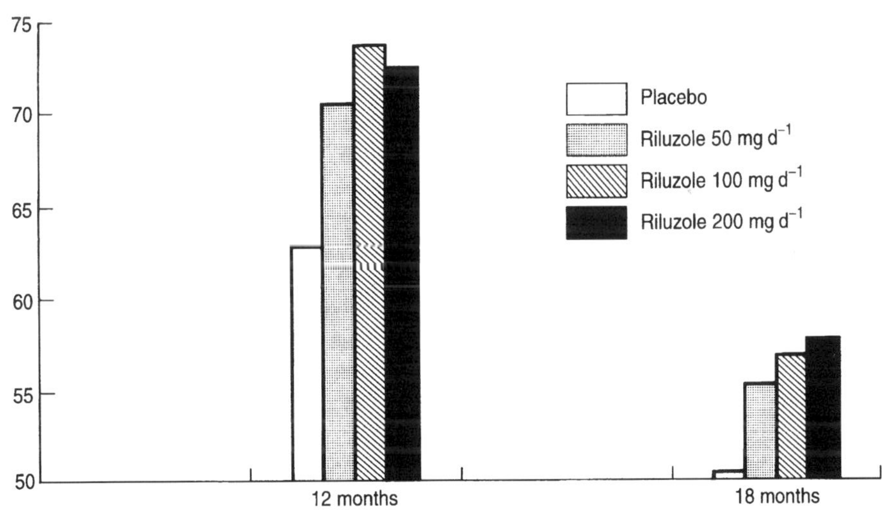

FIG. 6 Second pivotal study in ALS – survival rates at 12 months and at study completion.

involved 31 centers and eight countries. It might be possible that the variability of functional scale cotations hides a therapeutic effect.

7.3.6 *Safety*

Adverse events were recorded in the same way as in the first pivotal study. Most of the adverse events reported were related to the disease. However, some of them appeared to demonstrate a dose–response relationship and appeared to be drug related. The percentages of patients reporting these adverse effects are shown in Table 3. The most frequent side-effects were asthenia, nausea, and ALT elevations.

7.3.7 *Optimal daily dose*

A dose–effect relationship has been demonstrated for efficacy. Doses of 100 and 200 mg day^{-1} brought a significant improvement, but the dose of 200 mg day^{-1} did not bring a clear additional benefit in comparison with 100 mg day^{-1}. There was also a dose relationship in the occurrence of asthenia, nausea, and ALT increase. Therefore, the 100 mg daily dose appeared to have the best risk–benefit ratio.

8 CONCLUSION

The demonstration of the efficacy of riluzole in the treatment of ALS rests on the results of two independent double-blind placebo-controlled studies. In both studies, the primary efficacy parameter was survival, a strong and reliable end-point.

In the first study, the benefit was observed in the whole population, but mainly for the patients with the bulbar onset form of the disease. In the second study, the benefit was clearly not limited to these patients but observed in both groups (both the bulbar onset form and the limb onset form).

Survival analyses have demonstrated that riluzole is able to delay the progression of the disease and increases the survival of ALS patients. The second study has confirmed that the optimal daily dose is 100 mg day.$^{-1}$

Three conclusions arise from these two studies. Firstly, a neuroprotective drug can slow the course of ALS. Secondly, a product blocking glutamate transmission can bring therapeutic benefit to a neurodegenerative disease with an acceptable benefit–risk ratio. Thirdly, riluzole efficacy in patients suffering from ALS supports an involvement of EAAs, as a primary or secondary event, in ALS etiology.

10 Preclinical and Clinical Aspects of Remacemide Hydrochloride*

GENE C. PALMER[1] AND JOHN B. HUTCHISON[2]

[1]Department of Biology, Astra Arcus USA, PO Box 20890, Rochester, NY 14602, USA
[2]Department of Medical Affairs, Astra Charnwood, Bakewell Road, Loughborough, Leicestershire LE11 OHR, UK

1 PRECLINICAL EFFICACY STUDIES

1.1 Background and antiepileptic potential

Remacemide hydrochloride or (±)-2-amino-*N*-(1-methyl-1,2-diphenyl)acetamide hydrochloride (Fig. 1) was discovered† from a chemical series employing a rational synthetic approach based on molecular modeling. Initial preclinical profiling established upon acute efficacy/safety

* This chapter is dedicated to the memory of Dr Eric W. Lothman formerly from the Department of Neurology, University of Virginia, an outstanding scientist, teacher, collaborator, and friend.

† By Dr R. Griffith, and synthesized by Chris Becker.

EXCITATORY AMINO ACIDS – CLINICAL RESULTS WITH ANTAGONISTS
ISBN 0-12-546820-2

REMACEMIDE HYDROCHLORIDE

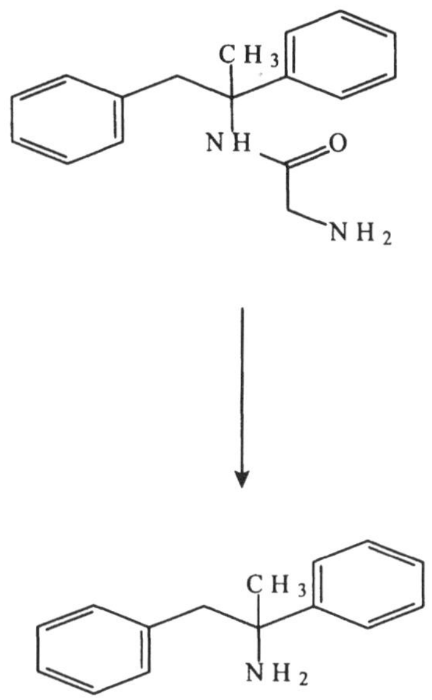

ARL 12495AA

FIG. 1 Chemical structures of remacemide and its desglycine metabolite ARL12495AA.

ratios in mice led to the identification of remacemide hydrochloride as the lead compound for further development (Griffith *et al.*, 1996). Detailed preclinical work generally followed the guidelines of the Antiepilepsy Drug Development Program (ADD Program), Epilepsy Branch, National Institutes of Health, who also provided assistance (Kupferberg, 1989). Subsequently a desglycinyl metabolite, ARL12495AA (former designation was FPL12495AA) (Fig. 1), was discovered in all species examined and was found to exert greater efficacy than the parent compound in most preclinical testing paradigms (Palmer *et al.*, 1992). Thus, remacemide may in part serve as a prodrug. The remaining metabolites are of pharmacological inconsequence and have been discussed previously (Palmer *et al*, 1992, 1993; Clark *et al.*, 1995).

1.2 Anticonvulsant profile

1.2.1 Maximal electroshock, auditory, and chemically induced seizures

Remacemide hydrochloride was effective orally, regarding prevention of seizures evoked by maximal electroshock (MES) in rodents (Table 1) (Stagnitto *et al.*, 1990; Garske *et al.*, 1991). Seizures produced by pentylenetetrazol, picrotoxin, bicuculline, BAY K 8644, or strychnine were only marginally inhibited (Stagnitto *et al.*, 1990; Palmer *et*

Table 1 Comparison of anti-seizure properties between remacemide hydrochloride and ARL12495AA

Experimental conditions	Remacemide (mg kg^{-1} body weight)	ARL12495AA (mg kg^{-1} body weight)
Mice		
ED_{50} MES test, p.o.[a]	48	40
ED_{50} MES test, i.v.[a,b]	15	5
ED_{50} MES test, s.c.[c]	21	8
ED_{50} MES test, i.p.[a]	22	17
ED_{50} NMDA seizures, i.p.[b]	57	32
ED_{50} NMDA lethality, i.p.	22	17
ED_{50} kainate seizures, i.p.[d]	60	Inactive
ED_{50} kainate lethality, i.p.	28	Inactive
ED_{50} 4-aminopyridine tonic seizures, i.p.[d]	18	18
ED_{50} audiogenic seizures in DBA2 mice, i.p.[e]	20	12
Rat		
ED_{50} MES test, p.o.[f]	22	10
ED_{50} MES test, s.c.[c]	12	4
ED_{50} MES test, i.v.[b]	6.6	3.5
Bicorneal kindling, p.o.[b]		
Established seizures	Inactive	Inactive
Epileptogenesis	Inactive	Inactive
Rapid hippocampal kindling[g]	Active	
Genetic absence seizure model[h]	Active	Active

NMDA, *N*-methyl-D-aspartate.

[a] Stagnitto *et al.* (1990).
[b] Palmer *et al.* (1992).
[c] Clark *et al.* (1995).
[d] Cramer *et al.* (1994).
[e] M. Tricklebank (unpublished data).
[f] Garske *et al.* (1991).
[g] E. Lothman (unpublished data).
[h] Van Luijtelaar and Coenen (1995).

al., 1993). Subsequently, remacemide hydrochloride was shown to inhibit auditory-induced seizures (DBA2 mice) and 4-aminopyridine-induced seizures (CF1 mice) (Stagnitto *et al.*, 1990; Cramer *et al.*, 1994). The desglycinyl metabolite normally exhibited greater potency in these tests for anticonvulsant efficacy (Table 1).

1.2.2 Kindling seizures

Neither remacemide hydrochloride nor ARL 12495AA was capable of preventing established kindled seizures or development of kindled seizures evoked by bicorneal application of subthreshold electroshock to rats (Palmer *et al.*, 1992). However, E. Lothman (unpublished data), employing the technique of rapid hippocampal kindling in rats (Lothman *et al.*, 1988), demonstrated the effectiveness of remacemide hydrochloride regarding shortening of electrographic afterdischarges, as well as attenuation of behavioral seizures.

1.2.3 Absence seizures

Remacemide hydrochloride and ARL12495AA were examined electrophysiologically in a genetic model of generalized absence seizures (WAG/Rij rats). Oral treatment with remacemide hydrochloride (20–80 mg kg^{-1}) or ARL12495AA (1–40 mg kg^{-1}) decreased spike-wave discharges. In addition, the highest dose of ARL12495AA produced electroencephalogram (EEG) changes consisting of decreased spindle frequencies and an increased amplitude of δ and high β wave frequencies (Van Luijtelaar and Coenen, 1995).

1.2.4 Excitatory amino acids

Remacemide hydrochloride and ARL12495AA inhibit seizures induced by *N*-methyl-D-aspartate (NMDA), whereas remacemide hydrochloride alone is active against kainic acid-induced seizures (see Table 1 and Section 1.3).

1.2.5 Isomers

Preclinical evaluations of the pharmacological properties for the enantiomers of remacemide hydrochloride and ARL12495AA indicated that the differences were not marked (Palmer *et al.*, 1991; Garske *et al.*, 1991).

1.3 Mechanism of action

Initial attempts to discover a mechanism of action for remacemide hydrochloride involved standard receptor-binding studies (GABAa, glutamate, adenosine, benzodiazepine, muscarinic, the broad-based NovaScreen (Nova Pharmaceutical Corp., Baltimore, Maryland, USA). The outcome was unsuccessful. The compound was next evaluated using *in vitro* electrophysiological techniques (penicillin-induced neuronal bursting, veratridine-elicited depolarization, synaptically evoked population spikes, recurrent inhibition, and induction of long-term potentiation). Again the results were unremarkable (Stagnitto *et al.*, 1990; Garske *et al.*, 1991; Harris *et al.*, 1992; Palmer *et al.*, 1993).

1.3.1 NMDA receptor

The role of the excitatory amino acid receptors, particularly NMDA and its associated subunits, as a molecular entity associated with the pathological sequelae of epilepsy and hypoxic/ischemic cell death has been established (Lehmann, 1989; Porsche-Wiebking, 1989). Since the molecular structure of remacemide includes a glycine moiety, a possible link to the strychnine-insensitive glycine subsite within the NMDA receptor complex was considered initially. Ligand-binding experiments demonstrated only weak affinity for remacemide hydrochloride and ARL12495AA at the glycine-binding site (Palmer *et al.*, 1993) (Table 2). The ionic channel subsite on the NMDA receptor was next considered because we observed prevention by both remacemide hydrochloride and ARL12495AA of convulsions and the subsequent mortality after intravenous administration of the NMDA analog (NMDLA) into mice (see Table 1) (Palmer *et al.*, 1992; M. Tricklebank, unpublished data). Parallel biochemical experiments (displacement of labelled MK-801 from rat synaptosomal membrane preparations) revealed that remacemide hydrochloride displayed weak (IC_{50} = 68 μM), and ARL12495AA possessed moderate (IC_{50} = 0.48 μM), affinities as uncompetitive NMDA receptor antagonists (Table 2). In further work using quantitative autoradiography, Porter and Greenamyre (1995) demonstrated that lower-affinity uncompetitive NMDA antagonists (remacemide hydrochloride, ARL12495AA, amantadine, budipine, and memantine) would bind more potently to the cerebellar subtype(s) of the NMDA receptor than the higher-affinity antagonists, namely MK-801, phencyclidine (PCP), etc. The converse was true for forebrain structures (Table 2).

Using whole-cell preparations, NMDA-induced depolarizations in neurons were inhibited by ARL12495AA (rat hippocampal slice, EC_{50} = 2 μM, compared to MK-801 EC_{50} = 0.02 μM; mouse cortical wedge, lowest effective concentration = 12.5 μM). Remacemide hydrochloride was relatively inactive in the hippocampal slice model (EC_{50} >30 μM) (Table 2) (Harris *et al.*, 1992; Hu and Davies, 1995). Recently, Subramaniam *et al.* (1996) using voltage-clamp recordings in cultured rat hippocampal neurons, demonstrated rapid and reversible inhibition of NMDA currents by remacemide hydrochloride and its isomers. In contrast, inhibition by enantiomers of ARL12495AA was more potent, slower in onset, and stereoselective (see Table 2 for details). The block by (+)-desglycine was strongly use- and voltage-dependent and was prevented by Mg^{2+}, suggesting interaction via an

Table 2 Comparison of *in vitro* mechanisms of action between remacemide hydrochloride and ARL12495AA[a]

Mechanism	Remacemide	ARL12495AA
Rat synaptic membranes		
IC_{50} MK-801 binding[c]	68 μM	0.48 μM
IC_{50} glycine binding[d]	274 μM	264 μM
IC_{50} MK-801 binding—quantitative autoradiography regional affinity[e]		
Frontal cortex	797 μM	1.8–14 μM[b]
Striatum	798 μM	2.2–15 μM[b]
Hippocampal CA1	968 μM	1.5–18 μM[b]
Cerebellum—granule cell	296 μM	0.5–4.4 μM[b]
Polyamine site—influence of spermine[f]	Stereoselective antagonism	No activity
IC_{50} NMDA currents[f] (hippocampal neurons)	67–75 μM[b] (rapid)	0.6–4 μM[b] (slower)
IC_{50} NMDA depolarization[g] (hippocampal slices)	>30 μM	2 μM
Mouse cortex – glutamate release by veratridine[h]		25–100 μM
Mouse spinal neurons IC_{50} sustained repetitive firing[i]	8 μM	0.8 μM
Rat synaptic membranes K_d batachrotoxin binding[j]	15.6 μM	7.9 μM
Dissociated rat cortical neurons patch clamp – high-frequency block of use dependent Na^+ channels[k]	Active at 30 μM	Active at 30 μM

[a]*In vivo* data expressed as mg (kg body weight)$^{-1}$.
[b]Data for enantiomers.
[c]Palmer *et al.* (1992).
[d]Palmer *et al.* (1993).
[e]Porter and Greenamyre (1995).
[f]Subramaniam *et al.* (1996)
[g]Harris *et al.* (1992).
[h]Srinivasan *et al.* (1995).
[i]Wamil *et al.* (1996).
[j]D. Kamp (unpublished data).
[k]D. Sanchez (unpublished data).

open channel mechanism. In more detailed investigation using only (+)-remacemide, the block of NMDA currents was partially voltage-dependent, indicating action at allosteric sites on the receptor, as well as via an open channel. In binding studies, (+)-remacemide slowed the dissociation of labelled MK-801 whereas (+)-desglycine did not. Inhibition of MK-801 binding by (+)-remacemide was reduced by spermine. Thus, (+)-remacemide appears to act at both the voltage-dependent site and the polyamine facilitatory site of the NMDA receptor.

Neuronal protective properties of remacemide hydrochloride and ARL12495AA were examined in primary cultures from rat cerebral cortex. Exposure of the neurons to 100 mM NMDA for 15 min led to 85% cell death within 24 h, an event prevented in a dose-dependent fashion by ARL12495AA (5–20 μM), but not remacemide hydrochloride (up to 100 μM). Treatment of cells with ARL12495AA also reduced the NMDA-elicited loss in membrane-associated protein kinase C activity and the magnitude in the NMDA-triggered Ca^{2+} surge into the cells. Inhibition of the NMDA-triggered Ca^{2+} surge by ARL12495AA was more rapid in onset and more prolonged at a subthreshold response than the slow-onset, total block observed by the high-affinity NMDA antagonists MK-801 and PCP (Black *et al.*, 1996).

1.3.2 Glutamate release

Veratridine-evoked release of glutamate from mouse cortical slices was inhibited by ARL12495AA (25 μM) (Table 2) (Srinivasan *et al.*, 1995). Veratridine depolarizes the Na^+ channel, which is a major site of action for remacemide hydrochloride and ARL12495AA (see below). Antagonism of Na^+ channel-coupled release of glutamate may be an initial step toward ultimate prevention against the toxicity produced by the excitatory amino acid neurotransmitters (Rataud *et al.*, 1994).

1.3.3 AMPA/kainate

Under *in vitro* conditions neither remacemide hydrochloride nor ARL12495AA has been shown to influence receptor ligand binding to 2-animo-3-(5-methyl-3-hydroxy-isoxazol-4-yl) propanoic acid (AMPA) or kainate receptors or provide protection against kainate/AMPA-induced currents or lethality in neuronal cultures (Hu and Davies, 1995; Black *et al.*, 1996; Subramanian *et al.*, 1996;

unpublished results from Novascreen). The data, although preliminary, suggest the primary neuroprotective actions of ARL12495AA are restricted to uncompetitive antagonism of the NMDA receptor. Alternatively, kainate-induced seizures/mortality were prevented in mice following pretreatment with remacemide hydrochloride, but not with ARL12495AA (see Table 1) (Cramer *et al.*, 1994).

1.3.4 *Activity at Na⁺ channels*

The following studies indicate a major site of action for remacemide hydrochloride and ARL12495AA to be the neuronal Na^+ channel. Evidence indicates the neuronal Na^+ channel to play a major role linked to the phenomenon of seizure spread, a characteristic of MES and generalized tonic/clonic seizures. The potency of antiepileptic drugs in the MES test is postulated to correlate with prevention of sustained repetitive firing, an activity linked to the fast Na^+ channel (Macdonald, 1989). In spinal neurons isolated from fetal mice, remacemide hydrochloride and ARL12495AA were shown to limit sustained repetitive firing (IC_{50} = 8 and 0.8 μM, respectively) (Table 2). The effective *in vitro* concentration of ARL12495AA roughly correlated to the plasma and cerebrospinal concentrations measured during efficacy (MES) testing in rats (Harris *et al.*, 1992; Wamil *et al.*, 1996).

Batrachotoxin-b binds to the open/activated site on the Na^+ channel, another locus of action for drugs thought effective in the MES test (Catterall *et al.*, 1981). Remacemide hydrochloride and ARL12495AA were active regarding displacement of batrachotoxin-b binding from synaptosomal membranes prepared from rat brain. The metabolite exhibited twofold greater potency (Table 2) (D. Kamp, unpublished data).

A patch clamp electrophysiology study using dissociated rat cortical neurons showed that remacemide hydrochloride, ARL12495AA, and phenytoin in a dose-related manner produced a use-dependent block of Na^+ channels by enhancing the process of Na^+ channel slow inactivation (D. Y. Sanchez and E. W. Harris, unpublished data).

1.3.5 *K⁺ channel activity*

Preliminary experiments indicate a possible site of action for remacemide hydrochloride/ARL12495AA to be linked to the K^+ channel. 4-Aminopyridine is postulated to cause seizures by inhibition of neuronal transient (A type) or sustained voltage-dependent K^+ channels. The resultant tonic seizures are prevented by drugs active in the MES test (Yamaguchi and Rogawski, 1992), including remacemide hydrochloride and ARL12495AA (see Table 1) (Cramer *et al.*, 1994). In preliminary patch clamp work with dissociated rat cortical neurons, remacemide hydrochloride and ARL12495AA were found to have an activating effect on voltage-dependent K^+ efflux (D. Sanchez, unpublished data).

1.3.6 *Conclusions*

From the available data the principal mechanisms of action of remacemide hydrochloride and the desglycinyl metabolite are twofold: (1) remacemide hydrochloride exhibits weak and ARL12495AA exhibits moderate potency as uncompetitive inhibitors of the ionic/channel subsite on the NMDA receptor; and (2) remacemide hydrochloride is a moderate and ARL12495AA is a potent antagonist of the Na^+ channel.

Other major differences between the two compounds include: (1) remacemide hydrochloride possesses more rapid on/off kinetics at the NMDA ionic channel site and acts allosterically at the facilitatory polyamine site to limit NMDA receptor gating, whereas ARL12495AA acts purely as a channel blocker (Subramaniam *et al.*, 1996); and (2) remacemide hydrochloride, unlike ARL12495AA, does not limit NMDA-induced Ca^{2+} surges nor protect neurons from NMDA-induced injury (Black *et al.*, 1996). Most likely the more prominent NMDA receptor properties of remacemide hydrochloride are reflected via its desglycinyl metabolite. Of principal interest, these lower-affinity NMDA receptor antagonists do not elicit the troublesome side-effects associated with the higher-affinity compounds, a factor limiting their potential for therapy in patients (Rogawski, 1992; Gee, 1994; Muir and Lees, 1995a,b; Palmer and Miller, 1996). Rogawski (1992) has argued that the more rapid association/dissociation kinetics of the lower-affinity NMDA antagonists contribute to their better safety profiles. The slower onset of NMDA receptor inhibition by MK-801-like compounds leads to a complete shut-down of receptor function including disruption of Ca^{2+} homeostasis, which may contribute to the severe side-effects encountered with this drug class (Black *et al.*, 1996). Moreover, the lower-affinity compounds reveal a preference for different regional NMDA receptor subtypes (Porter and Greenamyre, 1995). A sustained conversion *in vivo* of remacemide to its more potent metabolite likewise contributes to its overall safety profile. As an antiepileptic, remacemide hydrochloride/desglycine could inhibit seizure initiation via use-dependent blockade of the NMDA receptor. On the other hand, seizure spread would be prevented via inhibition at the Na^+ channel, an action possibly linked with K^+ channel activation. The explanatory hypotheses, though tenuous, await further experimental verification (Fig. 2).

1.4 Neuroprotective properties

Numerous studies have shown that antagonists of the excitatory amino acid receptors also protect the brain from the consequences of ischemia, anoxia, stroke, and trauma (Olney *et al.*, 1989; Lipton and Rosenberg, 1994; Vender *et al.*, 1995). A compound has yet to demonstrate effectiveness in the clinic; however, several, including remacemide hydrochloride, are currently under investigation (Gee, 1994; Lipton and Rosenberg, 1994). In the

"I had", said Sherlock Holmes, "Come to an entirely erroneous conclusion which shows, my dear Watson, how dangerous it always is to reason from insufficient data".

from THE ADVENTURE OF THE SPECKLED BAND by A. Conan Doyle

FIG. 2 Pitfalls of hypotheses with insufficient data.

subsequently described experiments with remacemide hydrochloride involving anoxia/ischemia, meticulous care was taken to maintain brain and body temperature.

1.4.1 Global ischemia/anoxia models (Table 3)

Potential neuroprotective properties for remacemide hydrochloride were demonstrated when pretreatment of mice exposed to anoxic conditions led to an extension of the time to death (Palmer *et al.*, 1991). Subsequent work showed protection of CA1 hippocampal neurons in rats and dogs following global ischemia. For example, remacemide hydrochloride treatment (20 mg kg^{-1} i.p., either s.i.d. or b.i.d. from 7 to 28 days) initiated after four-vessel occlusion (15 or 30 min) or two-vessel occlusion coupled to hypotension (10 min at 50 mmHg) not only protected the rat CA1 neurons but, in addition, memory loss was prevented and CA1 electrophysiological responses were partially preserved (Harris *et al.*, 1992; Ordy *et al.*, 1992; Palmer *et al.*, 1993; Lesiuk *et al.*, 1995). Eight minutes of global ischemia achieved by clamping the ascending aorta in the dog not only protected the CA1 neurons but also resulted in better neurology scores following a 7 day intravenous treatment regimen with remacemide hydrochloride (Palmer *et al.*, 1995). In the gerbil model of global forebrain ischemia, however, remacemide hydrochloride was ineffective in view of protection of the CA1 neurons and prevention of subsequent behavioral hyperactivity (D. Corbett, unpublished data). The plasma levels of ARL12495AA were barely detectable following administration of the parent compound to this species, suggestive of a metabolic difference between gerbils and other rodent species for remacemide (Palmer *et al.*, 1995). The effectiveness of remacemide hydrochloride in global ischemia indicates potential therapeutic utility in conditions of stroke, coronary artery bypass graft, drowning, asphyxiation, etc. (Fisher *et al.*, 1994).

1.4.2 Focal ischemia models (Table 4)

Focal models of ischemia are thought to mimic closely the majority of strokes seen in the clinic. Thus, the effectiveness of a compound herein would more likely predict therapeutic utility (Buchan, 1990; Hunter *et al.*, 1995). Ipsilateral clamping of the carotid arteries of spontaneously hypertensive rats (SHRs) prior to a 2 h occlusion of the middle cerebral artery (MCA) followed by 22 h recirculation results in large volume of infarction (>200 mm^3) restricted principally to cerebral tissue. Thirty minutes of pretreatment (20 mg kg^{-1}) followed by post-treatment (10 mg kg^{-1} at 4 and 12 h) with remacemide hydrochloride reduced the volume of cortical infarction to ~140 mm^3. Cortical infarction in WKY rats was, however, not reduced significantly by remacemide hydrochloride (Xue *et al.*, 1993; Palmer *et al.*, 1995), presumably because of greater collateral cerebral blood flow in this strain. In support of this work, Bannan *et al.* (1994), using a model of permanent MCA occlusion (6 h) in the cat, demonstrated a similar reduction in cortical infarct volume following 90 min pretreatment with remacemide hydrochloride (infusion 278 μg min^{-1}). Recently, L. Turski (Schering AG, unpublished data) looked at the influence of

Table 3 Neuroprotective actions of remacemide hydrochloride in animal models of global ischemia

Experimental situation[a]	Observations
Antihypoxia – mice[b,c]	
ED_{50} i.v. dosing	14.3 mg kg^{-1}
ED_{50} p.o. dosing	55.5 mg kg^{-1}
Rat four-vessel occlusion for 15 min[b,d]	
Histology assessment of CA1 (20 mg kg^{-1} i.p. at 1 h postischemia and s.i.d. for 14 days)	Modest protection
Rat four-vessel occlusion for 30 min[b,e]	
Histology assessment of CA1	Protection
Electrophysiology of CA1 (20 mg kg^{-1} i.p. at reflow and b.i.d. for 7 days)	Modest protection*
Rat four-vessel occlusion for 30 min[b,d]	
Effect on memory (T maze) (20 mg kg^{-1} i.p. at 1 h postischemia and s.i.d. for 28 days)	Prevents memory loss
Rat two-vessel occlusion for 10 min[f]	
Histology of CA1 at 7 days (20 mg kg^{-1} i.p. at reperfusion + 10 mg at 4 and 24 h)	Modest protection
Canine – 8 min clamp of ascending aorta[b]	
Histology assessment of CA1	Marked protection
Neurological scores (7.5 mg kg^{-1} i.v., b.i.d. for 3 days, s.i.d. for 4 days)	Improvement
Gerbil – 3 or 5 min bilateral carotid occlusion[b],	
Histology assessment of CA1	No protection
Open field motor behavior (40 mg kg^{-1} i.p. at time of reperfusion or 4 h postreperfusion, or 20 mg kg^{-1} for 3 days after 40 mg kg^{-1} i.p.)	No effect

*Not significant, $p = 0.06$.
[a]Rigid temperature control was in effect for all experiments.
[b]Palmer *et al.* (1995). [e]Harris *et al.* (1992).
[c]Palmer *et al.* (1991). [f]Lesiuk *et al.* (1995).
[d]Ordy *et al.* (1992).

intravenous infusion of remacemide hydrochloride in the mouse permanent MCA occlusion (24 h) model. When administered as a single dose of either 10 or 20 mg kg^{-1} h^{-1} at the time of MCA occlusion or 10 mg kg^{-1} h^{-1} at 30 min after MCA occlusion, remacemide hydrochloride significantly reduced the volume of cortical infarct. These findings formed the basis to evaluate tolerability of remacemide hydrochloride in patients with focal ischemia (Muir and Lees, 1995a,b).

1.4.3 *Parkinsonism* (Table 4)

Excessive excitatory amino acid stimulation at critical extrapyramidal motor connections in the brain has been postulated to worsen the symptoms of Parkinson's and Huntington's diseases (Beal, 1992). Based on this premise, Greenamyre and O'Brien (1991) assumed that low doses of excitatory amino acid antagonists could act in synergy with subthreshold doses of L-DOPA to alleviate the symptoms of Parkinson's disease. Akinesia in rats following reserpine pretreatment was reversed with L-DOPA. Reversal of akinesia was further magnified when L-DOPA was given with small doses of remacemide hydrochloride (10–40 mg kg^{-1} p.o.). In a second study, rhesus monkeys with parkinsonic indications induced by MPTP (1-methyl-4-phenyl-1,2,3,6-tetrahydropyridine) likewise exhibited a marked reduction in symptoms when a subthreshold dose of Sinemet (carbidopa + L-DOPA) was given in conjunction with 5 mg kg^{-1} p.o. of remacemide hydrochloride (Greenamyre *et al.*, 1994).

1.4.4 *Subarachnoid hemorrhage-induced edema and basilar arteriospasm* (Table 4)

Pretreatment of rats with remacemide hydrochloride reduces the amount of Evans blue dye extravasation into the ipsilateral side of the brain following topical application of blood or $FeCl_2$ to the corresponding cerebral cortex (D. Anderson and M. Zuccarello, unpublished data; see Palmer *et al.*, 1995). In support of these findings, NMDA receptors have been reported to be present in cerebral microvessels and possibly to mediate trauma-induced breakdown of the blood–brain barrier (Koenig *et al.*, 1992).

In a second study, vasospasm of the rabbit basilar artery was observed (using angiovist 282 coupled to serial digital subtraction angiography) at 72 h after injection of autologous blood into the cisterna magna. The degree of vasospasm was significantly reduced by 27% after treatment by remacemide hydrochloride (15 mg kg^{-1} i.p., 30 min after injection of autologous blood, 6 h later and b.i.d. for 3 days) (Zuccarello *et al.*, 1994). These results, though preliminary, are interesting, and suggest further inquiry into the possible use of remacemide hydrochloride for treatment of the sequelae of subarachnoid hemorrhage.

1.5 Acute and chronic safety considerations

Extensive acute safety/behavioral studies, as well as the initial 30 day and up to 1 year toxicology evaluations, revealed that relatively high doses of remacemide hydrochloride were required to produce overt behavioral–motor symptoms/mortality, and other than acute gastric/intestinal irritation, chronic drug effects on major organ systems were generally unremarkable. This topic has been covered in recent reviews (Palmer *et al.*, 1993; Clark *et al.*, 1995).

1.5.1 *Acute studies*

As a means to establish limits of tolerability, acute safety, effects on behavior, etc., remacemide hydrochloride was

Table 4 Neuroprotective actions of remacemide hydrochloride in animal models of focal ischemia, parkinsonism, cerebral edema, and basilar arteriospasm

Experimental situation	Observation
Rat: 2 h MCAO ischemia[a,b]	Cortical infarct volume:
SHR	28% reduction
Wistar	24% reduction*
(treatment: 20 mg kg^{-1} 30 min before and 10 mg kg^{-1} at 4 and 12 h post-MCA)	
Mouse: permanent MCAO[c]	Cortical infarct volume:
At 10 mg kg^{-1} h^{-1} at time of MCAO	25% reduction
At 20 mg kg^{-1} h^{-1} at time of MCAO	29% reduction
10 mg kg^{-1} h^{-1} at 30 min post-MCAO	24% reduction
10 mg kg^{-1} h^{-1} at 60 min post-MCAO	12% reduction*
(treatment: remacemide i.v. infusion over 4 h)	
Cat: 6 h permanent MCAO[a,d]	
(treatment: 278 μg min^{-1} for 90 min, i.v. prior to MCAO)	57% reduction in cortical infarct volume
Rat: reserpine-induced dyskinesia[a,e]	Dose-dependent potentiation of
(treatment: remacemide 5–40 mg kg^{-1} p.o.)	L-DOPA-induced motor activity
Monkey: MPTP-induced parkinsonism[a,e]	Potentiation of L-DOPA
(treatment: remacemide 5 mg kg^{-1}, carbidopa 25/250 mg p.o.)	amelioration of symptoms
Rat: subarachnoid-induced edema[a]	
Edema induced with topical blood	70% reduction in edema
Edema induced with topical $FeCl_2$	45% reduction in edema
(treatment: 10 mg kg^{-1} i.p. 30 min before and 20 mg kg^{-1} 60 min after SAH, measure Evans blue at 3 h after SAH)	
Rabbit: reduction in basilar arteriospasm[f]	27% reduction in vasospasm
(treatment: autologous blood into cisterna magna, monitor vasospasm at 72 h, remacemide 15 mg kg^{-1} i.p. 30 min after SAH, 6 h and b.i.d. for 3 days)	

MCAO, middle cerebral artery occlusion; MPTP, 1-methyl-4-phenyl-1,2,3,6-tetrahydropyridine; SAH, subarachnoid hemorrhage.
*Not significant.
[a]Palmer *et al.* (1995).
[b]Xue *et al.* (1993).
[c]L. Turski (unpublished data).
[d]Bannan *et al.* (1994).
[e]Greenamyre *et al.* (1994).
[f]Zuccarello *et al.* (1994).

examined extensively in mice, rats, and dogs. A summary of these findings is presented in Table 5 and published elsewhere in greater detail (Palmer *et al.*, 1993; Clark *et al.*, 1995). Tolerability ratios were more favorable for remacemide hydrochloride after oral dosing than via the other routes of administration. At high doses, proconvulsant properties of remacemide hydrochloride could, however, be experimentally demonstrated in mice and dogs (Palmer *et al.*, 1991, 1993).

A degree of tolerance to a 4 day treatment regimen with high doses of remacemide hydrochloride was observed using the MES test in mice (Palmer *et al.*, 1991). Tolerance was not shown, however, in the rat employing either a similar study design (Garske *et al.*, 1991) or with chronic experiments involving hippocampal kindled seizures (E. Lothman, unpublished data).

Disruption of learning and memory and PCP-like abuse liability are characteristic adverse behaviors seen in both laboratory animals and humans following treatment with high-affinity NMDA receptor antagonists (Koek *et al.*, 1987; Willets *et al.*, 1990; Gee, 1994; Hudzik and Palmer, 1995). Acute or subchronic doses of remacemide hydrochloride (up to 10 × the ED_{50} for the MES test) did not affect the ability of adult and adolescent rats to learn a bar-pressing task to obtain food, perform a delayed nonmatching to position paradigm, or remember the win-shift foraging strategy in the T maze (Ordy *et al.*, 1992; Hudzik and Palmer, 1995; D. Widzowski, unpublished data; Hudzik, unpublished data for adolescent rats). In four separate rat investigations (open field observation, drug discrimination, intravenous self-administration, or substitution for pleasurable intracranial self-stimulation), remacemide hydrochloride did not exhibit evidence of PCP-like behavior or abuse liability (Palmer *et al.*, 1993; Corbett *et al.*, 1996; Hudzik *et al.*, 1996).

Table 5 Acute–subchronic safety/behavioral pharmacology investigations with remacemide hydrochloride[a]

Test conditions	Observed activity
Mice	
TD_{50} (therapeutic index), p.o.[c,d]	370 (7.7) and 581 (17.6[b])
HD_{50} (hypnotic index), p.o.[c,d]	1172 (35.5)
MELD (safety margin), p.o.[c,d]	877 (18.2) and 407 (38.4[b])
MELD (safety margin), i.v.[c,e]	56 (3.8)
Tolerance to MES	Yes
Behavioral Symptoms, p.o.[f]	>400 mg kg^{-1}
Metrazol proconvulsant test[f]	Proconvulsant actions
Rat	
TD_{50} (therapeutic index), p.o.[g]	847 (39.6)
TD_{50} (therapeutic index), s.c.[h]	273 (22.4)
MELD (safety margin), p.o.[g]	983 (45.9)
MELD (safety margin), i.v.[g]	47 (7.2)
Tolerance to MES[c,g]	No
Spontaneous motor activity[g]	Increased activity at 200 mg kg^{-1}
Abuse liability—PCP behavior	
Open field observation, p.o.[c]	None up to 630 mg kg^{-1}
Drug discrimination, i.p.[c]	None up to 30 mg kg^{-1}
Substitution for intracranial self-stimulation, p.o.[c,i]	None up to 200 mg kg^{-1}
Self-administration, substitution for cocaine, i.v.[j]	None up to 84 mg kg^{-1} total
Acoustic startle reflex[c]	No effect up to 120 mg kg^{-1}
Learning/memory	
Win-shift strategy, T maze, i.p.[c,k]	No effect up to 20 mg kg^{-1}
Learning acquisition, p.o.[l]	Acute and subchronic – no effect at
Adult rats	120 or 500 mg kg^{-1}, respectively
Adolescent rats (subchronic)[h]	No effect to 60 mg kg^{-1}
Conscious dogs – symptoms, i.v.[c]	14 mg kg^{-1} – symptoms 44 mg kg^{-1} – convulsions

[a]TD_{50} for mice is determined from the inverted screen test and for rats it is the gang plank escape test. MELD is the median estimated lethal dose. Therapeutic indices, safety margins, and hypnotic indices are calculated by dividing the appropriate ED_{50}s from the MES test (Table 1) by the TD_{50}, HD_{50}, MELD, etc.
[b]Values for the Charles River strain of CF1 mice; all other data from the Harlan strain of CF1 mice.
[c]Palmer *et al.* (1993). [h]L. Freedman (unpublished data).
[d]Stagnitto *et al.* (1990). [i]Corbett *et al.* (1996).
[e]Palmer *et al.* (1992). [j]Hudzik *et al.* (1996).
[f]Palmer *et al.* (1991). [k]Ordy *et al.* (1992).
[g]Garske *et al.* (1991). [l]Hudzik and Palmer (1995).

1.5.2 Central nervous system vacuoles

No central nervous system pathology was observed using conventional brain techniques in any study. A detailed neuropathological examination with remacemide hydrochloride indicated a low propensity to produce the transient neuronal vacuoles in the retrosplenial and cingulate cortices (Palmer *et al.*, 1993; Clark *et al.*, 1995), a characteristic property of NMDA receptor antagonists (Olney *et al.*, 1989).

1.5.3 Reproductive toxicity study

In mice there were no effects of large doses of remacemide hydrochloride on time to mating, fertility index, number of implantation sites, or male reproductive organ histology. Male and female fertility in rats was likewise unaffected. In doses producing minimal to severe toxicity in rats there was lack of evidence of fetal toxicity or teratogenicity, although some skeletal retardation was apparent. Likewise, in rabbits no teratogenic effects of remacemide hydrochloride were observed (Palmer *et al.*, 1993; Clark *et al.*, 1995).

1.5.4 Mutagenic potential

Remacemide hydrochloride appeared free of genotoxicity when evaluated in two *in vitro* tests for mutagenicity (the Ames test and L5178Y mouse lymphoma cells) plus two tests for chromosome damage (cultured human lymphocytes or mouse bone marrow, i.e. the micronucleus test) (Palmer *et al.*, 1993; Clark *et al.*, 1995).

2 CLINICAL STUDIES

Remacemide hydrochloride has been studied in the following human subjects: human volunteers, patients with epilepsy, patients with acute ischemic stroke, patients undergoing coronary artery bypass surgery, and patients with Huntington's disease.

2.1 Human volunteers

A total of 169 young and elderly volunteers have been exposed to remacemide hydrochloride in a variety of studies. Single oral doses up to 400 mg were generally well tolerated, with evidence of dizziness, nausea, and dyspepsia at the highest doses. Repeated oral dosing (up to 600 mg day^{-1}) for periods of 2 weeks to a month were generally well tolerated by normal volunteers. Higher doses were associated with mood changes. There were no significant changes in vital sign measurements, laboratory tests, or psychometric performance following continued administration for up to 1 month. Similarly, single intravenous administrations by infusion of doses up to 300 mg (6 mg ml^{-1}) were not associated with specific treatment effects.

Pharmacokinetic evaluations of remacemide and ARL12495AA have been conducted. Remacemide is rapidly and extensively absorbed by man. In volunteers, maximal plasma concentrations of remacemide were achieved within 1–2 h of oral dosing (Fig. 3), with at least 90% of the administered dose absorbed. The desglycinyl metabolite appeared more slowly in plasma, with maximal concentrations achieved between 4 and 6 h. The metabolite persisted longer, with a half-life of elimination of around 12–15 h. Remacemide is 77% bound to plasma proteins in man, whilst 90% of the desglycinyl metabolite is subject to plasma protein binding.

Administration of remacemide hydrochloride with food has no effect on the extent of absorption but, as might be expected, delayed the time to peak plasma concentration. There was evidence for a slightly greater than proportional increase in C_{max} and AUC with increasing dose, but the half-life was independent of dose. The pharmacokinetic parameters for remacemide and ARL12495AA are presented in Table 6.

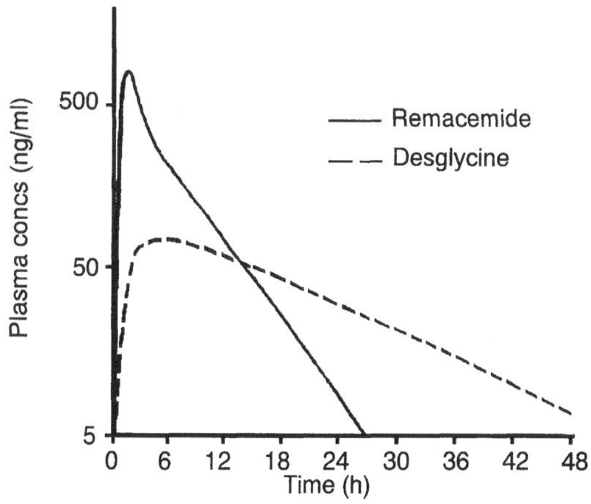

FIG. 3 Plasma concentrations of remacemide and the desglycinyl metabolite following a single oral dose (200 mg) to human volunteers.

Table 6 Pharmacokinetic parameters of remacemide and its desglycinyl metabolite in human volunteers[a]

Parameter	Remacemide	Desglycinyl metabolite
C_{max} (ng ml^{-1})	1069	143
C_{min} (ng ml^{-1})	124	90
AUC (ng ml^{-1} h^{-1})	4429	1348
$t_{1/2}$ (h)	3.3	12.9

[a]Multiple doses of remacemide hydrochloride were given orally (300 mg b.i.d.) to human volunteers (n = 7). C_{max} is the steady state maximal plasma concentration, C_{min} is the steady state predose plasma concentration, and AUC is the area under the plasma concentration versus time curve over one dose cycle.

[^{14}C]Remacemide hydrochloride was administered to volunteers to investigate its metabolism in man. Radioactivity amounting to 89% of the administered dose was recovered in the urine (9% in feces), whilst oral bioavailability was estimated at 60%. This indicates a degree of first-pass metabolism. Approximately 53% of radioactivity in the urine appeared within the first 9 h following administration of an oral dose. The metabolic profile of remacemide is complex, and a variety of metabolic products have been identified (Palmer *et al.*, 1992, 1993; Clark *et al.*, 1995). In addition to the desglycinyl moiety, significant amounts of a carbamoyl glucuronide are present in human plasma. Ring-hydroxylated metabolites of remacemide and the desglycinyl metabolite are also present in plasma. Remacemide itself is subject to hydroxylation of a terminal amine side-chain. Many of these primary metabolites are present as conjugates. Of the metabolites of remacemide, only the desglycinyl moiety and its hydroxylated forms possess significant pharmacological activity (Palmer *et al.*, 1992, 1993).

2.2 Epilepsy

Around 800 patients with medically refractory epilepsy have been exposed to remacemide hydrochloride in clinical trials to date. The majority have been enrolled into double-blind, placebo-controlled group comparative trials of efficacy, which are ongoing. The remaining patients have participated in a variety of smaller trials including double-blind cross-over trials of efficacy and pharmacokinetics, open and blinded trials of tolerability, and open efficacy trials involving assessments of safety, tolerability, effects on the EEG, and of co-medication withdrawal. All patients participating in remacemide hydrochloride trials have the opportunity to receive long-term treatment in an open fashion. To date, around 350 patients have entered long-term, open continuation trials, representing around 80% of those eligible.

Tolerability and safety have been assessed in patients receiving remacemide hydrochloride continuously for periods of up to 3 years. In general, these patients have been receiving concomitant medications which include carbamazepine, phenytoin, sodium valproate, lamotrigine, vigabatrin, gabapentin, and phenobarbital. Patients receiving enzyme-inducing co-medication tolerate higher daily doses of remacemide hydrochloride compared to volunteers since the plasma levels of remacemide and the desglycinyl metabolite are reduced in the presence of enzyme induction (see below). Consequently, doses up to 800 mg day^{-1} may be administered in a twice daily regimen, and up to 1200 mg day^{-1} in a three or four times daily regimen in enzyme-induced patients (Chadwick *et al.*, 1994). Patients not receiving enzyme-inducing co-medications, which represents a minority of refractory epileptics at present, have been shown to tolerate total daily doses up to 600 mg day^{-1}.

Dose-limiting side-effects in patients are similar to those in volunteers, and are generally benign and reversible. The most common effect is dizziness, which may resolve despite continued dosing. Other side-effects include abdominal symptoms (dyspepsia, nausea, abdominal pain, and vomiting), fatigue, somnolence, and visual abnormalities. Headache has been reported commonly in clinical trials, but the rates were similar between active and placebo-treated patients. Some symptoms were not observed in volunteers (e.g. visual abnormalities), and may be reflective of interaction with co-medicants. No consistent effects have been observed on vital signs, electrocardiogram (ECG) recording, psychometric evaluation, or laboratory tests in patients receiving remacemide hydrochloride.

Remacemide hydrochloride has been demonstrated to have the potential for interactions with enzyme-inducing anticonvulsants including carbamazepine and phenytoin (Scheyer *et al.*, 1992). No interactions have been demonstrated with sodium valproate. In a series of studies designed to quantify these interactions (Leach *et al.*, 1994), it was shown that the exposures to remacemide and the desglycinyl metabolite were 50–75% and 25–35%, respectively, compared to noninduced patients and volunteers for a given dose. Conversely, plasma concentrations of carbamazepine (but not its epoxide) showed a tendency to rise by an average of 28% in patients receiving carbamazepine monotherapy as remacemide hydrochloride (300 mg b.i.d.) was added. Similarly, some patients receiving phenytoin experienced a rise in plasma concentrations (12–22% on average) upon addition of remacemide hydrochloride (300 mg b.i.d.). These changes were not statistically significant possibly due to the intrinsic variability of phenytoin plasma concentrations in these patients.

In vitro studies using human liver microsomes have been performed to investigate the potential mechanism for the observed effects in patients. Carbamazepine is metabolized principally via 10, 11-epoxidation followed by slow hydrolysis of the epoxide and conjugation of phenolic metabolites. Phenytoin clearance is also dependent on oxidative metabolism. Both of these oxidative biotransformations are catalyzed by cytochrome P450 (CYP) enzymes: CYP 3A4 in the case of carbamazepine and CYP 2C9 for phenytoin. In human liver microsomes, remacemide hydrochloride was shown to be a modest inhibitor of CYP 2C9 and a more potent inhibitor of CYP 3A4. The desglycinyl metabolite is also an inhibitor of these systems, but is less potent. For comparison, the inhibitory potency of remacemide hydrochloride with respect to CYP 3A4 is considerably less than ketoconazole, and is considerably less potent than sulfonamides with respect to CYP 2C9. Importantly, the inhibitory interactions of remacemide hydrochloride appear to be predictable and quantitative, enabling them to be managed effectively.

The efficacy of remacemide hydrochloride was demonstrated in a double-blind, placebo-controlled cross-over trial as adjunctive treatment in patients with refractory epilepsy (Crawford *et al.*, 1992). Twenty-eight patients were randomized to receive either remacemide hydrochloride (600 mg day^{-1}) or a placebo for 1 month, after which they were subject to a 1 month washout period. After washout, they then received the alternate treatment. At the beginning of the trial, all patients had a month of baseline assessment to establish their background seizure pattern. At entry all patients had partial seizures with or without secondary generalization with an average seizure frequency of 19.3 per month (range 3–90 seizures). The majority (24/28) were receiving carbamazepine as co-medication, often in combination with another marketed anticonvulsant. Twenty-three patients completed the trial to the point where they were evaluated for efficacy.

During the treatment period with remacemide hydrochloride, the median seizure frequency was reduced from nine in the placebo month to six ($p = 0.041$). Thirty percent of patients had a seizure reduction of 50% or more during remacemide hydrochloride treatment compared to 9% during the placebo month. Three patients (11%) were seizure-free during the active treatment, whilst no patients were seizure-free during placebo treatment.

Further trials to examine the efficacy of higher and lower daily doses of remacemide hydrochloride in medically refractory epileptic patients are currently underway, as are trials to examine the potential for remacemide hydrochloride to be effective as monotherapy.

2.3 Other patient groups

2.3.1 *Acute stroke*

A preliminary examination of the tolerability and safety of remacemide hydrochloride in patients following acute ischemic stroke is nearing completion. Patients who are within 24 h of an acute ischemic stroke have been randomized to receive remacemide hydrochloride (six patients) or placebo (two patients) for 3 days by b.i.d. intravenous infusion. Dose levels between 100 and 600 mg b.i.d. have been examined.

Preliminary data from this stroke trial, which remains blinded, suggest there are no specific effects of remacemide hydrochloride on vital signs, blood chemistry, hematology, or the ECG. Adverse effects thought to be treatment related include venous irritation (which has been substantially reduced by dilution of the infusate) and some CNS effects. The latter range from headache, hallucination, diplopia, agitation, tremor, confusion, and hyperreflexia, though these have generally been confined to one or two individuals per dose group. As yet, no specific adverse effects relating to the administration of remacemide hydrochloride have been observed (Muir and Lees, 1995a), and the concentrations in plasma of the desglycinyl metabolite are consistent with the potential for neuroprotection, as manifest in a variety of animal models (Palmer *et al.*, 1995).

2.3.2 *Coronary artery bypass surgery*

Tolerability and safety were examined in patients undergoing coronary artery bypass surgery (CABS). A second study was a double-blind, placebo-controlled trial to examine the potential for remacemide hydrochloride administered before and after surgery to provide prophylactic neuroprotection by reducing the observed neuropsychological consequences reported in a proportion of such patients (Blauth *et al.*, 1989; Pugsley *et al.*, 1994). Remacemide hydrochloride (600 mg day^{-1}) was safe and well tolerated when administered for 5 days prior to and 5 days following surgery. Treatment-related adverse events were similar to those experienced by volunteers. With respect to demonstrating neuroprotection, there was an unexpectedly low incidence of neuropsychological deficits in the placebo-treated group compared to published data from the same clinical investigators and others (Blauth *et al.*, 1989; Pugsley *et al.*, 1994). Under these circumstances, the trial was insufficiently powered to demonstrate treatment differences, which were therefore not observed. It is believed that methodological advances and careful attention to the surgical and anesthetic procedures can produce marked reduction in the incidence of this complication, which may ultimately negate the requirement for pharmacological intervention.

2.3.3 *Huntington's disease*

A preliminary investigation has been made of the tolerability and safety of remacemide hydrochloride in Huntington's disease (Kieburtz *et al.*, 1995). There are sound theoretical grounds to believe that glutamate toxicity may be implicated in the pathogenesis of Huntington's disease, an inherited condition characterized by movement disorder and progressive dementia (Beal, 1992). Safety data have been collected from this group of patients, who received up to 600 mg day^{-1} of remacemide hydrochloride for 6 weeks in a randomized placebo-controlled study. Twenty-nine out of 31 patients enrolled completed the study, with one withdrawal due to an adverse event (vomiting). Remacemide hydrochloride was well tolerated by the remainder of patients, with few adverse effects. There were no demonstrable effects on other safety variables or on cognitive function. Further evaluation of the potential efficacy of remacemide hydrochloride in this patient group is planned.

3 CONCLUSIONS

Remacemide hydrochloride has attracted interest since it appears to be an example of an orally active antiepileptic and neuroprotective agent working not only via low-affinity uncompetitive inhibition of the channel site on the NMDA receptor, but also by antagonism of fast sodium channels, thereby inhibiting glutamate release presynaptically. The activity of remacemide hydrochloride in animals and man appears distinct from other NMDA receptor antagonists (the majority of which are high-affinity compounds) in as much as the tolerability appears to be good at pharmacologically effective doses. This is true for long-term administration to patients with epilepsy, and although neuroprotection has yet to be demonstrated in man, it is predicted from the animal model data.

Dual anticonvulsant and neuroprotective properties, if confirmed by larger-scale clinical trials, would offer unique advantages to the management of epilepsy. Although the precise mechanism by which brain damage occurs in epilepsy has yet to be fully understood, it appears that excitotoxic mechanisms play a part in mediating both the direct damage occurring during a seizure, and in the hypoxic injury resulting from prolonged seizure activity (Fujikawa, 1995). Furthermore, there is evidence from animal studies that NMDA receptor antagonists may be useful in preventing the development of chronic seizure disorders by inhibiting the development of kindling (McNamara *et al.*, 1988). Clearly, these potential advances will require formal evaluation in extended clinical trials.

To date, the preclinical and clinical data confirm that remacemide hydrochloride has potential as an effective anticonvulsant in refractory epilepsy, when added to conventional therapy. Further work has been undertaken to establish its potential as monotherapy. As far as can be judged from animal data, remacemide hydrochloride appears to be an effective neuroprotectant in a variety of models of cerebral ischemia and has been well tolerated in stroke patients at doses which are potentially neuroprotective. The results of ongoing clinical trials will help clarify the value of remacemide hydrochloride in these and other organic brain diseases.

Glossary

Some of the terms and abbreviations commonly used in this field and found in this book are explained here. Where appropriate, the chapters using a given term are cited. In many cases, the chapter(s) cited will provide additional information. Appended to the Glossary is a table summarizing the common names/abbreviations, structures and an indication of the type of activity, of some of the most commonly used compounds that are active at excitatory amino acid receptors. For additional definitions, please refer to *Excitatory Amino Acids and Synaptic Transmission*, 2nd edn (H. Wheal & A. Thomson, eds; 1995. Academic Press, London).

Akinesia Inability to initiate movement. (Chapters 8, 10)

Allodynia Condition in which ordinarily non-painful stimuli evoke pain. (Chapter 3)

Amyotrophic lateral sclerosis A disease of the motor tracts of the lateral columns and anterior horns of the spinal cord, causing progressive muscular atrophy, increased reflexes, fibrillary twitching, and spastic irritability of muscles; associated with a defect in the enzyme superoxide dismutase. A number of cases are inherited as an autosomal dominant trait. This disorder affects adults, is 90–95% sporadic in nature, and is usually fatal within 2–4 years of onset. (Chapters 8, 9)

Analgesia Loss of the sense of pain as that produced by the injection of local anesthetic agents or by systemic drugs such as the opioids, or resulting from disease interrupting pain pathways in the central or peripheral nervous system. (Chapters 3, 5, 6)

Anathria Loss of the power of articulate speech. (Chapter 2)

Anesthetic A compound that reversibly depresses neuronal function, producing loss of ability to perceive pain and/or other sensations. (Chapters 3, 4, 8, 10)

Angiomatous malformation Malformation relating to benign vascular tumour composed of blood or lymphatic vessels. (Chapter 2)

Anhedonia Inability to derive pleasure from day-to-day activities that the normal person finds pleasurable. It may appear as one of the earliest complaints in schizophrenia. (Chapter 5)

Anoxia Lack of oxygen in the circulating blood or in the tissues. (Chapters 6, 9, 10)

Anticonvulsant An agent acting to prevent or arrest seizures. (Chapters 1, 3, 4, 5, 6, 7, 8, 10)

Antidepressant An agent used to treat depression. (Chapters 5, 6)

Antinociceptive Tending to reduce the perception and behavioral effects of nociceptive stimuli. (Chapters 3, 6)

Anxiolytic Capable of preventing, reducing or eliminating anxiety. (Chapters 1, 5)

Aphasia A change in, or loss of, the ability to speak or to understand spoken or written speech, usually attributable to a cerebral lesion and occurring in the absence of any defect of phonation or articulation. (Chapter 4)

Ataxia Unsteadiness, incoordination or disorganization of movements in the absence of paralysis. In cerebral ataxia, the disorder is secondary to loss of the normal cerebellar influence upon motor activity. (Chapters 1, 2, 4, 5, 6, 7, 8)

AUC Area under the curve. (Chapters 1, 2, 4, 5, 6, 8, 10)

Bacteremia The presence of viable bacteria in the circulating blood. (Chapter 4)

Barthel index A scale used to assess physical function (so-called activities of daily living assessment), used widely in assessing progress in rehabilitation as well as static levels of ability. (Chapters 1, 4)

Brain Edema Diffuse or focal brain swelling due to an increase in intracellular but more particularly in extracellular fluid. (Chapters 2, 8)

Bulbar Onset Subtype of amyotrophic lateral sclerosis. (Chapter 9)

Cerebral Edema See brain edema. (Chapters 1, 4, 10)

Cerebral Ischemia Inadequate blood flow in the brain due to mechanical obstruction of the blood supply. (Chapters 1, 4, 6, 7, 8, 9, 10)

Cerebral Vasodilator An agent that causes dilation of the blood vessels in the brain. (Chapter 10)

Circadian Relating to a rhythm or recurring period of biological activity of roughly 24 hours. (Chapter 2)

Contralateral Turning Behavior Classical whole animal model for assessing dopaminergic agonist activity. The midbrain dopaminergic nuclei are unilaterally lesional, causing a supersensitivity of dopamine receptors in the basal ganglia. Systemic application of dopaminergic agents causes a turning behavior towards the side opposite from the lesion. (Chapter 6)

Decelerative Therapy Therapy to decrease heart rate. (Chapter 6)

Diplopia The condition in which a single object is perceived as two objects, also called double vision. (Chapters 2, 10)

Distal Paresthesia Any sensation such as pins and needles, burning, prickling, etc., which occurs spontaneously, without external cause occurring in the extremity or distant part of a limb. (Chapter 2)

Dysartria A disturbance of speech and language due to emotional stress, to brain injury, or to paralysis, incoordination, or spasticity of the muscles used for speaking. (Chapter 2)

Dyspepsia Impaired gastric function or 'upset stomach' due to some disorder of the stomach; characterized by epigastric pain, sometimes burning, nausea, and gaseous eructation. (Chapter 10)

Dysphoric Pertaining to a mood of general dissatisfaction, restlessness, depression, and anxiety; a feeling of unpleasantness or discomfort. (Chapter 6)

Dystonia Any abnormality of muscle tone. (Chapter 6)

Electroconvulsive Shock Therapy (ECT) Electroshock therapy: a form of treatment of mental disorders in which convulsions are produced by the passage of an electric current through the brain. (Chapter 5)

Electroencephalography A technique for recording the electrical activity of the brain, through the intact skull and scalp. An instrument is used which is made up of a number of channels, each of which comprises an amplifying and recording instrument which records the electrical events beneath and between a pair of recording electrodes. (Chapters 2, 5, 9)

Emesis Vomiting. (Chapter 8)

Emetic Response A response manifesting itself in vomiting. (Chapter 8)

Enantiomers Also called stereoisomers or optical isomers. One pair of compounds that differ only in being mirror images of each other in molecular structure. Characterized by the two arrangements of (the four different) groups around an asymmetric carbon atom which cannot be superimposed, being mirror images of each other. Known as the D and L forms or (*R*) and (*S*) forms. (Chapters 1, 2, 3, 10)

Epidural Overlying the dura mater. Filling the space between the dura and cranium. (Chapter 3)

Epilepsy A neurological disorder characterized by the tendency to suffer recurrent seizures or fits, whether minor or major, due to excessive neuronal discharges, and usually associated with some alteration of consciousness. The clinical manifestations of the attack may vary from complex abnormalities of behavior including generalized or focal convulsions to momentary spells of impaired consciousness. The condition may be idiopathic due to no obvious cause or symptomatic, resulting from a focal cerebral lesion, from diffuse brain disease, or from some metabolic disturbance affecting cerebral function. A genetic component is also likely in some forms of epilepsy. (Chapters 1, 2, 3, 5, 6, 8, 10)

EPMR Values Electron paramagnetic resonance values. (Chapter 2)

Facial/Conjunctival Hyperemia An increase in the volume of blood due to arterial or arteriolar dilatation in the face or in the mucous membrane investing the anterior surface of the eyeball and the posterior surface of the lids. (Chapter 2)

Fasciculations Spontaneous contraction of bundles of skeletal muscle fibres resulting in a localized twiching or flickering which can be seen under the skin or a mucuous membrane but does not produce movement at a joint. (Chapter 9)

Focal Ischemia A localized area of anemia due to mechanical obstruction (mainly arterial narrowing) of the blood supply. (Chapters 5, 6, 7, 8, 9, 10)

Geometric Isomerism A form of isomerism displayed by unsaturated or ring compounds where free rotation about a bond (usually a carbon–carbon bond) is restricted. (Chapter 2)

Gliosis Proliferation of the astrocytes in an area of damage in the brain or spinal cord. It is one means of scar formation in the central nervous system. (Chapter 1)

Global Ischemia Anemia due to mechanical obstruction (mainly arterial narrowing) of the blood supply affecting the whole brain. (Chapters 5, 6, 7, 8, 9, 10)

Guam Disease A disease showing the symptoms of both Parkinson's disease and amyotrophic lateral sclerosis, found in the people of Guam who ate cycas plant fruits during the war when other food was scarce. (Chapter 9)

Hemianopia Inability to see in half of the visual field. (Chapter 4)

Hyperacusis Abnormal acuteness of hearing due to increased irritability of the sensory neural mechanism. (Chapter 3)

Hyperalgesia Extreme sensitiveness to painful stimuli. (Chapters 3, 6, 8)

Hyperpathia Exaggerated subjective response to painful stimuli, with a continuing sensation of pain after the stimulation has ceased. (Chapter 3)

Hyperreflexia A condition in which deep tendon reflexes are exaggerated. (Chapter 10)

Hypertension High blood pressure. (Chapters 1, 4, 5, 6, 8, 10)

Hypobaric Hypoxia Decrease below normal levels of oxygen in inspired gases, arterial blood at less than normal atmospheric pressure. (Chapter 9)

Hypoglycemia An abnormally small concentration of glucose in the circulating blood, i.e. less than the minimum of the normal range. (Chapter 6)

Hypoglycemic Neuronal Death Neuronal death caused by hypoglycemia. (Chapter 6)

Hypoxia Decrease below normal levels of oxygen in inspired gases, arterial blood, or tissue. (Chapters 1, 4, 6, 9, 10)

Hypoxia–Ischemia Tissue hypoxia caused by arterial or arterioral obstruction or vasoconstriction. (Chapter 1)

Ischemic Penumbra The region of partial illumination or radiation caused by light or X-rays not originating from a point source, also called geometric unsharpness, related to or affected by ischemia. (Chapter 6)

Laminectomy Excision of the posterior arches and spinous processes of a vertebra. (Chapter 3)

Lathyrism A disease occurring in Ethopia, Algeria, and India, characterized by various nervous manifestations, tremors, spastic paraplegia, and paresthesias; resulting from the ingestion of peas of the genus *Lathyrus*. (Chapter 9)

Logorrhea Excessive, rapid speech, as in mania. (Chapter 2)

MCA Occlusion Medial cerebral artery occlusion used experimentally as an obstructive stroke model. (Chapters 1, 2, 6, 7, 8, 9, 10)

Metabolic Acidosis Acidosis that rises from a disorder of metabolism in which acid (excluding carbonic acid) accumulates in, or bicarbonate is lost from, extracellular fluid. It is distinguished from respiratory acidosis, in which the primary defect is retention of carbon dioxide. (Chapter 2)

Metabotropic Glutamate Receptors Glutamate receptors that do not form an ion channel, but exert their effect via an intracellular signalling mechanism. Separated now into three groups; group 1 being mGluR1 and 5, group 2 being mGluR2 and 3, and group 3 being mGluR4, 6 and 7. Group 1 receptors are coupled to phospho-inositol hydrolysis, and some may be coupled via different G-proteins (less pertussis toxin-sensitive) from those that are coupled to groups 2 and 3. Group 2 and group 3 receptors are negatively coupled to adenylate cyclase and are pertussis toxin-sensitive. (Chapters 2, 3, 7)

Myocardial Infarct (Global Ischemia) A discrete, usually wedge-shaped area of ischemic coagulative necrosis caused by interruption of blood flow relating to myocardium. (Chapters 4, 5)

Neuronal Plasticity The property of neurons of being able to modify their connection pattern, e.g. during learning or repair following injury. (Chapter 3)

Neuronal Vacuolization Formation of vacuoles in neurons. (Chapters 1, 3, 4, 5, 6)

Neuropathic Pertaining to, inducing, or causing disorders due to a neuronal dysfunction. (Chapter 3)

Neurotoxicity The quality of having a toxic effect upon the nervous system. (Chapters 1, 3, 5, 6, 7, 8, 9)

Nociceptive Denoting responsiveness or sensitivity to noxious stimuli capable of eliciting pain. (Chapter 3)

Noxious Harmful or injurious. (Chapter 3)

NSAIDs Nonsteroidal anti-inflammatory drugs. (Chapter 3)

Olney-Type Vacuoles Neuronal vacuoles seen in part of the cortical cingulate gyrus in rodents only after high doses of agents inhibiting the *N*-methyl-D-aspartate (NMDA) receptor. (Chapter 1)

Paraplegia Paralysis of both lower extremities and, generally, the lower trunk. (Chapters 1, 7, 9)

Parkinson's Disease A neurological syndrome usually resulting from deficiency of the neurotransmitter dopamine as the consequence of degenerative, vascular, or inflammatory changes in the basal ganglia; in particular the subotualia nigra, characterized by rhythmical muscular tremors, rigidity of movement, festination, droopy posture, and masklike facies. (Chapters 5, 6, 8, 9, 10)

Partial Agonists A compound that binds to the agonist-binding site and elicits in the receptor only a partial activation, i.e. the response will be less than the possible maximal response, no matter how high the concentration of the partial agonist is. (Chapter 5)

Photically Induced Myoclonus One or a series of shock-like contractions of a group of muscles, of variable regularity, synchrony, and symmetry, generally due to a central nervous system lesion, induced by rhythmic variation of a light source. (Chapter 8)

Photoepileptic Property making an individual susceptible to convulsions if stimulated by rhythmic variation of a light source. (Chapter 2)

Psychostimulation Antidepressant or mood-elevating agent, or also inducing euphoria and hallucinations. (Chapter 6)

Psychotomimetic Of a drug whose action(s) mimics psychosis. (Chapters 1, 2, 3, 4, 6, 8)

Ptosis Abnormal downward displacement of the upper lid, as may result from paralysis of the third cranial nerve, sympathetic denervation, or injury. (Chapters 5, 8)

Racemic Of a mixture of D and L stereoisomers of an optically active compound. (Chapters 1, 7)

Secondary Hyperalgesia Extreme sensitiveness to painful stimuli as a consequence of a primary cause. (Chapter 3)

Strychnine-Insensitive Glycine Site on the NMDA Receptor (SIGS) The NMDA receptor cannot be activated if no agonists interact with this site. (Chapter 2)

SOD-1 Superoxide dismutase: an enzyme that catalyzes the dismutation reaction, $2O_2^{-}\cdot + 2H^+ \rightarrow H_2O_2 + O_2$. There are three isozymes of SOD: an extracellular form (ECSOD) that contains copper and zinc, a cytoplasmic form that also contains copper and zinc, and a mitochondrial form that contains manganese; a deficiency of SOD is associated with amyotrophic lateral sclerosis. (Chapter 9)

Spasticity A state of increased muscular tone with exaggeration of the tendon reflexes. (Chapter 8)

Spatial Dissociation An unconscious process by which a group of mental processes is separated from the rest of the thinking processes, resulting in an independent functioning of these processes and a loss of the usual relationships; for example, a separation of affect from cognition. (Chapter 2)

Tachycardia Rapid beating of the heart, conventionally applied to rates over 100 beats per minute. (Chapters 1, 2, 6)

Tachyphylaxis Rapid appearance of progressive decrease in response following repetitive administration of a pharmacologically or physiologically active substance. (Chapter 2)

Tandem walk The act of walking by placing the heel of one foot in front of the toes of the opposite foot, while attempting to progress in a straight line. (Chapter 2)

Teratogenicity The property of producing fetal malformations. (Chapter 10)

Vacuolization Process of forming vacuoles (here in neurons). Indicates either incomplete histological fixation or overstimulation and/or injury to neurons. (Chapters 1, 3, 4, 5, 6, 7)

Vasoconstriction A narrowing of the lumen of the blood vessels. (Chapter 6)

Vasodilatation A widening of the lumen of the blood vessels. (Chapter 6)

Summary Table of Compounds and Their Clinical Status

P. L. HERRLING

Sandoz Pharma Ltd, CH-4002 Basle, Switzerland

Table 1

Name/Code	Structure	Mechanism	Route of application, human	Main indications	Main limiting side-effects in humans	Development status	Manufacturer
Selfotel (CGS 19755)	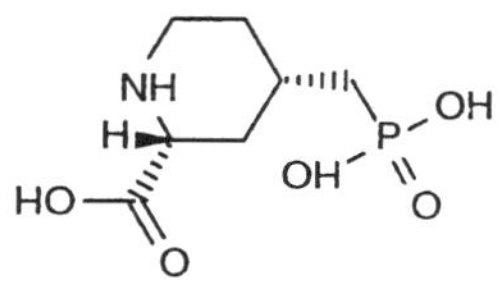	Competitive NMDA antagonist	Intravenous	Neuroprotection Head trauma Stroke	Hallucinations	Ph III on hold	Ciba
Trazec (SDZ EAA 494)	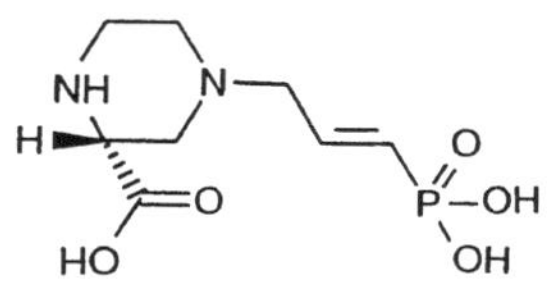	Competitive NMDA antagonist	Intravenous	Neuroprotection Head trauma	Hallucinations	PhIIa	Sandoz
CPP	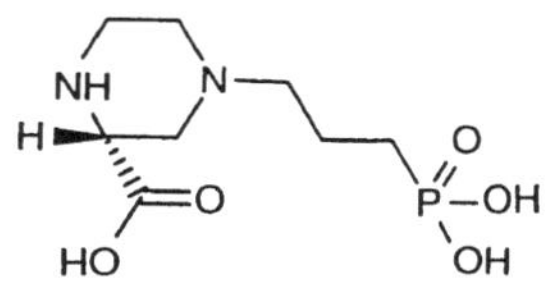	Competitive NMDA antagonist	Intrathecal	Neuropathic pain	Hallucinations	Exploratory human testing	Tocris
CERESTAT (CNS 1102)	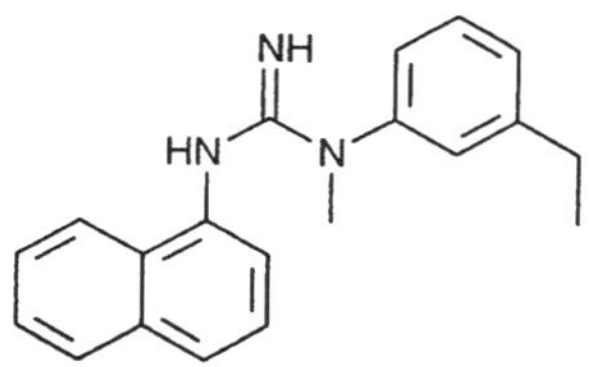	NMDA channel blocker	Intravenous	Neuroprotection Stroke	Disorientation Confusion Sedation Nausea	PhIII	Cambridge NeuroScience Inc
ACPC	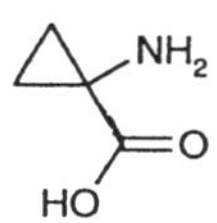	Partial agonist at glycine site of NMDA receptor	Intravenous	Neuroprotection Antidepression	None described	PhI	Symphony Pharmaceuticals
Eliprodil	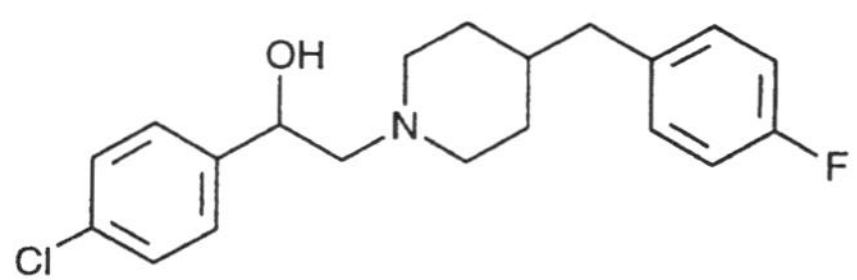	Polyamine site modulator NR2B antagonists Ca^{2+} antagonists	Oral Intravenous	Neuroprotection Stroke Head trauma	Headache Dizziness	PhII/III Terminated	Synthelabo

LY293558	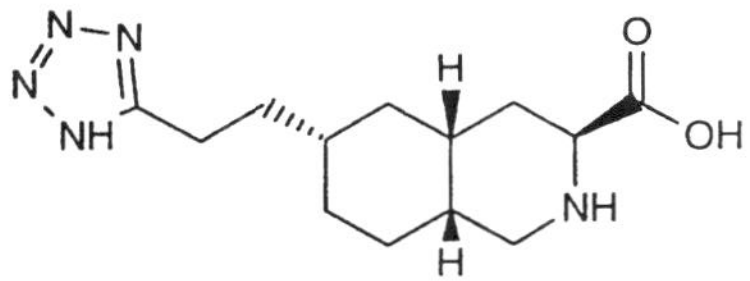	AMPA/KA Competitive antagonist	(Intravenous)	Neuroprotection	Unknown	Preclinical	Eli Lilly and Co
NBQX	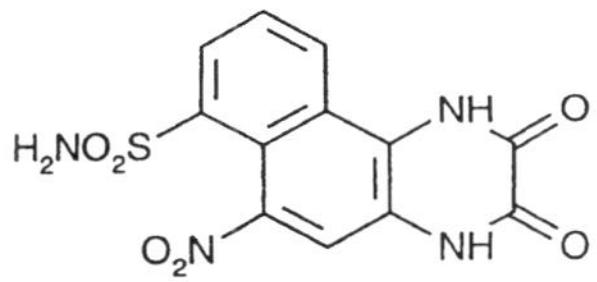	AMPA antagonist	Intravenous	Neuroprotection	Potential nephrotoxicity	PhI Terminated	Novo Nordisk
Riluzole	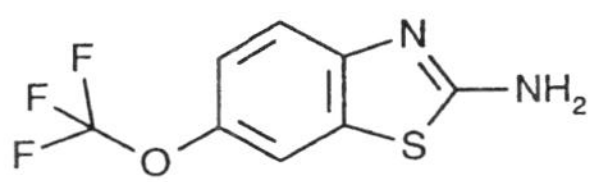	Non-competitive antagonist NMDA Glutamate-release inhibitor Na^+ channel blocker	Oral	Neuroprotection Amyotrophic lateral sclerosis (ALS)	Asthenia Nausea	Marketed	Rhone-Poulenc Rorer
Remacemide	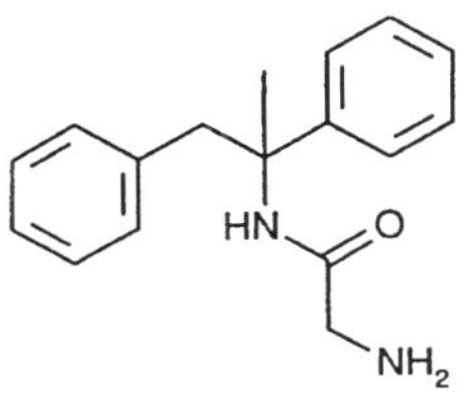	NMDA channel blocker Polyamine site modulator	Oral Intravenous	Neuroprotection Epilepsy Stroke Huntington's disease	Dizziness Nausea Dyspepsia Headache	PhII	Astra

References

Adeagbo, A.S. & Magbagbeola, A.O. (1985). Pharmacological actions of ifenprodil in the rat isolated anococcygeus muscle. *J. Pharm. Pharmacol.* **37**, 833–835. [*Chapter 6*]

Adeagbo, A.S. (1984). Vascular relaxation by ifenprodil in the isolated perfused rat mesenteric artery. *J. Cardiovasc. Pharmacol.* **6**, 1142–1147. [*Chapter 6*]

Aebischer, B., Frey, P., Haerter, H.P., Herrling, P.L., Mueller, W., Olverman, H.J. & Watkins, J.C. (1989). Synthesis and NMDA antagonistic properties of the enantiomers of 4-(3-phosphono-propyl)piperazine-2-carboxylic acid (CPP) and of the unsaturated analogue (*E*)-4-(3-phosphono-prop-2-enyl)piperazine-2-carboxylic acid (CPP-ene). *Helv. Chim. Acta* **72**, 1043–1051. [*Chapter 2*]

Aizenman, E. & Hartnett, K.A. (1992). The action of CGS 19755 on the redox enhancement of NMDA toxicity in rat cortical neurons *in vitro*. *Brain Res.* **585**, 28–34. [*Chapter 1*]

Aizenman, E., Lipton, S. & Loring, R. H. (1989). Selective modulation of NMDA responses by reduction and oxidation. *Neuron* **2**, 1257–1263. [*Chapter 6*]

Albers, G.W., Atkinson, R.P., Kelley, R.E. & Rosenbaum, D.M. (on behalf of the Dextrorphan Study Group) (1995). Safety, tolerability, and pharmacokinetics of the *N*-methyl-D-aspartate antagonist dextrorphan in patients with acute stroke. *Stroke* **26**, 254–258. [*Chapter 4*]

Albers, G.W., Goldberg, M.P. & Choi, D.W. (1989). *N*-Methyl-D-aspartate antagonists: ready for clinical trial in brain ischemia? *Ann. Neurol.* **25**, 398–404. [*Chapter 4*]

Allen, H.L. & Iversen, L.L. (1990). Phencyclidine, dizocilpine, and cerebrocortical neurons. *Science* **247**, 221. [*Chapter 1, 3*]

Anis, N.A., Berry, S.C., Burton, N.R. & Lodge, D. (1983). The dissociative anaesthetics, ketamine and phencyclidine, selectively reduce excitation of central mammalian neurones by *N*-methyl-aspartate. *Br. J. Pharmacol.* **79**, 565–575. [*Preface, Chapter 7*]

Aoki, M., Nomura, F., Stromski, M.E., Tsuji, M.K., Fackler, J.C., Hickey, P.R., Holtzman, D. & Jonas, R.A. (1994). Effects of MK-801 and NBQX on acute recovery of piglet cerebral metabolism after hypothermic circulatory arrest. *J. Cereb. Blood Flow Metab.* **14**, 156–165. [*Chapter 8*]

Arnér, S. & Meyerson, B.A. (1988). Lack of analgesic effect of opioids on neuropathic and idiopathic forms of pain. *Pain* **33**, 11–23. [*Chapter 3*]

Ascher, P. & Johnson, J.W. (1994). The NMDA receptor, its channel, and its modulation by glycine. In: 'The NMDA Receptor', 2nd edn (eds, G.L. Collingridge & J.C. Watkins), pp. 177–205. Oxford University Press: Oxford. [*Chapter 2*]

Auer, R.N. & Coulter, K.C. (1994). The nature and time course of neuronal vacuolation induced by the N-methyl-D-aspartate antagonist MK-801. *Acta Neuropathol. Berl.* **87**, 1–7. [*Chapter 3*]

Avenet, P., Leonardon, G., Besnard, F., Graham, D., Frost, J., Depoortere, H., Langer, S. Z. & Scatton, B. (1996). Antagonist properties of the stereoisomers of ifenprodil at NR1A/NR2B and NR1A/NR2B subtypes of the *N*-methyl-D-aspartate receptor expressed in *Xenopus* oocytes. *Eur. J. Pharmacol.* **296**, 209–213. [*Chapter 6*]

Backonja, M., Arndt, G., Gombar, K.A., Check, B. & M., Z. (1994). Response of chronic neuropathic pain syndromes to ketamine: a preliminary study. *Pain* **56**, 51–57. [*Chapter 3*]

Baker, A.J., Moulton, R.J., MacMillan, V.H. & Shedden P.M. (1993). Excitatory amino acids in cerebrospinal fluid following traumatic brain injury in humans. *J Neurosurg*. **79**, 369–372. [*Chapter 1*]

Balster, R.L., Nicholson, K.L. & Sanger, D.J. (1994). Evaluation of the reinforcing effects of eliprodil in rhesus monkeys and its discriminative stimulus effects in rats. *Drug Alcohol Depend.* **35**, 211–216. [*Chapter 6*]

Bannan, P.E., Graham, D.L., Lees, K.R. & McCulloch, J. (1994). Neuroprotective effect of remacemide hydrochloride in focal cerebral ischemia in the cat. *Brain Res.* **664**, 271–275. [*Chapter 10*]

Bastianetto, S., Perrault, G. & Sanger, D.J. (1995a). Pharmacological evidence for the involvement of sigma sites in DTG-induced contralateral circling in rats. *Neuropharmacology* **34**, 107–114. [*Chapter 6*]

Bastianetto, S., Rouquier, L., Perrault, G. & Sanger, D.J. (1995b). DTG-induced circling behaviour in rats may involve the interaction between σ sites and nigro-striatal dopaminergic pathways. *Neuropharmacology* **34**, 281–287. [*Chapter 6*]

Bath, C., Gilmore, J. & Bleakman, D. (1995). Inhibition of human N-type and rat P-type voltage-dependent calcium channels by ifenprodil and eliprodil. *Br. J. Pharmacol* Proceedings 5–7 Apr, **87P**. [*Chapter 6*]

Beal, M.F. (1992). Role of excitotoxicity in human neurological disease. *Current Opin. Neurobiol.* **2**, 657–662. [*Chapter 10*]

Ben-Ari, Y., Aniksztejn, L. & Bregestovski, P. (1992). Protein kinase C modulation of NMDA currents: an important link for LTP induction. *Trends Neurosci.* **15**, 333–339. [*Chapter 3*]

Benavides, J., Camelin, J.C., Mitrani, N., Flamand, F., Uzan, A., Legrand, J.J., Gueremy, C. & Le Fur, G. (1985). 2-Amino-6-trifluoromethoxy benzothiazole, a possible antagonist of excitatory amino acid neurotransmission II. Biochemical properties. *Neuropharmacology* **24**(11), 1085–1092. [*Chapter 9*]

Benavides, J., Peny, B., Allen, J. & Scatton, B. (1992). Pharmacological characterization of in vivo [^{3}H]Ifenprodil binding sites in the mouse brain. *J. Pharmacol. Exp. Ther.* **260**, 896–901. [*Chapter 6*]

Bennett, D.A., Bernard, P.S., Amrick, C.L., Wilson, D.E., Liebman, J.M. & Hutchison, A.J. (1989). Behavioral pharmacological profile of CGS 19755, a competitive antagonist at *N*-methyl-D-aspartate receptors. *J. Pharmacol. Exp. Ther.* **250**, 454–460. [*Chapter 1*]

Benoit, E. & Escande, D. (1991). Riluzole specifically blocks inactivated Na channels in myelinated nerve fibre. *Pflugers Arch.* **419**(6), 603–609. [*Chapter 9*]

Bensimon, G., Lacomblez, L. & Meininger, V. The ALS Study Group (1994). A controlled trial of riluzole in amyotrophic lateral sclerosis. *N. Engl. J. Med.* **330**(9), 585–591. [*Chapter 9*]

Benvenga, M.J. & Spaulding, T.C. (1988). Amnesic effect of the novel anticonvulsant MK-801. *Pharmacol. Biochem. Behav.* **30**, 205–207. [*Chapter 6*]

Benveniste, H. (1991). The excitotoxin hypothesis in relation to cerebral ischemia. *Cerebrovasc. Brain Metabol. Rev.* **3**, 213–245. [*Chapter 7*]

Benveniste, H., Drejer, J., Schousboe, A. & Diemer, N. H. (1984). Elevation of the extracellular concentrations of glutamate and aspartate in rat hippocampus during transient cerebral ischemia monitored by intracerebral microdialysis. *J. Neurochem.* **43**, 1369–1374. [*Chapter 6, 8*]

Benveniste, H., Jorgensen, M. B., Sandberg, M., Christensen, T., Hagberg, H. & Diemer, N. H. (1989). Ischemic damage in hippocampal CA1 is dependent on glutamate release and intact innervation from CA3. *J. Cereb. Blood Flow Metab.* **9**, 629–639. [*Chapter 6*]

Benveniste, M. & Mayer, M.L. (1991). Structure–activity analysis of binding kinetics for NMDA receptor competitive antagonists: the influence of conformational restrictions. *Br. J. Pharmacol.* **104**, 207–221. [*Chapter 2*]

Benveniste, M. & Mayer, M.L. (1993). Multiple effects of spermine on *N*-methyl-D-aspartic acid receptor responses of rat cultured hippocampal neurones. *J. Physiol.* **464**, 131–163. [*Chapter 6*]

Berger, P., Farrel, K., Sharp, F. & Skolnick, P. (1994). Drugs acting at the strychnine-insensitive glycine receptor do not induce HSP-70 protein in the cingulate cortex. *Neurosci. Lett.*, **168**, 147–150. [*Chapter 5*]

Bergeron, R., De Montigny, C. & Debonnel, G. (1995). Biphasic effects of sigma ligands on the neuronal response to *N*-methyl-D-aspartate. *Naunyn-Schmiedebergs Arch. Pharmacol.* **351**, 252–260. [*Chapter 6*]

Bernert, H., Löschmann, P.-A. & Turski L. (1992). NBQX protects against traumatic brain injury. *Soc. Neurosci. Abstr.* **18**, 457.2156–65. [*Chapter 8*]

Besnard, F., Renard, S., Langer, S. Z. & Graham, D. (1995). Expression and pharmacological characterization of NMDA receptor combinations NR1A-NR2A and NR1A-NR2B. *Soc. Neurosci. Abstr.* **21**, 41.8. [*Chapter 6*]

Biton, B., Granger, P., Carreau, A., Depoortere, H., Scatton, B. & Avenet, P. (1994). The NMDA receptor antagonist eliprodil (SL 82.0715) blocks voltage-operated Ca^{2+} channels in rat cultured cortical neurones. *Eur. J. Pharmacol.* **257**, 297–301. [*Chapter 6*]

Biton, B., Granger, P., Depoortere, H., Scatton, B. & Avenet, P. (1996). Block of P-type calcium channels by the NMDA receptor antagonist eliprodil in acutely dissociated rat Purkinje cells. *Eur. J. Pharmacol.* **294**, 91–100. [*Chapter 6*]

Björkman, R., Hallman, K.M., Hedner, J., Hedner, T. & Henning, M. (1994). Acetaminophen blocks spinal hyperalgesia induced by NMDA and substance P. *Pain* **57**, 259–265. [*Chapter 3*]

Black, M.A., Tremblay, R., Mealing, G.A-R., Durkin J.P., Whitfield, J.F. & Morely, P. (1996). The desglycinyl metabolite of remacemide hydrochloride is neuroprotective in cultured rat cortical neurons. *J. Neurochem.* **66**, 989–995. [*Chapter 10*]

Blauth, C., Griffin, S., Harrison, M., Klinger, L., Newman, S., Pugsley, W., Smith, P., Taylor, K., Treasure, T. & Venn, G. (1989). Neuropsychologic alterations after cardiac operation. *J. Thorac. Cardiovasc. Surg.* **98**, 454–455. [*Chapter 10*]

Block, G.A. (for the CNS 1102–003 Study Group) (1995). Final results from a dose-escalating safety and tolerance study of the non-competitive NMDA antagonist CNS 1102 in patients with acute cerebral ischemia. *Stroke* **26**, 185. [*Chapter 4*]

Bowes, M.P., Schoepp, D.D. & Zivin, J.A. (1994). The AMPA glutamate antagonist LY293558 improves neurological outcome following reversible spinal ischemia in rabbits. *Neurology* **44**, 190–191. [*Chapter 7*]

Branchereau, A. & Rouffy, J. (1995). Double-blind randomized controlled trial of ifenprodil tartrate versus placebo in chronic arterial occlusive disease of the legs at stage II of the Leriche and Fontaine classification. *J. Mal. Vasc.* **20**, 21–27. [*Chapter 6*]

Briggs, R.C., Bradley, W.G., Shahani B. (1991) Approach to the patient with neuromuscular disease. In 'Harrison's Principles of Internal Medicine' (eds, J.D. Wilson, E. Braunwald, K.J. Isselbacher, R.G. Petersdorf, J.B. Martin, A.S. Fauchi, R.K. Root), vol. 2, part 13, chapter 362, pp. 2088–2096. McGraw-Hill. [*Chapter 9*]

Browne, S.E. & McCulloch, J. (1994). AMPA receptor antagonists and local glucose utilization in the rat. *Brain Res.* **641**, 10–20. [*Chapter 7*]

Buchan, A.M. (1990). Do NMDA antagonists protect against cerebral ischemia: are clinical trials warranted? *Cebrovasc. Brain Metabolism Rev.* **2**, 1–26. [*Chapters 5, 10*]

Buchan, A.M., Li, H., Cho, S. & Pulsinelli, W.A. (1991a). Blockade of the AMPA receptor prevents CA1 hippocampal injury following severe but transient forebrain ischemia in adult rats. *Neurosci. Lett.* **132**, 255–258. [*Chapter 8*]

Buchan, A.M., Xue, D., Huang, Z.-G., Smith, H. & Lesiuk H. (1991b). Delayed AMPA receptor blockade reduces cerebral infarction induced by focal ischemia. *NeuroReport* **2**, 473–476. [*Chapter 8*]

Bullock, R. Graham, D.I., Chen, M.H., Lowe, D. & McCulloch, J. (1990a). Focal cerebral ischemia in the cat: pre-treatment with a competitive NMDA receptor antagonist, D-CPPene. *J. Cereb. Blood Flow Metab.* **10**, 668–674. [*Chapter 2*]

Bullock, R., Graham, D.I., Swanson, S. & McCulloch J. (1994). Neuroprotective effect of the AMPA receptor antagonist LY293558 in focal cerebral ischemia in the cat. *J. Cereb. Blood Flow Metab.* **14**, 466–471. [*Chapter 7*]

Bullock, R., McCulloch, J., Graham, D.I., Lowe, D., Chen, M.H. & Teasdale, G.M. (1990b). Focal ischaemic damage is reduced by CPP-ene. Studies in two animal models. *Stroke* **21**(Suppl. III), 32–36. [*Chapter 2*]

Burns, L.H., Sato, K., Willner, U. & Isacson O. (1993). Intra-nigra infusion of AMPA attenuates dopamine-dependent rotation in the rat. *NeuroReport* **4**, 1075–1078. [*Chapter 8*]

Cahusac, P.M., Evans, R.H., Hill, R.G., Rodriguez, R.E. & Smith, D.A. (1984). The behavioural effects of an N-methylaspartate receptor antagonist following application to the lumbar spinal cord of conscious rats. *Neuropharmacology* **23**, 719–724. [*Chapter 3*]

Capdevilla, J.L., Zapata, A., Viu, E., Faiman, C., Cherkofsky, S., Maccecchini, M.L. & Trullas, R. (1996). Effects of 1-Aminocyclopropanecarboxylic acid in the forced swim test: time-course and dose-response studies. Manuscript submitted. [*Chapter 5*]

Carpenter, M.P. (1981). Anatomy of the corpus striatum and brainstem integrating systems. In: 'Handbook of Physiology – The Nervous System, Part II' (eds, J.M. Brookhart & V.B. Mountcastle), pp. 947–995. American Physiological Society: Bethesda, MD. [*Chapter 8*]

Carron, C., Jullien, A. & Bucher, B. (1971). Synthesis and pharmacological properties of a series of 2-piperidino alkanol derivatives. *Arzneimittelforschung. Drug. Res.* **21**, 1992–1998. [*Chapter 6*]

Carter, A.J. (1994). Many agents that antagonize the NMDA receptor-channel complex in vivo also cause disturbances of motor coordination. *J. Pharm. Exp. Ther.* **269**, 573–580. [*Chapter 6*]

Carter, C., Benavides, J., Dana, C., Schoemaker, H., Perrault, G., Sanger, D. & Scatton, B. (1991). Non-competitive NMDA receptor antagonists acting on the polyamine site. In: 'Excitatory Amino Acid Antagonists' (ed. B. S. Meldrum), pp. 130–163. Blackwell: Oxford. [*Chapter 6*]

Carter, C., Benavides, J., Legendre, P., Vincent, J.D., Noel, F., Thuret, F., Lloyd, K. G., Arbilla, S., Zivkovic, B., Mackenzie, E.T., Scatton, B.S. & Langer, S.Z. (1988). Ifenprodil and SL 82.0715 as cerebral anti-ischemic agents. II. Evidence for *N*-methyl-D-aspartate receptor antagonist properties. *J. Pharmacol. Exp. Ther.* **247**, 1222–1232. [*Chapter 6*]

Carter, C., Minisclou, C. & Rivy, J.P. (1992). Glycine receptor status determines the effects of ifenprodil and the polyamines. *Br. J. Pharmacol.* **105**, 18P. [*Chapter 6*]

Carter, C., Rivy, J.P. & Scatton, B. (1989). Ifenprodil and SL 82.0715 are antagonists at the polyamine site of the N-methyl-D-aspartate (NMDA) receptor. *Eur. J. Pharmacol.* **164**, 611–612. [*Chapter 3*]

Carter, C.J., Lloyd, K.G., Zivkovic, B. & Scatton, B. (1990). Ifenprodil and SL 82.0715 as cerebral anti-ischemic agents. III. Evidence for antagonistic effects at the polyamine modulatory site within the *N*-methyl-D-aspartate receptor complex. *J. Pharmacol. Exp. Ther.* **253**, 475–482. [*Chapter 6*]

Catterall, W.A., Morrow, C.S., Daly, J.W. & Brown, G.B. (1981). Binding of batrachotoxin-A20-alpha-benzoate to a receptor site associated with sodium channels in synaptic nerve ending particles. *J. Biol. Chem.* **256**, 8922–8927. [*Chapter 10*]

Caudle, R.M. & Isaac, L. (1988). A novel interaction between dynorphin (1–13) and an N-methyl-D-asparatate site. *Brain Res.* **443**, 329–332. [*Chapter 3*]

Chadwick, D.W., Smith, D., Baker, G., Crawford, P., Anderson, W., Harrison, B. & Jamieson, V. (1994). Remacemide hydrochloride safety, tolerability, and pharmacokinetics. *Epilepsia* **35**(Suppl. 7), 61. [*Chapter 10*]

Chapman, A.G. & Meldrum, B.S. (1993). Excitatory amino acid antagonists and epilepsy. *Biochem. Soc. Trans.*, **21**, 106–110. [*Chapter 5*]

Chapman, A.G., Graham, J. & Meldrum, B.S. (1990). Potent oral anticonvulsant action of CPP and CPP-ene in DBA/2 mice. *Eur. J. Pharmacol.* **178**, 97–99. [*Chapter 2*]

Chapman, A.G., Meldrun B.S., Nanji, N. & Watkins J.C. (1987). Anticonvulsant action and biochemical effects in DBA/2 mice of CPP (3-((+/–)-2-carboxypiperazin-4-yl)-propyl-l-phosphonate), a novel N-methyl-D-aspartate antagonist. *Eur. J. Pharmacol.*, **139**, 91–96. [*Chapter 5*]

Charcot, J.M. (1874). De la sclérose latérale amyotrophique. *Progr. Mical* **23** (325–327); **24** (341–342); **29** (421–423); **31** (453–455). [*Chapter 9*]

Chazot, P.L., Coleman, S.K., Cik, M. & Stephenson, F.A. (1994). Molecular characterisation of *N*-methyl-D-aspartate receptors expressed in mammalian cells yields evidence for the existence of three subunit types within a discrete receptor molecule. *J. Biol. Chem.* **269**, 24403–24409. [*Chapter 6*]

Chen, M., Bullock, R., Graham, D.I., Frey, P., Lowe, D. & McCulloch, J. (1991). Evaluation of a competitive NMDA antagonist (D-CPPene) in feline focal cerebral ischemia. *Ann. Neurol.* **30**, 62–70. [*Chapter 2*]

Chenard, B.L., Shalaby, I. A., Koe, B. K., Ronau, R.T.,

Butler, T.W., Prochniak, M.A., Schmidt, A.W. & Fox, C. B. (1991). Separation of alpha 1 adrenergic and *N*-methyl-D-aspartate antagonist activity in a series of ifenprodil compounds. *J. Med. Chem.* **34**, 3085–3090. [*Chapter 6*]

Cheramy, A., Barbeito, L., Godeheu, G. & Glowinski, J. (1992). Riluzole inhibits the release of glutamate in the caudate nucleus of the cat in vivo. *Neurosci. Lett.* **147**(2), 209–212. [*Chapter 9*]

Cherkofsky, S.C. (1995). 1-Aminocyclopropane-carboxylic acid: mouse to man interspecies pharmacokinetic comparisons and allometric relationships. *J. Pharmaceutical Sci.*, **84**, 1231–1235. [*Chapter 5*]

Childs, A.M., Evans, R.H. & Watkins, J.C. (1988). The pharmacological selectivity of three NMDA antagonists. *Eur. J. Pharmacol.* **145**, 81–86. [*Preface*]

Choi D.W., Goldberg M.P., Monyer H., Giffard R.G. & Tecoma E.S. (1989). Competitive and non-competitive NMDA antagonists reduce acute neuronal injury *in vitro*. *J. Neurochem.* **52** (Suppl.), S24-D. [*Chapter 1*]

Choi, D. (1990). Methods for antagonizing glutamate neurotoxicity. *Cerebrovasc. Brain Metab. Rev.* **2**, 105–147. [*Chapter 4*]

Choi, D.W. (1985). Glutamate neurotoxicity in cortical cell culture is calcium dependent. *Neurosci. Lett.* **58**, 293–297. [*Chapter 9*]

Choi, D.W. (1988a). Glutamate neurotoxicity and diseases of the nervous system. *Neuron* **1**, 623–634. [*Chapter 6*]

Choi, D.W. (1988b). Calcium-mediated neurotoxicity: relationship to specific channel types and role in ischemic damage. *Trends Neurosci.* **11**, 465–469. [*Chapters 6, 7*]

Choi, D.W. (1992). Excitotoxic cell death. *J. Neurobiol.* **23**, 1261–1276. [*Chapter 6*]

Choi, S.C., Barnes, T.Y., Bullock, R., Germanson, T.A., Marmarou A. & Young, H.F. (1994). Temporal profile of outcomes in severe head injury. *J. Neurosurg.* **81**, 169–173. [*Chapters 1, 4*]

Church, J., Fletcher, E. J., Baxter, K. & Macdonald, J.F. (1994). Blockade by ifenprodil of high voltage-activated Ca^{2+} channels in rat and mouse cultured hippocampal pyramidal neurones: comparison with *N*-methyl-D-aspartate receptor antagonist actions. *Br. J. Pharmacol.* **113**, 499–507. [*Chapter 6*]

Ciabarra, A.M., Sullivan, J.M., Gahn, L.G., Pecht, G., Heinemann, S. & Sevarino, K. A. (1995). Cloning and characterisation of χ-1: a developmentally regulated member of a novel class of the ionotropic glutamate receptor family. *J. Neurosci.* **15**, 6498–6508. [*Chapter 6*]

Clark W.M. & Coull B.M. (for the CGS 19755 Study Group) (1994). Randomized trial of CGS 19755, a glutamate antagonist, in acute ischemic stroke treatment. *Neurology* **44** (Suppl.), A270. [*Chapter 1*]

Clark, B., Hutchison, J.B., Jamieson, V., Jones, T., Palmer, G.C. & Scheyer, R.D. (1995). Remacemide hydrochloride. In: 'Antiepileptic Drugs', 4th edn (eds, R.H. Levy, R.H. Mattson & B.S. Meldrum), Chap. 95, pp. 1035–1044. Raven Press; New York. [*Chapter 10*]

Coderre, T.J. & Van Empel, I. (1994). The utility of excitatory amino acid (EAA) antagonists as analgesic agents. I. Comparison of the antinociceptive activity of various classes of EAA antagonists in mechanical, thermal and chemical nociceptive tests. *Pain* **59**, 345–352. [*Chapter 6*]

Coderre, T.J. (1993). Potent analgesia induced in rats by combined action at PCP and polyamine recognition sites of the NMDA receptor complex. *Eur. J. Neurosci.* **5**, 390–393. [*Chapter 3*]

Collingridge, G.L. & Lester, R.A.J. (1989). Excitatory amino acid receptors in the central nervous system. *Pharmacol. Rev.* **40**, 143–210. [*Chapter 6*]

Collingridge, G.L. & Singer, W. (1990). Excitatory amino acid receptors and synaptic plasticity. *Trends Pharmacol. Sci.*, **11**, 290–296. [*Chapter 5*]

Contreras, P.C., Bremer, M.E. & Gray, N.M. (1990). Ifenprodil and SL 82.0715 potently inhibit binding of [3H](+)-3-PPP to sigma binding sites in rat brain. *Neurosci. Lett.* **116**, 190–193. [*Chapter 6*]

Corbett, D., McKay, K., Copeman, L. & Evans, S.J. (1996). Lack of abuse potential of the novel NMDA antagonist, remacemide HCl. *Pharmacol. Biochem. Behav.* in press. [*Chapter 10*]

Couratier, P., Hugon, J., Sindou, P., Vallat, J.M. & Dumas, M. (1993). Cell culture evidence for neuronal degeneration in amyotrophic lateral sclerosis being linked to glutamate AMPA/kainate receptors. *Lancet* **341**, 265–268. [*Chapters 8, 9*]

Couratier, P., Sindou, P., Esclaire, F., Louvel, E.& Hugon, J. (1994). Neuroprotective effects of riluzole in ALS CSF toxicity. *NeuroReport* **5**(8), 1012–1014. [*Chapter 9*]

Coyle, J.T., Bird, S.J., Evans, R.H., Gulley, R.L., Nadler, J.V., Nicklas, W.J. & Olney, J.W. (1981). Excitatory amino acid neurotoxins: selectivity, specificity, and mechanisms of action. *Neurosci. Res. Program Bull.* **19**, 1–427. [*Chapter 6*]

Cramer, C.L., Stagnitto, M.L., Knowles, M.A. & Palmer, G.C. (1994). Kainic acid and 4-aminopyridine seizure models in mice: Evaluation of efficacy of anti-epileptic agents and calcium antagonists. *Life Sci.* **54**, PL271–PL274. [*Chapter 10*]

Crawford, P., Richens, A., Mawer, G., Cooper, P. & Hutchison, J.B. (1992). A double blind placebo controlled cross-over study of remacemide hydrochloride on adjunctive therapy in patients with refractory epilepsy. *Epilepsy* **1**(Suppl. A), p7–13. [*Chapter 10*]

Cudennec, A., Duverger, D., Benavides, J., Scatton, B. & Nowicki, J.P. (1994). Effect of eliprodil, an NMDA receptor antagonist acting at the polyamine modulatory site, on local cerebral glucose use in the rat in the rat brain. *Brain Res.* **664**, 41–48. [*Chapter 6*]

Curtis D.R. & Watkins, J.C. (1965). The pharmacology of amino acids related to gamma-aminobutyric acid. *Pharm. Rev.* **17** 347–391. [*Chapter 2*]

Curtis, D.R. & Watkins, J.C. (1960). The excitation and depression of spinal neurones by structurally related amino acids. *J. Neurochem.* **6**, 117–141. [*Preface, Chapter 2*]

Dalgaard, L., Hjortkjær, R.K., Regnier, B. & Nordholm, L. (1994). Pharmacokinetics of the neuroprotective glutamate antagonist NBQX (6-nitro-7-sulfamoyl-benzo(F)quinoxaline-2,3-dione) in mice, rats, and dogs: Interactions with probenecid. *Drug Metab. Dispos.* **22**, 289–293. [*Chapter 8*]

Dall, V., Ørntoft, U., Schmidt, A. & Nordholm, L. (1993). Interaction of the competitive AMPA receptor antagonist NBQX with hexobarbital. *Pharmacol. Biochem. Behav.* **46**, 73–76. [*Chapter 8*]

Davies, J., Evans, R.H., Francis, A.A. & Watkins, J.C. (1979). Excitatory amino acid receptors and synaptic excitation in the mammalian central nervous system. *J. Physiol. (Paris)* **75**, 641–654. [*Preface*]

Davies, J., Evans, R.H., Francis, A.A., Jones, A.W., Smith, D.A.S. & Watkins, J.C. (1982). Conformational aspects of the actions of some piperidine dicarboxylic acids at excitatory amino acid receptors in the mammalian and amphibian spinal cord. *Neurochem. Res.* **7** 1119–1133. [*Chapter 2*]

Davies, J., Evans, R.H., Herrling, P.L., Jones, A.W., Olverman, H.J., Pook, P. & Watkins, J.C. (1986). CPP, a new potent and selective NMDA antagonist. Depression of central neuron responses, affinity for [^{3}H]D-AP5 binding sites on brain membranes and anticonvulsant activity. *Brain Res.* **382**, 169–173. [*Preface, Chapters 2, 3*]

Davies, J., Francis, A.A., Jones, A.W. & Watkins, J.C. (1981). 2-Amino-5- phosphonovalerate (2APV), a potent and selective antagonist of amino acid-induced and synaptic excitation. *Neurosci. Lett.* **21**, 77–81. [*Preface*]

Davies, S.N. & Lodge, D. (1987). Evidence for involvement of N-methylaspartate receptors in 'wind-up' of class 2 neurones in the dorsal horn of the rat. *Brain Res.* **424**, 402–406. [*Chapter 3*]

Debonnel, G. (1993). Current hypotheses on sigma receptors and their physiological role: Possible implications in psychiatry. *J. Psychiatr. Neurosci.* **18**, 157–170. [*Chapter 6*]

Debono, M.W., Le Guern, J., Canton, T., Doble, A. & Pradier, L. (1993). Inhibition by riluzole of electrophysiological responses modified by rat kainate and NMDA receptors expressed in Xenopus oocytes. *Eur. J. Pharmacol.* **235**(2–3), 283–289. [*Chapter 9*]

DeGraba, T.J., Ostrow, P., Hanson, S. & Grotta, J.C. (1994). Motor performance, histologic damage, and calcium influx in rats treated with NBQX after focal ischemia. *J Cereb. Blood Flow Metab.* **14**, 262–268. [*Chapter 8*]

Delage, I., Gotti, B. & Mackenzie, E.T. (1983). Flow/metabolism coupling in experimental focal cerebral ischaemia: reestablishment of coupling by pharmacotherapy. *Eur. Neurol.* **22**, 5. [*Chapter 6*]

Deren Wesolek, A. & Maj, J. (1993). Central effects of SL 82.0715, an antagonist of polyamine site of the NMDA receptor complex. *Pol. J. Pharmacol.* **45**, 467–480. [*Chapter 6*]

Desai, M.M., Burnett, J.P., Ornstein, P.L. & Schoepp, D.D. (1995). Cyclothiazide acts on a site on the α-amino-3-hydroxy-5-methyl-4-isoxazole propionic acid receptor complex that does not recognize competitive or non-competitive AMPA receptor antagonists. *J. Pharmacol. Exp. Ther.* **272**, 38–43. [*Chapter 7*]

DeSarro, G.B. & DeSarro, A. (1993). Anticonvulsant properties of non-competitive antagonists of the *N*-methyl-D-aspartate receptor in genetically epilepsy-prone rats: comparison with CPPene. *Neuropharmacology* **32**, 51–58. [*Chapters 2, 6*]

Dessi, F., Ben-Ari, Y. & Charriaut-Marlangue, C. (1993). Riluzole prevents anoxic injury in cultured cerebellar granule neurons. *Eur. J. Pharmacol.* **250**, 325–328. [*Chapter 9*]

Dich, N.J., Svendsen, L.B. & Berthelsen, P. (1992). Intramuscular low-dose ketamine versus pethidine for postoperative pain treatment after thoracic surgery. *Acta Anaesthesiol. Scand.* **36**, 583–587. [*Chapter 3*]

Dickenson, A.H. & Sullivan, A.F. (1987). Evidence for a role of the NMDA receptor in the frequency dependent potentiation of deep rat dorsal horn nociceptive neurones following C fibre stimulation. *Neuropharmacology* **26**, 1235–1238. [*Chapter 3*]

Diemer, N.H., Jørgensen, M.B., Johansen, F.F., Sheardown, M.J. & Honore T. (1992). Protection against ischemic hippocampal damage in the rat with a new non-NMDA antagonist, NBQX. *Acta Neurol. Scand.* **86**, 45–49. [*Chapter 8*]

Diemer, N.H., Johansen-F.F. & Jorgensen, M.B. (1990). *N*-methyl-D-aspartate and non-*N*-methyl-D-aspartate antagonists in global cerebral ischemia. *Stroke* (Suppl III) 21; 39–42. [*Chapter 5*]

Dingledine, R. (1983). *N*-Methyl aspartate activates voltage-dependent calcium conductance in rat hippocampal pyramidal cells. *J. Physiol. (Lond.)* **343**, 385. [*Preface*]

Doble, A. (1996). Pharmacology and mechanism of action of riluzole. *Neurology*, in press. [*Chapter 9*]

Doble, A., Hubert, J.P. & Blanchard, J.C. (1992). Pertussis toxin pretreatment abolishes the inhibitory effect of riluzole and carbachol on D-[3H]aspartate release from cultured cerebellar granule cells. *Neurosci. Lett.* **140**(2), 251–254. [*Chapter 9*]

Donevan, S.D. & Rogawski, M.A. (1993). GYKI 52466, a 2,3-benzodiazepine, is a highly selective, noncompetitive antagonist of AMPA/kainate receptor responses. *Neuron* **10**, 51–59. [*Preface*]

Drummond, J.C., Oh, Y-S. & Cole, D.J. (1991). The influence of phenylephrine-induced hypertension during focal cerebral ischemia on the formation of brain edema. *J. Neurosurg. Anesthesiol.* **1**, 4–11. [*Chapter 4*]

Dubner, R. & Ruda, M.A. (1992). Activity-dependent

neuronal plasticity following tissue injury and inflamation. *Trends Neurosci.* **15**, 96–103. [*Chapter 3*]

Duermueller, N., Craggs, M. & Meldrum, B.S. (1994). The effect of the non-NMDA receptor antagonist GYKI 52466 and NBQX and the competitive NMDA receptor antagonist D-CPPene on the development of amygdala kindling and on amygdala-kindled seizures. *Epilepsy Res.* **17**, 167–174. [*Chapter 2*]

Durand, G.M., Bennett, M.V. & Zukin, R.S. (1993). Splice variants of the *N*-methyl-D-aspartate receptor NR1 identify domains involved in regulation by polyamines and protein kinase C. *Proc. Natl Acad. Sci. USA* **90**, 6731–6735 [Published erratum: (1993) in *Proc. Natl Acad. Sci. USA* **90**, 9739.]. [*Chapter 6*]

Durand, G.M., Gregor, P., Zheng, X., Bennett, M.V.L., UHL, G.R. & Zukin, S. (1992). Cloning of an apparent splice variant of the rat *N*-methyl-D-aspartate receptor NMDAR1 with altered sensitivity to polyamines and activators of protein kinase C. *Proc. Natl Acad. Sci. USA* **89**, 9359–9363. [*Chapter 6*]

Duval, D., Roome, N., Gauffeny, C., Nowicki, J.P. & Scatton, B. (1992). SL 82.0715, an NMDA antagonist acting at the polyamine site, does not induce neurotoxic effects on rat cortical neurons. *Neurosci. Lett.* **137**, 193–197. [*Chapters 3, 6*]

Eide, P.K., Jørum, E., Stubhaug, A., Bremnes, J. & Breivik, H. (1994). Relief of post-herpetic neuralgia with the N-methyl-D-aspartic acid receptor antagonist ketamine: a double-blind, cross-over comparison with morphine and placebo. *Pain* **58**, 347–354. [*Chapters 3, 6*]

Eisenberg, E., Lacross, S. & Strassman, A.M. (1995). The clinically tested *N*-methyl-D-aspartate receptor antagonist memantine blocks and reverses thermal hyperalgesia in a rat model of painful mono-neuropathy. *Neurosci. Lett.* **187**, 17–20. [*Chapter 6*]

Elliott, K., Minami, N., Kolesnikov, Y.A. & Pasternak, G.W. (1994). The NMDA receptor antagonists, LY274614 and MK-801, and the nitric oxide synthase inhibitor, NG-nitro-L-arginine, attenuate analgesic tolerance to the mu-opioid morphine but not to kappa opioids. *Pain* **56**, 69–75. [*Chapter 3*]

Estevez, A.G., Stutzmann, J.M. & Barbeito, L. (1995). Protective effect of riluzole on excitatory amino acid-mediated neurotoxicity in motoneuron-enriched cultures. *Eur. J. Pharmacol.* **280**, 47–53. [*Chapter 9*]

Evans, R.H., Francis, A.A., Hunt, K., Oakes, D.J. & Watkins, J.C. (1979). Antagonism of excitatory amino acid-induced responses and of synaptic excitation in the isolated spinal cord of the frog. *Br. J. Pharmacol.* **67** 591–603. [*Chapter 2*]

Evans, R.H., Francis, A.A., Jones, A.W., Smith, D.A.S. & Watkins, J.C. (1982). The effect of a series of omega-phosphonic-alpha-carboxylic amino acids on electrically evoked and excitant amino acid-induced responses in isolated spinal cord preparations. *Br. J. Pharmacol.* **75**, 65–75. [*Preface, Chapter 2*]

Evoniuk, G.E., Hertzman, R.P. & Skolnick, P. (1991). A rapid method for evaluating the behavorial effects of phenocyclidine-like dissociative anesthetics in mice. *Psychopharmacol.*, **105**, 125–128. [*Chapter 5*]

Faden, A. & Salzman, S. (1992). Pharmacological strategies in CNS trauma. *Trends Pharmacol. Sci.* **13**, 29–35. [*Chapter 5*]

Fagg, G.E. (1987). Phencyclidine and related drugs bind to the activated N-methyl D-aspartate receptor-channel complex in rat brain membranes. *Neurosci. Lett.* **76**, 221–227. [*Preface*]

Fayad, P.B., Edwards, K., Hormes, J. & Lees, K.R. (1996). The safety and tolerability of non weight-adjusted doses of aptiganel HCl (CERESTAT®) in acute ischemic stroke patients. *American Heart Association* [abstract to be published in *Stroke*]. [*Chapter 4*]

Fink, K., Schmitz, V., Boing, C. & Gothert, M. (1995). Stimulation of serotonin release in the rat brain cortex by activation of ionotropic glutamate receptors and its modulation via α2-heteroreceptors. *Naunyn Schmiedebergs Arch. Pharmacol.* **352**, 394–401. [*Chapter 6*]

Fink-Jensen, A., Judge, M.E., Hansen, J.B., Jacobsen, P., Turski, L., Olney, J. & Honoré, T. (1992). Inhibition of cisplatin-induced emesis in ferrets by the non-NMDA receptor antagonists NBQX and CNQX. *Neurosci. Lett.* **137**, 173–177. [*Chapter 8*]

Fisher M. (1995). An Overview of Cytoprotective Therapy for Acute Ischemic Stroke. In: 'Proceedings of the 19th Princeton Conference' (eds, M.A. Moskowitz & L.R. Caplan), pp. 391–401. Butterworth-Heinemann: London. [*Chapter 4*]

Fisher, M., Jones, S. & Sacco, R.L. (1994). Prophylactic neuroprotection for cerebral ischemia. *Stroke* **25**, 1075–1080. [*Chapter 10*]

Fix, A.S. (1994). Pathomorphologic effects of N-methyl-D-aspartate antagonists in the rat posterior cingulate/retrosplenial cerebral cortex: a review. *Drug Dev. Res.* **32**, 147–152. [*Chapter 7*]

Fix, A.S., Horn, J.W., Wightman, K.A., Johnson, C.A., Long, C.G., Storts, R.W., Farber N., Wozniak D.F. & Olney J.W. (1993). Neuronal vacuolization and necrosis induced by the noncompetitive *N*-methyl-D-aspartate (NMDA) antagonist (+)MK801 (dizocilpine maleate): a light and electron microscopic evaluation of the rat retrosplenial cortex. *Exp. Neurol.* **123**, 204–215. [*Chapters 1, 6*]

Fossom, L.H., Von Lubitz, D.K.J.E., Lin, R.C.S. & Skolnick, P. (1995). Neuroprotective actions of 1-aminocyclopropanecarboxylic acid (ACPC): a partial agonist at strychnine-insensitive glycine sites. Neurol. Res. 17, 265–269. [*Chapter 5*]

Foster, A.C., Gill, R., Kemp, J.A. & Woodruff, G.N. (1987). Systemic administration of MK 801 prevents *N*-methyl-D-aspartate-induced neuronal degeneration in rat brain. *Neurosci. Lett.* **76**, 307–311. [*Chapter 2*]

Foutz, A.S., Pierrefiche, O. & Denavit-Saubié, M. (1994). Combined blockade of NMDA and non-NMDA receptors produces respiratory arrest in the adult cat. *NeuroReport* **5**, 481–484. [*Chapter 8*]

Francis, A.A., Jones, A.W. & Watkins, J.C. (1980). Dipeptide antagonists of amino acid-induced and synaptic excitation in the frog spinal cord. *J. Neurochem.* **35**, 1458–1460. [*Chapter 2*]

French E.D., Mura A. & Wang T. (1993). MK-801, phencyclidine (PCP), and PCP-like drugs increase burst firing in rat A10 dopamine neurons: comparison to competitive NMDA antagonists. *Synapse* **13**, 108–116. [*Chapter 1*]

Fujikawa, D.G. (1995). Neuroprotective effect of ketamine administered after status epilepticus onset. *Epilepsia* **36**, 186–195. [*Chapter 10*]

Gallagher, M.J., Lynch, D.R., Pritchett, D.B. & Robinson, M.B. (1995). Ifenprodil and polyamines bind to distinct regions of the NMDA receptor complex: a study using chimeric NMDAR2 subunits. *Soc. Neurosci. Abstr.* **21**, 41.4. [*Chapter 6*]

Gamzu, E.R. (1995). CERESTAT® in the treatment of acute cerebral ischemia and TBI. In: 'Ishemic Stroke: Recent Advances in Understanding and Therapy' (eds, J. Grotta, L.P. Miller & A. Buchanan), pp. 86–110. International Business Communications. [*Chapter 4*]

Gamzu, E.R. (for the CNS 1102-002 Study Group) (1994). CERESTAT™ (CNS 1102), an NMDA antagonist in severe traumatic brain injury (TBI) patients: a safety study. *American Neurology Association* Oct. [*Chapter 4*]

Garske, G.E., Palmer, G.C., Napier, J.J., Griffith, R.C., Freedman, L.R., Harris, E.W., Ray, R., McCreedy, S.A., Blosser, J.C., Woodhead, J.H., White, H.S. & Swinyard, E.A. (1991). Preclinical profile of the anticonvulsant remacemide and its enantiomers in the rat. *Epilepsy Res.* **9**, 161–174. [*Chapter 10*]

Gean, P.W. & Shinnick-Gallaher, P. (1988). Epileptiform activity induced by magnesium-free solution in slices of rat amygdala: antagonism by N-methyl-D-aspartate receptor antagonists. *Neuropharmacol.*, **27**, 557–562. [*Chapter 5*]

Gee, K.R. (1994). Therapeutic potential of PCP receptor ligands. *Exp. Opin. Invest. Drugs.* **3**, 1021–1030. [*Chapter 10*]

Giffard, R.G., Monyer, H., Christine, C.W. & Choi, D.W. (1990). Acidosis reduces NMDA receptor activation, glutamate neurotoxicity, and oxygen-glucose deprivation neuronal injury in cortical cultures. *Brain Res.* **506**, 339–342. [*Chapter 6*]

Gill, R. & Lodge, D. (1994). The neuroprotective effects of a novel AMPA/kainate antagonist (3*SR*,4a*RS*,6*RS*,8a*RS*)-6-[2-(1*H*-tetrazol-5-yl)-ethyl]-1,2,3,4,4a,5,6,7,8a-decahydroisoquinoline-3-carboxylic acid (LY215490) in focal ischemia. *Neuropharmacology* **33**, 1529–1536. [*Chapter 7*]

Gill, R. (1994). The pharmacology of α-amino-3-hydroxy-5-methyl-4-isoxazole propionate (AMPA)/kainate antagonists and their role in cerebral ischemia. *Cerebrovasc. Brain Metab. Rev.* **6**, 225–256. [*Chapter 7*]

Gill, R., Brazell, C., Woodruff, G.N. & Kemp, J.A. (1991). The neuroprotectant action of dizocilpine (MK801) in the rat middle cerebral artery occlusion model of focal ischaemia. *Br. J. Pharmacol.* **103**, 2030–2036. [*Chapter 7*]

Gill, R., Foster, A.C. & Woodruff, G.N. (1987). Systemic administration of MK-801 protects against ischemia-induced hippocampal neurodegeneration in the gerbil. *J. Neurosci.* **7**, 3343–3349. [*Chapter 7*]

Gill, R., Nordholm, L. & Lodge, D. (1992). The neuroprotective actions of 2,3-dihydroxy-6-nitro-7-sulfamoyl-benzo(F)quinoxaline (NBQX) in a rat focal ischaemia model. *Brain Res.* **580**, 35–43. [*Chapter 8*]

Ginsberg, M. (1995a). Neuroprotection in brain ischemia: an update Part I. *Neuroscientist* **1**, 95–103. [*Chapter 4*]

Ginsberg, M. (1995b). Neuroprotection in brain ischemia: an update Part II. *Neuroscientist* **1**, 164–175. [*Chapter 4*]

Ginski, M.J. & Witkin, J.M. (1994). Sensitive and rapid behavioural differentiation of *N*-methyl-D-aspartate receptor antagonists. *Psychopharmacology* **114**, 573–582. [*Chapter 6*]

Giroux, C., Rosen, P. & Scatton, B. (1994). Preclinical pharmacology and clinical safety profile of eliprodil, an atypical NMDA receptor antagonist. In 'Pharmacology of Cerebral Ischemia' (eds, J. Krieglstein & H. Oberpischler-Schwenk), pp. 643–651, Wissenschaftliche: Stuttgart. [*Chapter 6*]

Goldberg, M.P., Monyer, H. & Choi, D.W. (1988a). Hypoxic neuronal injury in vitro depends on extracellular glutamine. *Neurosci. Lett.* **94**, 52–57. [*Chapter 6*]

Goldberg, M.P., Pham, P.C. & Choi, D.W. (1987a). Dextrorphan and dextromethorphan attenuate hypoxic injury in neuronal culture. *Neurosci. Lett.* **80**, 11–15. [*Chapter 6*]

Goldberg, M.P., Viseskul, V. & Choi, D.W. (1988b). Phencyclidine receptor ligands attenuate cortical neuronal injury after *N*-methyl-D-aspartate exposure or hypoxia. *J. Pharmacol. Exp. Ther.* **245**, 1081–1087. [*Chapter 6*]

Goldberg, M.P., Weiss, J.H., Pham, P.C. & Choi, D.W. (1987b). *N*-Methyl-D-aspartate receptors mediate hypoxic neuronal injury in cortical culture. *J. Pharmacol. Exp. Ther.* **243**, 784–791. [*Chapter 6*]

Gordh, T., Karlsten, R. & Kristensen, J. (1995). Intervention with spinal NMDA, adenosine, and NO systems for pain modulation. *Ann. Med.* **27**, 229–234. [*Chapter 2*]

Gorelick, D.A. & Balster, R.L. (1995). Phencyclidine (PCP). In: 'Psychopharmacology: The Fourth Generation of Progress' (eds, F.E. Bloom & D.J. Kupfer), pp. 1767–1776, Raven Press: New York. [*Chapter 6*]

Gotti B., Benavides J., MacKenzie E.T. & Scatton B. (1990). The pharmacotherapy of focal cortical ischaemia in the mouse. *Brain Res.* **522**, 290–307. [*Chapters 1, 6*]

Gotti, B., Duverger, D., Bertin, J., Carter, C., Dupont, R., Frost, J., Gaudilliere, B., Mackenzie, E.T., Rousseau, J.

& Scatton, B. (1988). Ifenprodil and SL 82.0715 as cerebral anti-ischemic agents. I. Evidence for efficacy in models of focal cerebral ischemia. *J. Pharmacol. Exp. Ther.* **247**, 1211–1221. [*Chapter 6*]

Gotti, B., Mackenzie, E.T. & Young, A.R. (1984). The pharmacotherapy of focal cerebral ischaemia: development of an experimental model. *J. Chron. Dis. Ther. Res.* **8**, 44–61. [*Chapter 6*]

Graham, D., Darles, G. & Langer, S.Z. (1992). The neuroprotective properties of ifenprodil, a novel NMDA receptor antagonist, in neuronal cell culture toxicity studies. *Eur. J. Pharmacol.* **226**, 373–376. [*Chapter 6*]

Graybiel, A.M., (1990). Neurotransmitters and neuromodulators in the basal ganglia. *Trends Neurosci.* **13**, 244. [*Chapter 8*]

Greenamyre, J.T. & O'Brien, C.F. (1991). N-Methyl-D-aspartate antagonists in the treatment of Parkinson's disease. *Arch. Neurol.* **48**, 977–981. [*Chapter 10*]

Greenamyre, J.T. & Porter, R.H. (1994). Anatomy and physiology of glutamate in the CNS. Neurology, 44 (Suppl. 8), S7-S13. [*Chapter 5*]

Greenamyre, J.T., Eller, R.V., Zhang, A., Olvadia, A., Kurlan, R. & Gash, D.M. (1994). Antiparkinsonian effects of remacemide hydrochloride, a glutamate antagonist, in rodent and primate models of Parkinson's disease. *Ann. Neurol.* **35**, 655–661. [*Chapter 10*]

Griffith, R.C., Murray, R.J., Napier, J.J., Gentile, R.J., Becker, C.W., Smith, C., Schmiesing, R.J., Balestra, M., Frankenheim, J., Garske, G.E., Stagnitto, M.L. & Palmer G.C. (1996). Synthesis and anticonvulsant activity of 1,2-diarylalkylamine amino acid conjugates. *J. Med. Chem.*, submitted. [*Chapter 10*]

Grimwood, S., Moseley, A.M., Carling, R.W., Leeson, P.D. & Foster, A.C. (1992). Characterization of the binding of [^{3}H]L-689,560, an antagonist for the glycine site on the *N*-methyl-D-aspartate receptor, to rat brain membranes. *Mol. Pharmacol.* **41**, 923–930. [*Chapter 6*]

Grimwood, S., Struthers, L. & Foster, A.C. (1994). Polyamines modulate [3H]L-689,560 binding to the glycine site of the NMDA receptor from rat brain. *Eur. J. Pharmacol.* **266**, 43–50. [*Chapter 6*]

Grotta J., Clark W., Coull B., Pettigrew L.C., Mackay B., Goldstein L.B., Meissner I., Murphy D. & LaRue L. (1995). Safety and tolerability of the glutamate antagonist CGS 19755 (Selfotel) in patients with acute ischemic stroke. *Stroke* **26**, 602–605. [*Chapters 1, 4, 6*]

Gurney, M.E., Cutting, F.B., Zhai, P., Doble, A., Taylor, C.P., Andous, P.K. & Hall, E.D. (1996). Benefit of vitamin E, riluzole, and gabapentin in a transgenic model of familial amyotrophic lateral sclerosis. *Ann. Neurol.* **2**, 147–157. [*Chapter 9*]

Gurney, M.E., Pu, H., Chiu, A.Y., Dal Canto, M.C., Polehow, C.Y., Alexander, D.D., Caliendo, J., Hentati, A., Kwon, Y.W. & Deng, H.X. (1994). Motor neuron degeneration in mice that express a human Cu, Zn superoxide dismutase mutation. *Science* **264**, 1772–1775. [*Chapter 9*]

Hagberg, H., Gilland, E., Diemer, N.-H. & Andiné P. (1994). Hypoxia–ischemia in the neonatal rat brain: histopathology after post-treatment with NMDA and non-NMDA receptor antagonists. *Biol. Neonate* **66**, 205–213. [*Chapter 8*]

Hagelin, A. & Lundberg, D. (1981). Ketamine for postoperative analgesia after upper abdominal surgery. *Clin. Ther.* **4**, 229–233. [*Chapter 3*]

Hanner, M., Moebius, F.F., Weber, F., Grabner, M., Striessnig, J. & Glossmann, H. (1995). Phenylalkylamine Ca^{2+} binding protein: molecular cloning, tissue distribution and heterologous expression. *J. Biol. Chem.* **270**, 7551–7557. [*Chapter 6*]

Hargreaves, R.J., Rigby, M., Smith, D. & Hill, R.G. (1993a). Lack of effect of L-687,414 ((+)-cis-4-methyl-HA-966), an NMDA receptor antagonist acting at the glycine site, on cerebral glucose metabolism and cortical neuronal morphology. *Br. J. Pharmacol.* **110**, 36–42. [*Chapter 3*]

Hargreaves, R.J., Rigby, M., Smith, D., Hill, R.G. & Iversen, L.L. (1993b). Competitive as well as uncompetitive *N*-methyl-D-aspartate receptor antagonists affect cortical neuronal morphology and cerebral glucose metabolism. *Neurochem. Res.* **18**, 1263–1269. [*Chapter 6*]

Harris, E., Stagnitto, M., Garske, G., Cregan, E., Ray, R., Julien, R., Wilson, T., Machulskis, G., Bialobok, P., White, J. & Palmer, G. (1992). Neuroprotection in animals by remacemide: a novel anticonvulsant effective against maximal electroshock seizures in mice. In: 'Multiple Sigma and PCP Receptor Ligands: Mechanisms for Neuromodulation and Neuroprotection?' (eds, J-M. Kamenka & E.F. Domino), pp. 643–653. NPP Books: Ann Arbor, MI. [*Chapter 10*]

Harrison, N.L. & Simmonds, H.A. (1985). Quantitative studies on some antagonists of *N*-methyl-D-aspartate in slices of cerebral cortex. *Br. J. Pharmacol.* **87**, 381–391. [*Chapters 2, 7*]

Hartvig, P., Valtysson, J., Lindner, C.J., Kristensen, J.D., Karlsten, R., Gustafsson, L., Persson, J., Svensson, J.O., Øye, I., Antoni, G., Westerberg, G. & Långström, B. (1995). CNS effects of subdissociative doses of (S)-ketamine are related to plasma and brain concentrations measured with positron emission tomography in healthy volunteers. *Clin. Pharmacol. Ther.* **58**, 165–173. [*Chapter 3*]

Hashimoto, K. & London, E.D. (1993). Further characterization of [^{3}H]ifenprodil binding to sigma receptors in rat brain. *Eur. J. Pharmacol.* **236**, 159–163. [*Chapter 6*]

Hashimoto, K. & London, E.D. (1995). Interactions of erythro-ifenprodil, threo-ifenprodil, erythro-iodoifenprodil, and eliprodil with subtypes of sigma receptors. *Eur. J. Pharmacol.* **273**, 307–310. [*Chapter 6*]

Hashimoto, K., Mantione, C.R., Spada, M.R., Neumeyer, J.L. & London, E.D. (1994). Further characterization of [3H]ifenprodil binding in rat brain. *Eur. J. Pharmacol.* **266**, 67–77. [*Chapter 6*]

Hashimoto, K., Mantione, C.R., Spadda, M., Neumeyer,

J.L. & London, E. (1992). Subcellular distribution of [^{3}H]ifenprodil binding in rat brain. *Soc. Neurosci. Abstr.* **18**, 1115. [*Chapter 6*]

Hays, S.J., Bigge, C.F., Novak, P.M., Drummond, J.T., Bobovski, T.P., Rice, M.J., Johnson, G., Brahce, L.J. & Coughenour, L.L. (1990). New and versatile approaches to the synthesis of CPP-related competitive NMDA antagonists. Preliminary structure-activity relationships and pharmacological evaluation. *J. Med. Chem.* **33**, 2916–2924. [*Chapter 3*]

Hebert, T., Drapeau, P., Pradier, L. & Dunn, R.J. (1994). Block of the rat brain IIA sodium channel alpha subunit by the neuroprotective drug riluzole. *Mol. Pharmacol.* **45**, 1055–1060. [*Chapter 9*]

Hedner, T., Qian-Ling, G., Hedner, J. & Samuelsson, H. (1991). Involvement of nitric oxide (NO) in nociceptive processing in the rat (abstract). Scandinavian Association for the study of Pain, 15th Annual Meeting. [*Chapter 3*]

Herrling, P.L., Morris, R. & Salt, T.E. (1983). Effects of excitatory amino acids and their antagonists on membrane and action potentials of cat caudate neurones. *J. Physiol.* **339**, 207–222. [*Chapter 2*]

Herrling, P.L. (1985). Pharmacology of the corticocaudate excitatory postsynaptic potential in the cat: evidence for mediation by quisqualate- or kainate receptors. *Neuroscience* **14**, 417–426. [*Chapter 2*]

Hicks, T.P., Hall, J.G. & McLennan, H. (1978). Ranking of excitatory amino acids by the antagonists glutamic acid diethylester and D-alpha-amino adipic acid. *Can. J. Physiol. Pharmacol.* **56**, 901–907. [*Preface*]

Hoffmann, V., Coppejans, H., Vercauteren, M. & Adriaensen, H. (1994). Successful treatment of post-herpetic neuralgia with oral ketamine. *Clin. J. Pain* **10**, 240–242. [*Chapter 3*]

Hollmann, M. & Heinemann, S. (1994). Cloned glutamate receptors. *Annu. Rev. Neurosci.* **17**, 31–108. [*Chapter 2*]

Hollmann, M., Boulter, J., Maron, C., Beasley, L., Pecht, G. & Heinemann, S. (1993). Zinc potentiates agonist-induced currents at certain splice variants of the NMDA receptor. *Neuron* **10**, 943–954. [*Chapter 6*]

Hollmann, M., O'Shea-Greenfield, A., Rogers, S.W. & Heinemann, S. (1989). Cloning by functional expression of a member of the glutamate receptor family. *Nature* **343**, 643–648. [*Preface*]

Holmes, K.H., Bilkey, D.K., Laverty, R. & Goddard, G.V. (1990). The N-methyl-D-aspartate antagonists aminophosphonovalerate and carboxypiperazinephosphonate retard the development and expression of kindled seizures. *Brain Res.* **506**, 227–235. [*Chapter 3*]

Honda, H. & Sakai, Y. (1987). The mode of action of ifenprodil tartrate in isolated canine cerebral and femoral arteries. *Arch. Int. Pharmacodyn. Ther.* **285**, 211–225. [*Chapter 6*]

Honda, H., Sakai, Y., Iwata, T., Ohba, S., Kanezuka, T. & Irino, O. (1988). Effects of ifenprodil tartrate on alpha-adrenoceptors and Ca2+ movement in isolated canine saphenous veins. *Arch. Int. Pharmacodyn. Ther.* **292**, 112–121 [*Chapter 6*]

Honda, H., Shibuya, T. & Salafsky, B. (1989a). Effects of ifenprodil tartrate on calcium flux in arteries and brain synaptosomes. *Proc. West. Pharmacol. Soc.* **32**, 155–158. [*Chapter 6*]

Honda, H., Shibuya, T. & Salafsky, B. (1989b). Effects of Ca2+ antagonists on $^{45}Ca^{2+}$ uptake by rat whole brain P1 and P2 fractions. *Res. Commun. Chem. Pathol. Pharmacol.* **64**, 463–470. [*Chapter 6*]

Honore, T., Daviees, S.N., Drejer, J., Fletcher, Jacobsen, P., Lodge, D. & Nielsen, F.E. (1988). Quinoxalinediones: potent competitive non-NMDA glutamate receptor antagonists. *Science* **241** 701–703. [*Chapter 8*]

Houamed, K.M., Kuijper, J.L., Gilbert, T.L., Haldeman, B.A., O'Hara, P.J., Mulvihill, E.R., Almers, W. & Hagen, F.S. (1991). Cloning, expression, and gene structure of a G protein-coupled glutamate receptor from rat brain. *Science* **252**, 1318–1321. [*Preface*]

Howell, S.E., Miller, S.R., McCallister, J.D., Cherkofsky, S.C. & Patrick, K.S. (1995). Gas chromatographic-mass spectrometric determination of urinary 1-aminocyclopropanecarboxylic acid in mice using deuterated internal standard. *J. Chromatography*, **B 663**, 148–152. [*Chapter 5*]

Hu, R.Q. & Davies, J.A. (1995). The effect of the desglycinyl metabolite of remacemide on cortical wedges prepared from DBA/2 mice. *Eur. J. Pharmacol.* **287**, 251–256. [*Chapter 10*]

Hubert, J.P. & Doble, A. (1989). Ibotenic acid stimulates D-[3H]aspartate release from cultured cerebellar granule cells. *Neurosci. Lett.* **96**(3), 345–350. [*Chapter 9*]

Hubert, J.P., Delumeau, J.C., Glowinski, J., Premont, J. & Doble, A. (1994). Antagonism by riluzole of entry of calcium evoked by NMDA and veratridine in rat cultured granule cells: evidence for a dual mechanism of action. *Br. J. Pharmacol.* **113**, 261–267. [*Chapter 9*]

Hudzik, T.J. & Palmer, G.C. (1995). Effects of anticonvulsants in a novel operant learning paradigm in rats: Comparison of remacemide hydrochloride and FPL 15896AR to other anticonvulsant agents. *Epilepsy Res.* **21**, 183–193. [*Chapter 10*]

Hudzik, T.J., Freedman, L. & Palmer, G.C. (1996). Remacemide hydrochloride and ARL 15896AR lack abuse potential: Additional differences from other uncompetitive NMDA antagonists. *Epilepsia* **37**, 544–550. [*Chapter 10*]

Hugon, J., Vallat, J.M., Spencer, P.S., Leboutet, J. & Barthe, D. (1989). Kainic acid induces early and delayed degenerative neuronal changes in rat spinal cord. *Neurosci. Lett.* **104**, 258–262. [*Chapter 9*]

Hunter, A.J., Green, A.R. & Cross, A.J. (1995). Animal models of acute ischaemic stroke: can they predict clinically successful neuroprotective drugs? *TIPS* **16**, 123–128. [*Chapter 10*]

Hutchison, A.J., Williams, M., Angst, C., de Jesus, R.,

Blanchard, L., Jackson, R.H., Wilusz, E.J., Murphy, D.E., Bernard, P.S., Schneider, J., Campbell, T., Guida, W. & Sills, M.A. (1989). 4-(Phosphonoalkyl)- and 4-(phosphonoalkenyl)-2-piperidinecarboxylic acids: synthesis, activity at *N*-methyl-D-aspartic acid receptors, and anticonvulsant activity. *J. Med. Chem.* **32** 2171–2178. [*Chapter 2*]

Imperato, A., Scrocco, M.G., Bacchi, S. & Angelucci, L. (1990). NMDA receptors and in vivo dopamine release in the nucleus accumbens and caudatus. *Eur. J. Pharmacol.* **187**, 555–556. [*Chapter 2*]

Ingwersen, S.H., Ørstrøm, J.K., Petersen, P., Drustrup, J., Bruno, L. & Nordholm, L. (1994). Human Pharmacokinetics of the neuroprotective agent NBQX. *Am. J. Therap.* **1**, 1–8. [*Chapter 8*]

Irwin, S. (1968). Comprehensive observational assessment: Ia. A systematic, quantitative procedure for assessing the behavioral and physiological state of the mouse. *Psychopharmacologia* **13**, 222–257. [*Chapter 2*]

Izquierdo, I. (1993). Long-term potentiation and the mechanisms of memory. *Drug Dev. Res.* **30**, 1–17. [*Chapter 6*]

Jackson, A. & Sanger, D.J. (1988). Is the discriminative stimulus produced by phencyclidine due to an interaction with *N*-methyl-D-aspartate receptors? *Psychopharmacology* **96**, 87–92. [*Chapter 6*]

Jahangir, S.M., Islam, F. & Aziz, L. (1993). Ketamine infusion for postoperative analgesia in asthmatics: a comparison with intermittent meperidine. *Anesth. Analg.* **76**, 45–49. [*Chapter 3*]

Jane, D.E., Olverman, H.J. & Watkins, J.C. (1994). Agonists and competitive antagonists: Structure-activity and molecular modelling studies. In: 'The NMDA Receptor', 2nd edn (eds, G.L. Collingridge and J.C. Watkins), pp. 31–104. Oxford University Press: Oxford. [*Chapter 2*]

Jasper, J.R. (1992). Evolving concepts of partial agonism. the beta-adrenergic receptor as a paradigm. *Biochem. Pharmacol.*, **43**, 119–130. [*Chapter 5*]

Javitt, D.C. & Zukin, S.R. (1995). Biexponential kinetics of [^{3}H]MK-801 binding: evidence for access to closed and open *N*-methyl-D-aspartate receptor channels. *Mol. Pharmacol.* **35**, 387–393. [*Chapter 6*]

Joachimsson, P.-O., Hedstrand, U. & Eklund, A. (1986). Low-dose ketamine infusion for analgesia during postoperative ventilator treatment. *Acta Anaesthesiol. Scand.* **30**, 697–702. [*Chapter 3*]

Johnson, J.W. & Ascher, P. (1987). Glycine potentiates the NMDA response in cultured mouse brain neurons. *Nature* **325**, 529–531. [*Preface, Chapter 6*]

Johnson, J.W. & Ascher, P. (1990). Voltage-dependent block by intracellular Mg^{++} of *N*-methyl-D-aspartate activated channels. *Biophys. J.* **57**, 1085–1090. [*Chapter 2*]

Jones, A.W., Smith, D.A.S. & Watkins, J.C. (1984). Structure–activity relations of dipeptide antagonists of excitatory amino acids. *Neuroscience* **13** 573–581. [*Chapter 2*]

Judge, M.E., Sheardown, M.J., Jacobsen, P. & Honoré, T. (1991). Protection against post-ischemic behavioral pathology by the α-amino-3-hydroxy-5-methyl-4-isoxazolepropionic acid (AMPA) antagonist 2,3-dihydroxy-6-nitro-7-sulfamoyl-benzo(f)quinoxaline (NBQX) in the gerbil. *Neurosci. Lett.* **133**, 291–294. [*Chapter 8*]

Jurna, I., Spohrer, B. & Bock, R. (1992). Intrathecal injection of acetylsalisylic acid, salicylic acid and indometacin depress C fibre evoked activity in the rat thalamus and spinal cord. *Pain* **49**, 249–256. [*Chapter 3*]

Kaiser, F., Hansen, A.J., Laursen, H. & Sheardown, M.J. (1992). NBQX protects against delayed cell death following complete forebrain ischaemia in the rat. *Mol Neuropharmacol.* **2**, 219–220. [*Chapter 8*]

Kaku D.A., Giffard R.G. & Choi D.W. (1993). Neuroprotective effects of glutamate antagonists and extracellular acidity. *Science* **260**, 1516–1518. [*Chapter 1*]

Karbon, E.W., Patch, R.J., Pontecorvo, M.J. & Ferkany, J.W. (1990). Ifenprodil potently interacts with [^{3}H](+)-3-PPP-labeled sigma binding sites in guinea pig brain membranes. *Eur. J. Pharmacol.* **176**, 247–248. [*Chapter 6*]

Karlsten, R. & Gordh, T. (1996). An A1-selective adenosine agonist abolishes allodynia elicited by vibration and touch after intrathecal injection. *Anesth. Analg.* (in press) [*Chapter 3*]

Karp, S.J., Masu, M., Eki, T., Ozawa, K. & Nakanishi, S. (1993). Molecular cloning and chromosomal localization of the key subunit of the human *N*-methyl-D-aspartate receptor. *J. Biol. Chem.* **268**, 3728–3733. [*Chapter 6*]

Keinänen, K., Wisden, W., Sommer, B., Werner, P., Herb, A., Verdoorn, T.A., Sakmann, B. & Seeburg, P.H. (1990). A family of AMPA-selective glutamate receptors. *Science* **249**, 556–560. [*Preface*]

Kemp, J.A. & Leeson, P.D. (1993). The glycine site of the NMDA receptor – five years on. *Trends Pharmacol. Sci.* **14**, 20–25. [*Chapter 6*]

Kessler, J.A., Fenstermacher, J.D. & Owens, E.S. (1976). Spinal subarachnoid perfusion of rhesus monkeys. *Am. J. Physiol.* **230**, 614–618. [*Chapter 3*]

Kieburtz, K., McDermott, M., Marshall, F.J., Feigine, A., Como, P., Abwender, D., Zimmerman, C., Hickey, C., Orme, C., Bordwell, K., Greenamyre, J.T., Dunn, C. & Shoulson, I. (1995). Evidence of the glutamate antagonist remacemide hydrochloride in Huntington's disease. *Neurology* **45**(Suppl. 4), 335P. [*Chapter 10*]

Kirk, C.J., Reddy, N.L., Fischer, J.B., Wolcott, T.C., Knapp, A.G. & McBurney, R.N. (1994). *In vitro* neuroprotection by substituted guanidines with varying affinities for the *N*-methyl-D-aspartate receptor ionophore and for sigma sites. *J. Pharm. Exp. Ther.* **271**, 1080–1085. [*Chapter 4*]

Klepstad, P., Maurset, A., Moberg, E.R. & Oye, I. (1990). Evidence of a role for NMDA receptors in pain perception. *Eur. J. Pharmacol.* **187**, 513–518. [*Chapter 3*]

Klockgether, T. & Turski, L. (1989). Excitatory amino

acids and the basal ganglia: Implications for the therapy of Parkinson's disease. *Trends Neurosci.* **12**, 285–286. [*Chapter 8*]

Kloog, Y., Nadler, V. & Sokolovsky, M. (1988). Mode of binding of [^{3}H]dibenzocycloalklenimine (MK-801) to the *N*-methyl-D-aspartate (NMDA) receptor and its therapeutic implication. *Febs Lett.* **230**, 167–170. [*Chapter 6*]

Knöpfel, T., Kuhn, R. & Allgeier, H. (1995). Metabotropic glutamate receptors: novel targets for drug development. *J. Med. Chem.* **38** 1417–1426. [*Chapter 2*]

Kocklar, A., Zivin, J.A., Lyden, P.D. & Mazzarella, V. (1988). Glutamate antagonist therapy reduces neurological deficits produced by focal central nervous system ischemia. *Arch. Neurol.* **45**, 148–153. [*Chapter 7*]

Koek W. & Colpaert F.C. (1991). Use of a conflict procedure in pigeons to characterize anxiolytic drug activity: evaluation of *N*-methyl-D-aspartate antagonists. *Life Sci.* **49**, PL37-PL42. [*Chapter 1*]

Koek, W. & Colpaert, F.C. (1992). *N*-Methyl-D-aspartate antagonism and phencyclidine like activity: behavorial effects of glycine site ligands. In ''Multiple Sigma and PCP Receptor Ligands: Mechanisms for Neurodilation and Neuroprotection?'' (eds. J.M. Kamenka and F.F. Domino), pp. 655–671. NPP Books, Ann Arbor, MI. [*Chapter 5*]

Koek, W., Woods, J.H. & Ornstein, P. (1987). A simple and rapid method for assessing similarities among directly observable behavioral effects of drugs: PCP-like effects of 2-amino-5-phosphonovalerate in rats. *Psychopharmacology* **91**, 297–304. [*Chapter 10*]

Koenig, H., Trout, J.J., Gladstone, A.D. & Lu, C.Y. (1992). Capillary NMDA receptors regulate blood-brain barrier function and breakdown. *Brain Res.* **588**, 297–303. [*Chapter 10*]

Kolesnikov, Y.A., Maccecchini, M.L. & Pasternak, G.W. (1994). l-Aminocyclopropane carboxylic acid (ACPC) prevents mu and delta opioid tolerance. *Life Sci.* **55**, 1393–1398. [*Chapter 5*]

Kolesnikov, Y.A., Pick, C.G. & Pasternak, G.W. (1992). NG-nitro-L-arginine prevents morphine tolerance. *Eur. J. Pharmacol.* **221**, 399–400. [*Chapters 3, 5*]

Koltchine, V., Anantharam, V., Wilson, A., Bayley, H. & Treistman, S.N. (1993). Homomeric assemblies of NMDAR1 splice variants are sensitive to ethanol. *Neurosci. Lett.* **152**, 13–16. [*Chapter 6*]

Kraaier, V., Van Huffelen, A.C. & Wieneke, G.H. (1988). Quantitative EEG changes due to hypobaric hypoxia in normal subjects. *Electroenceph. Clin. Neurophysiol.* **69**, 303–312. [*Chapter 9*]

Kristensen, J.D. (1994). Spinal NMDA receptors and pain: experimental studies on behaviour, antinociception, spinal cord blood flow and neurotoxicity, pharmacokinetics and a clinical application. Doctorial Thesis No. 459, Uppsala University. [*Chapter 3*]

Kristensen, J.D., Hartvig, P., Karlsten, R., Gordh, T. & Halldin, M. (1995). CSF and plasma pharmacokinetics of the NMDA receptor antagonist CPP after intrathecal, extradural and i.v. administration in anaesthetized pigs. *Br. J. Anaesth.* **74**, 193–200. [*Chapter 3*]

Kristensen, J.D., Karlsten, R. & Gordh, T. (1994a). Laser-Doppler evaluation of spinal cord blood flow after intrathecal administration of an NMDA antagonist in rats. *Anesth. Analg.* **78**, 925–931. [*Chapter 3*]

Kristensen, J.D., Karlsten, R., Gordh, T. & Berge, O.-G. (1994b). The NMDA antagonist 3-(2-carboxypiperazin-4-yl)propyl-1-phosphonic acid (CPP) has antinociceptive effect after intrathecal injection in the rat. *Pain* **56**, 59–68. [*Chapter 3*]

Kristensen, J.D., Karlsten, R., Gordh, T. & Holtz, A. (1993a). Spinal cord blood flow after intrathecal injection of a NMDA-antagonist or an adenosine-agonist in rats. *Anesth. Analg.* **76**, 1279–1283. [*Chapter 3*]

Kristensen, J.D., Post, C., Gordh, T. & Svensson, B.A. (1993b). Spinal cord morphology and antinociception after chronic intrathecal administration of excitatory amino acid antagonists in the rat. *Pain* **54**, 309–316. [*Chapter 3*]

Kristensen, J.D., Svensson, B. & Gordh, T.J. (1992). The NMDA receptor antagonist CPP abolishes neurologic 'wind-up pain' after intrathecal administration in humans. *Pain* **51**, 249–253. [*Chapters 2, 3*]

Krnjevic, K. & Phillis, J.W. (1963). lontophoretic studies of neurones in the mammalian cerebral cortex. *J. Physiol. (Lond.)* **165**, 274–304. [*Preface*]

Kucharczyk, J., Mintorovitch, J., Moseley, M.E., Asgari, H.S., Sevick, R.J., Derugin, N. & Norman, D. (1991). Ischaemic brain damage: reduction by sodium-calcium ion channel modulator RS-87476. *Radiology* **179**, 221–227. [*Chapter 6*]

Kupferberg, H.J. (1989). Antiepileptic drug development program: a cooperative effort of government and industry. *Epilepsia* **30**(Suppl. 1), S51–S56. [*Chapter 10*]

Kusiak, J.W. & Norton, D.D. (1993). A splice variant of the *N*-methyl-D-aspartate (NMDAR1) receptor. *Mol. Brain Res.* **20**, 64–70. [*Chapter 6*]

Kutsuwada, T., Kashiwabuchi, N., Mori, H., Sakimura, K., Kushiya, E., Araki, K., Meguro, H., Masaki, H., Kumanishi, T. & Arakawa, M. (1992). Molecular diversity of the NMDA receptor channel [see comments]. *Nature* **358**, 36–41. [*Chapter 3*]

Lacomblez, L., Bensimon, P., Leigh, G., Guillet, N. & Meininger, V. The ALS Study Group II (1996). A controlled trial of riluzole in amyotrophic lateral sclerosis. *Lancet* **347**, 1425–1431. [*Chapter 9*]

Lacomblez, L., Bouche, P., Bensimon, Meininger, V. (1989). A double-blind, placebo-controlled trial of high doses of gangliosides in amyotrophic lateral sclerosis. *Neurology* **39** 1635–1637. [*Chapter 9*]

Lafon-Casal, M., Pietri, S., Culcasi, M. & Bocksert, J. (1993). NMDA-dependent superoxide production and neurotoxicity. *Nature* **364**, 535–537. [*Chapter 9*]

Lahti, A.C., Holcomb, H.H., Medoff, D.R. & Tamminga, C.A. (1995). Ketamine activates psychosis and alters limbic blood flow in schizophrenia. *Neuroreport* **6**, 869–872. [*Chapter 6*]

Lanier, W.L., Christopherson, T.J., Scheithauer, B.W., Mude, J.H., Hofer, R.E. & Michenfelder, J.D. (1993). Post-ischemic treatment with NBQX does not improve outcome in a canine model of complete cerebral ischemia. *J. Cereb. Blood Flow Metab.* **13** (Suppl. 1), S701. [*Chapter 8*]

Laurie, D.L. & Seeburg, P.H. (1994). Ligand affinities at recombinant *N*-methyl-D-aspartate receptors depend on subunit composition. *Eur. J. Pharmacology. (Mol. Pharmac. Sect.)* **268**, 335–345. [*Chapters 2, 6*]

Layer, R.T., Popik, P., Olds, T. & Skolnick, P. (1996). Antidepressant-like actions of the polyamine site NMDA antagonist, Eliprodil (SL82.0715). *Pharmacol. Biochem. Behav.* **52**, 621–627. [*Chapter 6*]

Le Bourdelles, B., Wafford, K.A., Kemp, J.A., Marshall, G., Bain, C., Wilcox, A.S., Sikela, J.M. & Whiting, P.J. (1994). Cloning, functional coexpression, and pharmacological characterisation of human cDNAs encoding NMDA receptor NR1 and NR2A subunits. *J. Neurochem.* **62**, 2091–2098. [*Chapter 6*]

Le Peillet, E., Arvin, B., Moncada, C. & Meldrum, B.S. (1992). The non-NMDA antagonists, NBQX and GYKI 52466, protect against cortical and striatal cell loss following transient global ischaemia in the rat. *Brain Res.* **571**, 115–120. [*Chapter 8*]

Leach, J.P., Blacklaw, J., Stewart, M., Jones, T, Jamieson, V., Richens, A. & Brodie, M.J. (1994). Mutual pharmacokinetic interactions between remacemide and carbamazepine and phenytoin but not sodium valproate in epileptic patients. *Epilepsia* **35**(Suppl. 8), 52. [*Chapter 10*]

Leach, M.J., Marden, C.M. & Miller, A.A. (1986). Pharmacological studies of lamotrigine, a novel potential antiepileptic drug: II. Neurochemical studies on the mechanism of action. *Epilepsia* **27**, 490–497. [*Chapter 3*]

Leander, J.D., Parli, C.J., Potts, B. & Lodge, D. (1992). Relation of plasma and brain concentrations of the anticonvulsant ameltolide to its pharmacological effects. *Epilepsia.* **33**, 696–704. [*Chapter 7*]

Legendre, P. & Westbrook, G.L. (1991). Ifenprodil blocks *N*-methyl-D-aspartate receptors by a two-component mechanism. *Mol. Pharmacol.* **40**, 289–298. [*Chapter 6*]

Lehmann J., Chapman A.G., Meldrum B.S., Hutchison A., Tsai C. & Wood P.L. (1988a). CGS 19755 is a potent and competitive antagonist at NMDA-type receptors. *Eur. J. Pharmacol.* **154**, 89–93. [*Chapter 1*]

Lehmann J., Hutchison A.J., McPherson S.E., Mondadori C., Schmutz M., Sinton C.M., Tsai C., Murphy D.E., Steel D.J., Williams M., Cheney D.L. & Wood P.L. (1988b). CGS 19755, a selective and competitive *N*-methyl-D-aspartate type excitatory amino acid receptor antagonist. *J. Pharmacol. Exp. Ther.* **246**, 65–75. [*Chapter 1*]

Lehmann, J. (1989). The NMDA receptor. *Drugs Future* **14**, 1059–1071. [*Chapter 10*]

Lehmann, J., Schneider, J., McPherson, S., Tsai, C., Bennett, D.A., Pastor, G., Steel, D.J., Boehm, C., Cheney, D.L., Liebman, J.M., Williams, M. & Wood, P.L. (1987). CPP, a selective N-methyl-D-aspartate (NMDA)-type receptor antagonist: Characterization in vitro and in vivo. *J. Pharmacol. Exp. Ther.* **240**, 737–746. [*Preface, Chapter 3*]

Lerma, J. (1992). Spermine regulates *N*-methyl-D-aspartate receptor desensitization. *Neuron* **8**, 343–352. [*Chapter 6*]

Lesiuk, H.J., Barnes, K., Todd, M.L. & Hewitt, K. (1995). NMDA antagonists: A reappraisal–protection against transient forebrain ischemia in rodents by the competitive NMDA antagonist remacemide. *Can. J. Neurol. Sci.* **22**(Suppl. 1), S14. [*Chapter 10*]

Lewis, S.J., Barres, C., Jacob, H.J., Ohta, H. & Brody, M.J. (1989). Cardiovascular effects of the *N*-methyl-D-aspartate receptor antagonist MK-801 in conscious rats. *Hypertension* **73**, 759–765. [*Chapter 6*]

Li, H. & Buchan, A.M. (1993). Treatment with an AMPA antagonist 12 hours following severe normothermic forebrain ischemia prevents CA1 neuronal injury. *J. Cereb. Blood Flow Metab.* **13**, 933–939. [*Chapter 8*]

Li, Y., Kawamura, S., Yasui, N., Shirazawa, M. & Fukasawa, H. (1994). Therapeutic effects of nilvadipine on rat focal cerebral ischaemia. *Exp. Brain. Res.* **99**, 1–6. [*Chapter 6*]

Lippert, K., Welsch, M. & Krieglstein, J. (1994). Over-additive protective effect of dizocilpine and NBQX against neuronal damage. *Eur. J. Pharmacol.* **253**, 207–213. [*Chapter 8*]

Lipton, S.A. & Rosenberg, P.A. (1994). Excitatory amino acids as a final common pathway for neurologic disorders. *N. Engl. J. Med.* **330**, 613–622. [*Chapters 4, 9, 10*]

Lipton, S.A. (1993). Prospects for clinically tolerated NMDA antagonists: open-channel blockers and alternative redox states of nitric oxide. *Trends Neurosci.* **12**, 527–532. [*Chapter 3*]

Lockhart, B.P., Soulard, P., Benicourt, C., Privat, A. & Junien, J.L. (1995). Distinct neuroprotective profiles for sigma ligands against *N*-methyl-D-aspartate (NMDA) and hypoxia-mediated neurotoxicity in neuronal culture toxicity studies. *Brain Res.* **675**, 110–120. [*Chapter 6*]

Lockhart, B.P., Vila, J., Hamedi Sangsari, F., Privat, A. & Vignon, J. (1993). Neurotoxic effect of the anti-HIV drug D-aspartate beta-hydroxamate for rat primary neuronal cultures: attenuation by *N*-methyl-D-aspartate (NMDA) antagonists. *Brain Res.* **630**, 32–40. [*Chapter 6*]

Lodge, D., Jones, M. & Fletcher, E. (1994). Non-competitive antagonists of *N*-methyl-D-aspartate. In: 'The NMDA Receptor', 2nd edn (eds, G.L. Collingridge and J.C. Watkins), pp. 105–131. Oxford University Press: Oxford. [*Chapter 2*]

Loescher, W. & Honack, D. (1991). Anticonvulsant and behavioral effects of two novel competitive *N*-methyl-D-aspartic acid receptor antagonists, CGP 37849 and CGP 39551, in the kindling model of epilepsy. Comparison with MK 801 and carbamazepin. *J. Pharmacol. Exp. Ther.* **256**, 432–440. [*Chapter 2*]

Loescher, W., Rundfeldt, C. & Honack, D. (1993). Low doses of NMDA receptor antagonists synergistically increase the anticonvulsant effect of the AMPA receptor antagonist NBQX in the kindling model of epilepsy. *Eur. J. Neurosci.* **5**, 1545–1550. [*Chapters 2, 8*]

Long, J.B. & Skolnick, P. (1994). 1-Aminocyclopropanecarboxylic acid protects against dynorphin A-induced spinal injury. *Eur. J. Pharmacol.* **261**, 295–301. [*Chapter 5*]

Lopez, O.T. & Lanthorn, T.H. (1996). The glycine site partial agonist, ACPC, reduces infarct volume following middle cerebral artery occlusion in the mouse. *Eur. J. Pharmacol.*, in press. [*Chapter 5*]

Lothman, E.W., Salerno, R.A., Perlin, J.B. & Kaiser, D.L. (1988). Screening and characterization of antiepileptic drugs with rapidly recurring hippocampal seizures in rats. *Epilepsy Res.* **2**, 367–379. [*Chapter 10*]

Louvel, E., Hugon J. & Doble, A. (1996). Clinical trends in amyotrophic lateral sclerosis. *TIPS* in press. [*Chapter 9*]

Lowe, D.A., Emre, M., Frey, P., Kelly, P.H., Malanowski, J., McAllister, K.H., Neijt, H.C., Ruedeberg, C., Urwyler, S., White, T.G. & Herrling, P.L. (1994). The pharmacology of SDZ EAA 494, a competitive NMDA antagonist. *Neurochem. Int.* **25**, 583–600. [*Chapter 2*]

Lowe, D.A., Neijt, H.C. & Aebischer, B. (1990). D-CPPene (SDZ EAA 494), a potent and competitive *N*-methyl-D-aspartate (NMDA) antagonist: effect on spontaneous activity and NMDA-induced depolarizations in rat neocortical slice preparation compared with other CPP derivatives and MK-801. *Neurosci. Lett.* **113**, 315–321. [*Chapter 2*]

Lucas, D.R. & Newhouse, J.P. (1957). The toxic effect of sodium L-glutamate on the inner layers of the retina. *AMA Arch. Ophthalmol.* **58**, 193–201. [*Chapter 9*]

Luquin, M.R., Obeso, J.A., Laguna, J., Guillin, J. & Martinez-Lage, J.M. (1993). The AMPA receptor antagonist NBQX does not alter the motor response induced by selective dopamine agonists in MPTP-treated monkeys. *Eur. J. Pharmacol.* **235**, 297–300. [*Chapter 8*]

MacDermott, A.B., Mayer, M.L., Westbrook, G.L., Smith, S.J. & Barjer, J.L. (1986). NMDA-receptor activation increase cytoplasmic calcium concentration in cultured spinal cord neurones. *Nature* **323**, 519. [*Preface, Chapter 6*]

Macdonald, R.L. (1989). Antiepileptic drug action. *Epilepsia* **30**(Suppl. 1), S19–S28. [*Chapter 10*]

MacKenzie, E.T., Gotti, B., Nowicki, J.P. & Young, A.R. (1984). Adrenergic blockers as cerebral antiischaemic agents. In: 'Neurotransmitters and the Cerebral Circulation, LERS Monograph Series' (eds, G. Bartholini, J.C. Friedmann, S.Z. Langer, P.L. Morselli & A. Wick), Vol. 2, pp. 219–244. Raven Press: New York. [*Chapter 6*]

Madden K.P., Clark W.M. & Zivin J.A. (1993). Delayed therapy of experimental ischemia with competitive *N*-methyl-D-aspartate antagonists in rabbits. *Stroke* **24**, 1068–1071. [*Chapter 1*]

Madden, K.P., Clark, W.M., Kochhar, A. & Zivin, J.A. (1992). Efficacy of LY 233053, a competitive glutamate antagonist, in experimental central nervous system ischemia. *J. Neurosurg.* **76**, 106–110. [*Chapter 8*]

Malgouris, C., Bardot, F., Daniel, M., Pellis, F., Rataud, J., Uzan, J., Blanchard, J.C. & Laduron, P.M. (1989). Riluzole, a novel antiglutamate, prevents memory loss and hippocampal neuronal damage in ischemic gerbils. *J. Neurosci.* **9**(11), 3720–3727. [*Chapter 9*]

Mankowitz, E., Brock-Utne, J.G., Cosnett, J.E. & Green-Thompson, R. (1982). Epidural ketamine. *S. Afr. Med. J.* **61**, 441–442. [*Chapter 3*]

Markabi S. (1994). Selfotel (CGS 19755). The preliminary clinical experience. In 'Pharmacology of Cerebral Ischemia 1994' (eds J. Krieglstein, H. Oberpichler-Schwenk), pp. 635–642. Wissenschaftliche Verlagsgesellschaft mbH, Stuttgart. [*Chapter 1*]

Martin, D, Thompson, M.A. & Nadler, J.V. (1993). The neuroprotective agent riluzole inhibits release of glutamate and aspartate from slices of hippocampal area CAI. *Eur. J. Pharmacol.* **250**(3), 473–476. [*Chapter 9*]

Marvizon, J.C. & Baudry, M. (1994). [^{3}H]dizocilpine association kinetics distinguish stimulatory and inhibitory polyamine sites of *N*-methyl-D-aspartate receptors. *J. Neurochem.* **63**, 963–971. [*Chapter 6*]

Massieu, L. Thedinga, K.H., McVey, M. & Fagg, G.E. (1993). A comparative analysis of the neuroprotective properties of competitive and uncompetitive *N*-methyl-D-aspartate receptor antagonists *in vivo*: implications for the process of excitotoxic degeneration and its therapy. *Neuroscience*. **5**, 883–892. [*Chapter 2*]

Masu, M., Tanabe, Y., Tsuchida, K., Shigemoto, R. & Nakanishi, S. (1991). Sequence and expression of a metabotropic glutamate receptor. *Nature* **349**, 760–765. [*Preface*]

Matsumoto, R.R., Hemstreet, M.K., Lai, N.L., Thurkauf, A., De Costa, B.R., Rice, K.C., Hellewell, S.B., Bowen, W.D. & Walker, J.M. (1990). Drug specificity of pharmacological dystonia. *Pharmacol. Biochem. Behav.* **36**, 151–155. [*Chapter 6*]

Maurset, A., Skoglund, L.A., Hustveit, O. & Oye, I. (1989). Comparison of ketamine and pethidine in experimental and postoperative pain. *Pain* **36**, 37–41. [*Chapter 3*]

Mayer, M.L. & Westbrook, G.L. (1987a). The physiology of excitatory amino acids in the vertebrate central nervous system. *Prog. Neurobiol.* **28**, 197–276. [*Chapter 6*]

Mayer, M.L. & Westbrook, G.W. (1987b). Permeation and block of *N*-methyl-D-aspartic receptor channels by divalent cations in mouse cultured central neurones. *J. Physiol.* **394**, 501–527. [*Chapter 6*]

Mayer, M.L., Vyklicky, L.J. & Clements, J. (1989). Regulation of NMDA receptor desensitization in mouse hippocampal neurons by glycine. *Nature* 338, 425–427. [*Chapter 3*]

Mayer, M.L., Westbrook, G.W. & Guthrie, P.B. (1984). Voltage-dependent block by Mg^{++} of NMDA responses in spinal cord neurones. *Nature* **309**, 261–263. [*Chapter 6*]

McAllister, K. (1992). *N*-Methyl-D-aspartate receptor antagonists and channel blockers have different effects upon a spinal seizure model in mice. *Eur. J. Pharmacol.* **231**, 309–312. [*Chapter 2*]

McBurney, R.N. (1994). Therapeutic potential of NMDA antagonists in neurodegenerative diseases. *Neurobiol. Aging* **15**, 271–273. [*Chapter 4*]

McCabe, R.T., Wasterlain, G.G., Kucharczyk, N., Sofia, R.D. & Vogel, J.R. (1993). Evidence for anticonvulsant and neuroprotectant action of felbamate mediated by strychnine-insensitive glycine receptors. *J. Pharmacol. Exp. Ther.* **264**, 1248–1252. [*Chapter 3*]

McCool, B.A. & Lovinger, D.M. (1995). Ifenprodil inhibition of the 5-Hydroxytryptamine$_3$ receptor. *Neuropharmacology* **34**, 621–629. [*Chapter 6*]

McCulloch, J. & Iversen, L.L. (1995). Autoradiographic assessment of the effects of *N*-methyl-D-aspartate (NMDA) receptor antagonists *in vivo. Neurochem. Res.* **16**, 951–963. [*Chapter 6*]

McCulloch, J., Bullock, R. & Teasdale, G.M. (1991). Excitatory amino acid antagonists: Opportunitites for the treatment of ischaemic brain damage in man. In: 'Excitatory Amino Acid Antagonist' (Ed., B. Meldrum), pp. 287–326. Blackwell: Oxford. [*Chapter 4*]

McFarlane, C., Warner, D.S., Todd, M.M. & Nordholm, L. (1992). AMPA receptor competitive antagonism reduces halothane MAC in rats. *Anesthesiology* **77**, 1165–1170. [*Chapter 8*]

McIntosh T.K., Smith D.H., Hayes R.L., Vink R. & Simon R.P. (1992). Role of excitatory amino acid neurotransmitters in the pathogenesis of traumatic brain injury. In: 'Excitatory amino acids. Fidia Research Foundation Symposium Series', Vol. 9 (ed., R.P. Simon), pp. 247–253. Thieme: New York. [*Chapter 1*]

McLennan, H. & Liu, J. (1982). The action of six antagonists of the excitatory amino acids on neurones of the rat spinal cord. *Exp. Brain. Res.* **45**, 151–156. [*Preface*]

McNamara, J.O., Russell, R.D., Rigsbee, L. & Bonhaus, D.W. (1988). Anticonvulsant and antiepileptogenic actions of MK-801 in the kindling and electroshock models. *Neuropharmacology* **27**, 563–568. [*Chapter 10*]

McQuay, H.J. & Dickenson, A.H. (1990). Implications of nervous system plasticity for pain management. *Anaesthesia* **45**, 110–112. [*Chapter 3*]

Meadows, M-E., Fisher, M. & Minematsu, K. (1994). Delayed treatment with a noncompetitive NMDA antagonist, CNS-1102, reduces infarct size in rats. *Cerebrovasc. Disease* **4**, 26–31. [*Chapter 4*]

Meden, P., Overgaard, K., Sereghy T. & Boysen G. (1993). Enhancing the efficacy of thrombolysis by AMPA receptor blockade with NBQX in a rat embolic stroke model. *J. Neurol. Sci.* **119**, 209–216. [*Chapter 8*]

Meldrum, B. & Garthwaite, J. (1990). Excitatory amino acid neurotoxicity and neurodegenerative disease. *Trends Pharmacol. Sci.* **11**, 379–387. [*Chapter 5*]

Meller, S.T. & Gebhart, G.F. (1993). Nitric oxide (NO) and nociceptive processing in the spinal cord. *Pain* **52**, 127–136. [*Chapter 3*]

Millan, M.J. & Seguin, L. (1994). Chemically-diverse ligands at the glycine B site coupled to *N*-methyl-D-aspartate (NMDA) receptors selectively block the late phase of formalin-induced pain in mice. *Neurosci. Lett.* **178**, 139–143. [*Chapter 6*]

Miller, J.D., Bullock, R., Graham, D.I., Chen, M.H. & Teasdale, G.M. (1990). Ischemic brain damage in a model of acute subdural hematoma. *Neurosurgery* **27**, 148–153. [*Chapter 2*]

Miller, R., LaGrone, J., Skolnick, P. & Boje, K.M. (1992). High performance liquid chormatographic assay for 1-aminocyclopropanecarboxylic acid from plasma and brain. *J. Chromatogr.* **578**, 103–108. [*Chapter 5*]

Miller, W.C. & DeLong, M.R. (1987). Altered tonic activity of neurones in the globus pallidus and subthalamic nucleus in the primate MTPT model of Parkinsonism. In: 'The Basal Ganglia', (eds, M.B. Carpenter & A. Jayaraman) Vol. II, pp. 415–427. Plenum Press: New York. [*Chapter 8*]

Minematsu, K., Fisher, M., Li, L., Davis, M.A., Knapp, A.G., Cotter, R.E., McBurney, R.N. & Sotak, C.H. (1993a). Effects of a novel NMDA antagonist on experimental stroke rapidly and quantitatively assessed by diffusion-weighted MRI. *Neurology* **43**, 397–403. [*Chapter 4*]

Minematsu, K., Fisher, M., Li, L. & Sotak, C. (1993b). Diffusion and perfusion MRI studies to evaluate a non-competitive NMDA antagonist and reperfusion in experimental stroke in rats. *Stroke* **24**, 2074–2081. [*Chapter 4*]

Mitchell, I.J., Cross, A.J., Sambrock, M.A. & Crossman, A. (1986). Neural mechanisms mediating 1-methyl-4-phenyl-1,2,3,6-tetrahydropyridine-induced parkinsonism in the monkey: relative contributions of the striatopallidal and striatonigral pathways as suggested by 2-deoxyglucose uptake. *Neurosci. Lett.* **63**, 61. [*Chapter 8*]

Mitchell, I.J., Hughes, N., Carroll, C.B. & Brotchie, J.M. (1995). Reversal of parkinsonian symptoms by intra-striatal and systemic manipulations of excitatory amino acid and dopamine transmission in the bilateral 6-OHDA lesioned marmoset. *Behav. Pharmacol.* **6**, 492–507. [*Chapter 6*]

Mizoule, J., Meldrum, B., Mazadier, M., Croucher, M., Ollat, C., Uzan, A., Legrand, J.J. Gueremy, G. & Le Fur, G. (1985). 2-Amino-6-trifluoromethoxy benzothiazole, a possible antagonist of excitatory amino acid neurotransmission I. Anticonvulsant properties. *Neuropharmacology* **24**(8), 767–773. [*Chapter 9*]

Mocanda, S., Palmer, R.J.M. & Higgs, E.A. (1991). Nitric oxide: physiology, pathophysiology and pharmacology. *Pharmacol Rev* **43**, 109–142. [*Chapter 3*]

Moebius, F.F., Burrows, G.G., Hanner, M., Schmid, E.,

Striessnig, J. & Glossmann, H. (1993). Identification of a 27-kDa high affinity phenylalkylamine-binding polypeptide as the σ_1 binding site by photoaffinity labelling and ligand-directed antibodies. *Mol. Pharmacol.* **44**, 966–971. [*Chapter 6*]

Moebius, F.F., Hanner, M., Knaus, H., Weber, F., Striessnig, J. & Glossmann, H. (1994). Purification and amino-terminal sequencing of the high-affinity phenylalkylamine Ca^{2+} antagonist binding protein from guinea pig liver endoplasmic reticulum. *J. Biol. Chem.* **269**, 29314–29320. [*Chapter 6*]

Monaghan, D.T., Bridges, R.J. & Cotman, C.W. (1989). The excitatory amino acid receptors: their classes, pharmacology and distinct properties in the function of the central nervous system. *Annu. Rev. Pharmacol. Toxicol.* **29** 365–402. [*Chapters 2, 5*]

Mondadori, C. & Weiskrantz, L. (1993). NMDA receptor blockers facilitate and impair learning via different mechanisms. *Behav. Neural Biol.* **60**, 205–210. [*Chapter 2*]

Monnet, F.P., Debonnel, G. & De Montigny, C. (1992). In vivo electrophysiological evidence for a selective modulation of *N*-methyl-D-aspartate-induced neuronal activation in rat CA_3 dorsal hippocampus by sigma ligands. *J. Pharm. Exp. Ther.* **261**, 123–130. [*Chapter 6*]

Monnet, F.P., Debonnel, G., Bergeron, R., Gronier, B. & De Montigny, C. (1994). The effects of sigma ligands and of neuropeptide Y on N-methyl-D-aspartate-induced neuronal activation of CA_3 dorsal hippocampla neurones are differentially affected by pertussis toxin. *B. J. Pharmacol.* **112**, 709–715. [*Chapter 6*]

Montastruc, J.L., Rascol, O., Senard, J.M. & Rascol, A. (1992). A pilot study of *N*-methyl-D-aspartate (NMDA) antagonist in Parkinson's disease *J. Neurol. Neurosurg. Psychiatry* **55**, 630–631. [*Chapter 6*]

Monyer, H. & Choi, D.W. (1988). Morphinans attenuate cortical neuronal injury induced by glucose deprivation *in vitro. Brain Res.* **446**, 144–148. [*Chapter 6*]

Monyer, H., Goldberg, M.P. & Choi, D.W. (1989). Glucose deprivation neuronal injury in cortical culture. *Brain Res.* **483**, 347–354. [*Chapter 6*]

Monyer, H., Sprengel, R., Schoepfer, R., Herb, A., Higuchi, M., Lomeli, H., Burnashev, N., Sakmann, B. & Seeburg, P.H. (1992). Heteromeric NMDA receptors: Molecular and functional distinction of subtypes. *Science* **256**, 1217–1221. [*Preface, Chapter 3, 6*]

Moon, I.S., Apperson, M.L. & Kennedy, M.B. (1994). The major tyrosine-phosphorylated protein in the postsynaptic density fraction is *N*-methyl-D-aspartate receptor subunit 2B. *Proc. Natl Acad. Sci. USA* **91**, 3954–3958. [*Chapter 6*]

Morimoto K., Katayama K., Inoue K. & Sato K. (1991). Effects of competitive and noncompetitive NMDA receptor antagonists on kindling and LTP. *Pharmacol Biochem. Behav.* **40**, 893–899. [*Chapter 1*]

Moriyoshi, K., Masu, M., Ishii, T., Shigemoto, R., Mizuno, N. & Nakanishi, S. (1991). Molecular cloning and characterization of the rat NMDA receptor. *Nature* **354**, 31–37. [*Chapters 3, 6*]

Morris, R.G.M. (1989). Synaptic plasticity and learning: selective impairment of learning in rats and blockade of long-term potentiation *in vivo* by the *N*-methyl-D-aspartate receptor antagonist AP5. *J. Neurosci* **9**, 3040–3057. [*Chapter 6*]

Muir, K.W. & Lees, K.R. (1995b). Clinical experience with excitatory amino acid antagonist drugs. *Stroke* **26**, 503–513. [*Chapters 2, 4, 6, 10*]

Muir, K.W. & Lees, K.R. (1995a). Initial experience with remacemide hydrochloride in patients with acute ischaemic stroke. *Ann. NY Acad. Sci.* **765**, 322–323. [*Chapter 10*]

Muir, K.W., Grosset, D.G., Gamzu, E. & Lees, K.R. (1994). Pharmacological effects of the non-competitive NMDA antagonist CNS 1102 in normal volunteers. *Br. J. Clin. Pharmacol.* **38**, 33–38. [*Chapters 2, 4*]

Murphy D.E., Hutchison A.J., Hurt S. D., Williams M. & Sills M.A. (1988). Characterization of the binding of [^{3}H]-CGS 19755: a novel *N*-methyl-D-aspartate antagonist with nanomolar affinity in rat brain. *Br. J. Pharmacol.* **95**, 932–938. [*Chapter 1*]

Muscat, R., Papp, M. & Willner, P. (1992). Reversal of stress-induced anhedonia by the atypical antidepressants, fluoxetine and maprotiline. *Psychopharmacol.* **109**, 443–438. [*Chapter 5*]

Nabeshima, T., Kozawa, T., Furukawa, H. & Kameyama, T. (1986). Phencyclidine-induced retrograde amnesia in mice. *Psychopharmacology* **89**, 334–337. [*Chapter 6*]

Nakamura-Craig, M. & Follenfant, R.L. (1994). Lamotrigine and analogs: a new treatment for chronic pain? In 'Proceedings of the 7th World Congress on Pain' (eds, G.F. Gebhart, D.L. Hammond and T.S. Jensen), pp. 725–730. IASP Press: Seattle. [*Chapter 3*]

Nakanishi S. (1992). Molecular diversity of glutamate receptors and implications for brain function. *Science* **258**, 597–603. [*Chapters 2, 5*]

Nakayama, H., Ginsberg, M.D. & Dietrich, W.D. (1988). (*S*)-Emopamil, a novel calcium channel blocker and serotonin S_2 antagonist, markedly reduces infarct size following middle cerebral artery occlusion in the rat. *Neurology* **38**, 1667–1673. [*Chapter 6*]

Nankai, M., Fage, D. & Carter, C. (1995). Striatal NMDA receptor subtypes: the pharmacology of NMDA-evoked dopamine, -aminobutyric acid, acetylcholine and spermidine release. *Eur. J. Pharmacol.* **286**, 61–70. [*Chapter 6*]

Nässtróm, J., Karlsson, U. & Post, C. (1992). Antinociceptive actions of different classes of excitatory amino acid receptor antagonists in mice. *Eur. J. Pharmacol.* **212**, 21–29. [*Chapter 3*]

Nellgård, B. & Wieloch, T. (1992). Postischemic blockade of AMPA but not NMDA receptors mitigates neuronal damage in the rat brain following transient severe cerebral ischemia. *J. Cereb. Blood Flow Metab.* **12**, 2–11. [*Chapter 8*]

Nicolas, C. & Carter, C. (1994). Autoradiographic distribution and characteristics of high- and low-affinity polyamine-sensitive [3H]ifenprodil sites in the rat brain: possible relationship to NMDAR2B receptors and calmodulin. *J. Neurochem.* **63**, 2248–2258. [*Chapter 6*]

Nicolas, C., Fage, D. & Carter, C. (1994). NMDA receptors with different sensitivities to magnesium and ifenprodil control the release of (14C)acetylcholine and [3H] spermidine from rat striatal slices *in vitro. J. Neurochem.* **62**, 1835–1839. [*Chapter 6*]

Nielsen-Kudsk, F., Mellemkjær, S. & Bang L. (1991). Cardiac uptake kinetics and possible dynamic effects of a new cerebral antiischaemic compound FG 9202 (NBQX) studied in the isolated rabbit heart. *Pharmacol. Toxicol.* **69**, 127–131. [*Chapter 8*]

NINDS rt-PA Stroke Study Group (1995). Tissue plasminogen activator for acute ischemic stroke. *N. Engl. J. Med.* **333**, 1581–1587. [*Chapter 4*]

Norris, F.H., Calanchini, P.R., Fallat, R.H., Panchari, S. & Jewett, B. (1974). The administration of guanidine in amyotrophic lateral sclerosis. *Neurology* **24**, 721–728. [*Chapter 9*]

Nowak, G., Trullas, R., Layer, R.T., Skolnick, P. & Paul, I.A. (1993). Adaptive changes in the *N*-Methyl-D-aspartate receptor complex after chronic treatment with imipramine and l-aminocyclopropanecarboxylic acid. *J. Pharmacol. Exper. Ther.* **265**, 1380–1386. [*Chapter 5*]

Nowak, L., Bregestowski, P., Ascher, P., Herbert, A. & Prochiantz, A. (1984). Magnesium gates glutamate-activated channels in mouse central neurones. *Nature* **307**, 462–465. [*Chapter 6*]

Nowicki, J.P., Jourdain, D. & Mackenzie, E.T. (1989). NMDA-induced changes in NADH fluorescence *in vivo*: antagonism by TCP, MK-801 and SL-82.0715. *J. Cereb. Blood Flow Metab.* **9** (Suppl 1): S300. [*Chapter 6*]

Oblin, A. & Schoemaker, H. (1994). Complex allosteric modulation of the binding of the NMDA receptor antagonist [3H]CGP39653. *Eur. J. Pharmacol.* **266**, 103–106. [*Chapter 3, 6*]

Oblin, A., Allen, J. & Schoemaker, H. (1994a). Further characterisation of the interaction of eliprodil with the NMDA receptor *in vitro. Can. J. Physiol. Pharmacol.* **72**, 331. [*Chapter 6*]

Oblin, A., Carter, C. & Schoemaker, H. (1994b). Pharmacological characterisation of the binding of the glycine antagonist [^{3}H]5,7-dichlorokynurenate and its modulation by spermine. *Br. J. Pharmacol.* **112**, 355P. [*Chapter 6*]

Ogita, K., Ohkawara, A., Suzuki, T., Ohgaki, T., Uchida, S., Meguri, H. & Yoneda, Y. (1992). Effects of ifenprodil on the *N*-methyl-D-aspartate receptor ionophore complex in rat brain. *Neurochem. Int.* **21**, 135–147. [*Chapter 6*]

Olney J.W., Labruyere J. & Price M.T. (1989). Pathological changes induced in cerebrocortical neurons by phencyclidine and related drugs. *Science* **244**, 1360–1362. [*Chapters 1, 4, 6, 10*]

Olney, J.W. & De Gubareff, T. (1978). Glutamate neurotoxicity and Huntington's chorea. *Nature* **271**, 557–559. [*Chapter 6*]

Olney, J.W. (1969). Brain lesions, obesity, and other disturbances in mice treated with monosodium glutamate. *Science* **164**, 719–721. [*Chapters 6, 9*]

Olney, J.W. (1981). Kainic acid and other excitotoxins: a comparative analysis. *Adv. Biochem. Psychopharmacol.* **27**, 375–384. [*Chapter 6*]

Olney, J.W. (1994). Neurotoxicity of NMDA receptor antagonists: an overview. *Psychopharmacol. Bull.* **30**, 533–540 [*Chapter 6*]

Olney, J.W., Ho, O.L. & Rhee, V. (1971). Cytotoxic effects of acidic and sulfur containing amino acids on the infant mouse central nervous system. *Exp. Brain Res.* **14**, 61–76. [*Chapter 9*]

Olney, J.W., Labruyere, J. & Price, M.T. (1990). Pathological changes induced in cerebrocortical neurons by phencyclidine and related drugs. *Science*, **244**, 1360–1362. [*Chapter 5*]

Olney, J.W., Labruyere, J., Wang, G., Wozniak, D.F., Price, M.T. & Sesma, M.A. (1991). NMDA antagonist neurotoxicity: mechanism and prevention. *Science* **254**, 1515–1518. [*Chapter 3*]

Olney, J.W., Rhee, V. & Ho, O.L. (1974). Kainic acid: a powerful neurotoxic analogue of glutamate. *Brain Res.* **77**, 507–512. [*Chapter 6*]

Olverman, H.J., Jones, A.W. & Watkins, J.C. (1984). L-Glutamate has higher affinity than other amino acids for [^{3}H]-D-AP5 binding sites in rat brain membranes. *Nature (Lond.)* **307** 460–462. [*Chapter 2*]

Olverman, H.J., Jones, A.W., Mewett, K.N. & Watkins, J.C. (1988b). Structure/activity relations of NMDA receptor ligands as studied by their inhibition of [^{3}H]D-AP5 binding in rat brain membranes. *Neuroscience* **26** 17–31. [*Chapter 2*]

Olverman, H.J., Jones. A.W. & Watkins, J.C. (1988a). [^{3}H]D-2-Amino-5-phosphonopentanoate as a ligand for *N*-methyl-D-aspartate receptors in the mammalian central nervous system. *Neuroscience* **26**, 1–15. [*Chapter 2*]

Olverman, H.J., Monaghan, D.T., Cotman, C.W. & Watkins, J.C. (1986). [^{3}H]CPP, a new competitive ligand for NMDA receptors. *Eur. J. Pharmacol.* **131** 161–162. [*Chapter 2*]

Oppenheim, D.R. (1984). In: '*Greenfield's Neuropathology*', 4th edn (eds, J. Hume, J.A.N. Corsellis, L.W. Duchen, pp. 699–743. Wiley. [*Chapter 9*]

Ordy, J.M., Volpe, B., Murray, R., Thomas, G., Bialobok, P., Wegenack, T.M. & Dunlap, W. (1992). Pharmacological effects of remacemide and MK-801 on memory and hippocampal CA1 damage in the rat four-vessel occlusion (4-VO) model of global ischemia. In: 'The Role of Neurotransmitters in Brain Injury' (eds, M.Y-T. Globus & W.D. Dietrich), pp. 83–92. Plenum Press: New York. [*Chapter 10*]

Ornstein, P.L., Arnold, M.B., Augenstein, N.K., Lodge, D., Leander, J.D. & Schoepp, D.D. (1993).

3*SR*,4a*RS*,6*RS*,8a*RS*-6-(2-(1*H*-Tetrazol-5-yl)ethyl)-decahydroisoquinoline3-carboxylic acid: a structurally novel, systemically active, competitive AMPA receptor antagonist. *J. Med. Chem.* **36**, 2046–2048. [*Chapter 7*]

Overgaard, K., Sereghy, T., Pedersen, H. & Boysen G. (1993). Neuroprotection with NBQX and thrombolysis with rt-PA in rat embolic stroke. *Neurol. Res.* **15**, 344–349. [*Chapter 8*]

Øye, I., Paulsen, O. & Mauerset, A. (1991). Effects of ketamine on sensory perception: evidence for a role of N-methyl-D-aspartate receptors. *J. Pharmacol. Exper. Ther.* **260**, 1209–1213. [*Chapter 3*]

Paleckova, V., Palecek, J., McAdoo, D.J. & Willis W.D. (1992). The non-NMDA antagonist CNQX prevents release of amino acids into the rat spinal cord dorsal horn evoked by sciatic nerve stimulation. *Neurosci. Lett.* **148**, 19–22. [*Chapter 8*]

Palmer, G.C. & Miller, J.A. (1996). The search for new antiepileptic drugs, Parts 1 & 2. *Pharmaceut. News* **3**, 7–11, 13–15. [*Chapter 10*]

Palmer, G.C., Clark, B. & Hutchison, J.B. (1993). Antiepileptic and neuroprotective potential of remacemide hydrochloride. *Drugs Future* **18**, 1021–1042. [*Chapter 10*]

Palmer, G.C., Cregan, E.F., Borrelli, A.R. & Willett, F. (1995). Neuroprotective properties of the uncompetitive NMDA receptor antagonist remacemide hydrochloride. *Ann. NY Acad. Sci.* **765**, 236–247. [*Chapter 10*]

Palmer, G.C., Murray, R.J., Wilson, T.C.M., Eisman, M.S., Ray, R.K., Griffith, R.C., Napier, J.J., Fedorchuk, M., Stagnitto, M.L. & Garske, G.E. (1992). Biological profile of the metabolites and potential metabolites of the anticonvulsant remacemide. *Epilepsy Res.* **12**, 9–20. [*Chapter 10*]

Palmer, G.C., Stagnitto, M.L., Ordy, J.M., Griffith, R.C., Napier, J.J., Gentile, R.J., Woodhead, J.H., White, H.S. & Swinyard, E.A. (1991). Preclinical profile of stereoisomers of the anticonvulsant remacemide in mice. *Epilepsy Res.* **8**, 36–48. [*Chapter 10*]

Panter S.S. & Faden A.I. (1992). Pretreatment with NMDA antagonists limits release of excitatory amino acids following traumatic brain injury. *Neurosci. Lett.* **136**, 165–168. [*Chapter 1*]

Papp, M. & Moryl, E. (1993b). Similar effect of chronic treatment with imipramine and the NMDA antagonists CGP 37849 and MK-801 in a chronic mild stress model of depression in rats. *Eur. J. Neuropsychopharmacol.* **3**, 348–349. [*Chapter 5*]

Papp, M. & Moryl, E. (1994). Antidepressant activity of non-competitive and competitive NMDA receptor antagonists in a chronic mild stress model of depression. *Eur. J. Pharmacol.* **263**, 1–7. [*Chapter 5*]

Papp, M. & Moryl, E. (1996). Differential effects of functional NMDA antagonists in the conditioned place preference test. *Eur. J. Pharmacology*, in press. [*Chapter 5*]

Papp, M., Lappas, S., Muscat, R. & Willner, P. (1992). Attenuation of place preference conditioning but not place aversion conditioning by chronic mild stress. *J. Psychopharmacol.* **6**, 352–356. [*Chapter 5*]

Papp, M., Willner, P. & Muscat, R. (1991). An animal model of anhedonia: attentuation of sucrose consumption and place preference conditioning by chronic unpredictable mild stress. *Psychopharmacol.* **104**, 255–259. [*Chapter 5*]

Papp, M., Willner, P. & Muscat, R. (1993a). Behavioral sensitization to a dopamine agonist is associated with reversal of stress-induced anhedonia. *Psychopharmacol.* **110**, 159–164. [*Chapter 5*]

Park, C.K., McBurney, R.N., Holt, W.F., Cotter, R.E., McCulloch, J., Kang, J.K. & Choi, C.C. (1993). The dose-dependency of the antischemic efficacy and of the side effects of a novel NMDA antagonist, CNS 1102. *J. Cerebr. Blood Flow Metab.* **13**, 641. [*Chapter 4*]

Park, C.K., McCulloch, J., Jung, D.S., Kang, J.K. & Choi, C.R. (1994a). Do *N*-methyl-D-aspartate antagonists have disproportionately greater effects on brain swelling than on ischemic damage in focal cerebral infarction? *Acta Neurochir.* **60** (Suppl.), 279–281. [*Chapter 2*]

Park, C.K., McCulloch, J., Kang, J.K. & Choi, C.R. (1994b). Pre-treatment with a competitive NMDA antagonist D-CPPene attenuates focal cerebral infarction and brain swelling in awake rats. *Acta Neurochir.* **127**, 220–226. [*Chapter 2*]

Park, C.K., Nehls, D.G., Graham, D.I., Teasdale G.M. & McCulloch, J. (1988). Focal cerebral ischemia in the cat: treatment with the glutamate antagonist MK-801 after induction of ischaemia. *J. Cereb. Blood Flow Metab.* **8**, 757–762. [*Chapter 7*]

Patat, A., Molinier, P., Hergueta, T., Brohier, S., Zieleniuk, I., Danjou, P., Warot, D. & Puech, A. (1994). Lack of amnestic, psychotomimetic or impairing effect on psychomotor performance of eliprodil, a new NMDA antagonist. *Int. Clin. Psychopharmacol.* **9**, 155–162. [*Chapter 6*]

Patel, S. Chapman, A.G., Graham, J.L., Meldrum, B.S. & Frey, P. (1990). Anticonvulsant activity of the NMDA antagonists, D(–)4-(3-phosphonopropyl)piperazine-2-carboxylic acid (D-CPP) and D(–)(E)4-(3-phosphonoprop-2-enyl)piperazine-carboxylic acid (D-CPPene) in a rodent and a primate model of reflex epilepsy. *Epilepsy Res.* **7**, 3–10. [*Chapters 2, 3*]

Paul, I.A., Layer, R.T., Skolnick, P. & Nowak, G. (1993). Adaptation of the NMDA receptor in rat cortex following chronic electroconvulsive shock or imipramine. *Eur. J. Pharmacol.* **247**, 305–311. [*Chapter 5*]

Paul, I.A., Nowak, G., Layer, R.T., Popik, P. & Skolnick, P. (1994). Adaptation of the N-Methyl-D-Aspartate receptor complex following chronic antidepressant treatments. *J. Pharmacol. Exper. Ther.* **269**, 95–102. [*Chapter 5*]

Perkins, M.N. & Stone, T.W. (1985). Action of kynurenic acid and quinolinic acid in the rat hippocampus in vivo. *Exp. Neurol.* **88**, 570–579. [*Chapter 3*]

Perkins, M.N., Collins, J.F. & Stone, T.W. (1982) Isomers

of 2-amino-7- phosphonoheptanoic acid as antagonists of neuronal excitants. *Neurosci.Lett.* **32**, 65–68. [*Preface, Chapter 2*]

Perrault, G., Morel, E., Joly, D., Sanger, D. & Zivkovic, B. (1990). Differential neuropsychopharmacological profiles of NMDA antagonists: ifenprodil-like, PCP-like and CPP-like compounds. *Eur. J. Pharmacol.* **183**, 942–948. [*Chapter 6*]

Perrault, G., Morel, E., Joly, D., Sanger, D.J. & Zivkovic, B. (1989). Comparison of the pharmacological profiles of four NMDA antagonists, ifenprodil, SL82.0715, MK-801 and CPP, in mice. *Br. J. Pharmacol.* **97**, 580p. [*Chapter 6*]

Perrier, M.L. & Benavides, J. (1995). Pharmacological heterogeneity of NMDA receptors in cerebellar granule cells in immature rat slices. A microfluorimetric study with the $[CA^{2+}]_i$ sensitive dye Indo-1. *Neuropharmacology* **34**, 35–42. [*Chapter 6*]

Perry, T.L., Hansen, S. & Jones, K., (1987). Brain glutamate deficiency in amyotrophic lateral sclerosis. *Neurology* **37**, 1845–1848. [*Chapter 9*]

Persson, J., Axelsson, G., Hallin, R.G. & Gustafsson, L.L. (1995). Beneficial effects of ketamine in a chronic pain state with allodynia, possibly due to central sensitization. *Pain* **60**, 217–222. [*Chapter 3*]

Pin J-P. & Duvoisin, R. (1995). Neurotransmitter receptors I. The metabotropic glutamate receptors: structure and functions. *Neuropharmacology* **34** 1–26. [*Chapter 2*]

Pingping, Z., Ogita, K., Han, D. & Yoneda, Y. (1993). Comparative studies on binding of 3 different ligands to the *N*-methyl-D-aspartate recognition domain in brain synaptic membranes treated with Triton X-100. *Brain Res.* **609**, 253–261. [*Chapter 6*]

Plaitakis, A. & Caroscio, J.T. (1987). Abnormal glutamate metabolism in amyotrophic lateral sclerosis. *Ann. Neurol.* **22**, 575–579. [*Chapter 9*]

Poignet, H., Nowicki, J.P. & Scatton, B. (1992). Lack of neuroprotective effect of sigma ligands in a model of focal cerebral ischaemia in the mouse. *Brain Res.* **596**, 320–324. [*Chapter 6*]

Porsche-Wiebking, E. (1989). New *N*-methyl-D-aspartate antagonists for the treatment of stroke. *Drug Develop. Res.* **17**, 367–375. [*Chapter 10*]

Porter, H.P. & Greenamyre, J.T. (1995). Regional variations in the pharmacology of NMDA receptor channel blockers: implications for therapeutic potential. *J. Neurochem.* **64**, 614–623. [*Chapter 10*]

Pratt, J., Rataud, J., Bardot, F., Roux, M., Blanchard, J.C. Laduron, P.M. & Stutzmann, J.M. (1992). Neuroprotective actions of riluzole in rodent models of global and focal cerebral ischaemia. *Neurosci. Lett.* **140**(2), 225–230. [*Chapter 9*]

Price, D., Mao, J., Frenk, H. & Mayer, D. (1994). The *N*-methyl-D-aspartate receptor antagonist dextromethorphan selectively reduces temporal summation of second pain in man. *Pain* **59**, 165–174. [*Chapter 6*]

Priestley, T., Ochu, E. & Kemp, J.A. (1994). Subtypes of NMDA receptor in neurones cultured from rat brain. *Neuroreport* **5**, 1763–1765. [*Chapter 6*]

Pugsley, W., Klinger, L., Paschalis, C., Treasure, T., Harrison, M. & Newman, S. (1994). The impact of microemboli during cardiopulmonary bypass on neuropsychological functioning. *Stroke* **25**, 1393–1399. [*Chapter 10*]

Pullan, L. & Powell, R.J. (1991). Spermine reciprocally changes the affinity of NMDA receptor agonists and antagonists. *Eur. J. Pharmacol.* **207**, 173–174. [*Chapter 6*]

Pullan, L.M., Stumpo, R.J., Powel, R.J., Paschetto, K.A. & Britt, M. (1992). Neomycin is an agonist at a polyamine site on the *N*-methyl-D-aspartate receptor. *J. Neurochem.* **59**, 2087–2093. [*Chapter 6*]

Ransom, R.W. & Stec, N.L. (1988). Cooperative modulation of [3H]MK-801 binding to the N-methyl-D-aspartate receptor-ion channel complex by L-glutamate, glycine, and polyamines. *J. Neurochem.* **51**, 830–836. [*Preface, Chapter 6*]

Ransom, R.W. (1991). Polyamine and ifenprodil interactions with the NMDA receptor's glycine site. *Eur. J. Pharmacol.* **208**, 67–71. [*Chapter 6*]

Rataud, J., Debarnot, F., Mary, V., Pratt, J. & Stutzmann J-M. (1994). Comparative study of voltage-sensitive sodium channel blockers in focal ischaemia and electric convulsions in rodents. *Neurosci. Lett.* **172**, 19–23. [*Chapter 10*]

Reddy, N.L., Hu, L-Y., Cotter, R.E., Fischer, J.B., Wong, W.J., McBurney, R.N., Weber, E., Holmes, D.L., Wong, S.T., Prasad, R. & Keana, J.F.W. (1994). Synthesis and structure-activity studies of *N,N'*-diarylguanidine derivatives. *N*-(1-Naphthyl)-*N'*-(3-ethylphenyl)-*N'*-methylguanidine: a new, selective non-competitive NMDA receptor antagonist. *J. Med. Chem.* **37**, 260–267. [*Chapter 4*]

Redmond J.M., Zehr K.J., Blue M.E., Lange, M.S., Gillinov, A.M., Troncoso, J.C., Cameron, D.E., Johnston, M.V. & Baumgartner W.A. (1995). AMPA glutamate receptor antagonism reduces neurologic injury after hypothermic circulatory arrest. *Ann. Thorac. Surg.* **59**, 579–584. [*Chapter 8*]

Ren, K., Hylden, J.L.K., Williams, G.M., Ruda, M.A. & Dubner, R. (1992). The effect of a non-competitive NMDA receptor antagonist, MK801, on behavioral hyperalgesia and dorsal horn neuronal activity in rats with unilateral inflamation. *Pain* **50**, 331–344. [*Chapter 3*]

Reynolds, I.J. & Miller, R.J. (1989). Ifenprodil is a novel type of *N*-methyl-D-aspartate receptor antagonist: interaction with polyamines. *Mol. Pharmacol.* **36**, 758–765. [*Chapter 6*]

Rho, J.M., Donevan, S.D. & Rogawski, M.A. (1994). Mechanism of action of the anticonvulsant felbamate: opposing effects on N-methyl-D-aspartate and gamma-aminobutyric acid receptors. *Ann. Neurol.* **35**, 229–234. [*Chapter 3*]

Rock, D.M. & Macdonald, R.L. (1992a). Spermine and related polyamines produce a voltage-dependent reduction of *N*-methyl-D-aspartate receptor single-channel conductance. *Mol. Pharmacol.* **42**, 157–164. [*Chapter 6*]

Rock, D.M. & Macdonald, R.L. (1992b). The polyamine spermine has multiple actions on *N*-methyl-D-aspartate receptor single-channel currents in cultured cortical neurons. *Mol. Pharmacol.* **41**, 83–88. [*Chapter 6*]

Rogawski, M.A. (1992). The NMDA receptor, NMDA antagonists and epilepsy therapy. *Drugs* **44**, 279–292. [*Chapter 10*]

Rosen, D.R., Siddique, T., Patterson, D., Figlewicz, D.A., Sapp, P., Hentati, A., Donaldson, D., Goto, J., O'Regan, J.P., Deng, H.X, *et al.* (1993). Mutations in Cu/Zn superoxide dismutase gene are associated with familial amyotrophic lateral sclerosis. *Nature* **362**, 59–62. [*Chapter 9*]

Rosner, M.J., Rosner, S.D. & Johnson, A.H. (1995). Cerebral perfusion pressure: management protocol and clinical results. *J. Neurosurg.* **83**, 949–962. [*Chapter 4*]

Rothman, S.M. & Olney, J.W. (1986). Glutamate and the pathophysiology of hypoxic-ischemic brain damage. *Ann. Neurol.* **19**, 105–111. [*Chapter 3*]

Rothstein, J.D. & Kluncl, R.W. (1995). Neuroprotective strategies in a model of chronic glutamate-mediated motor neuron toxicity. *J. Neurochem.* **65**, 643–651. [*Chapter 9*]

Rothstein, J.D., Jin, L., Dykes-Hoberg, M. & Kuncl, R.W. (1993). Chronic inhibition of glutamate uptake produces a model slow neurotoxicity. *Proc. Natl Acad. Sci. USA* **90**, 6591–6595. [*Chapter 9*]

Rothstein, J.D., Martin, L.J. & Kuncl, R.W. (1992). Decreased glutamate transport by the brain and spinal cord in amyotrophic lateral sclerosis. *N. Engl. J. Med.* **326**, 1464 1468. [*Chapter 9*]

Rothstein, J.D., Tsai, G. & Kuncl, R.W., Clawson, L., Cornblath, D.R., Drachman, D.B., Pestronk, A., Stauch, B.L., Coyle, J.T. (1990). Abnormal excitatory amino acid metabolism in amyotrophic lateral sclerosis. *Ann. Neurol.* **28**, 18–25. [*Chapter 9*]

Rothstein, J.D., Van Kammen, M., Levey, A.I., Martin, L.J. & Kuncl, R.W. (1995a). Selective loss of glial glutamate transporter GLT-1 in amyotrophic lateral sclerosis. *Ann. Neurol.* **38**, 1, 73–84. [*Chapter 9*]

Rowland, L.P. & McLeod, J.G. (1988). World Federation of Neurology Research Committee Research Group on Neuromuscular Diseases. *J. Neurol. Sci.* **86**, 333–360. [*Chapter 9*]

Roytblat, L., Korotkoruchko, A., Katz, J., Glazer, M., Greemberg, L. & Fisher, A. (1993). Postoperative pain: the effect of low-dose ketamine in addition to general anesthesia. *Anesth. Analg.* **77**, 1161–1165. [*Chapter 3*]

Sacaan, A.I. & Johnson, K.M. (1989). Spermine enhances binding to the glycine site associated with the *N*-methyl-D-aspartate complex. *J. Pharm. Exp. Ther.* **36**, 836–839. [*Chapter 6*]

Sanada T., Pitts L.H. & Nishimura M.C. (1990). The expression of heat shock protein after brain injury in the rat: combined effect of NMDA antagonist, voltage-dependent calcium blocker, and a free radical scavenger. *Soc Neurosci. Abstr.* **16**(2), 1340–552.12. [*Chapter 1*]

Sanger, D.J. & Jackson, A. (1989). Effects of phencyclidine and other *N*-methyl-D-aspartate antagonists on the schedule-controlled behavior of rats. *J. Pharmacol. Exp. Ther.* **248**, 1215–1221. [*Chapter 6*]

Sanger, D.J. & Joly, D. (1991). Effects of NMDA receptor antagonists and sigma ligands on the acquisition of conditioned fear in mice. *Psychopharmacology* **104**, 27–34. [*Chapter 6*]

Sanger, D.J., Perrault, G. & Audi, E. (1995). The anti-ischaemic drug, eliprodil, blocks hyperactivity induced by intra-accumbens injection of spermine. *Behav. Pharmacol.* **6**, 520–526. [*Chapter 6*]

Sauer D., Allegrini P.R., Cosenti A., Pataki A., Amacker H. & Fagg G.E. (1993). Characterization of the cerebro-protective efficacy of the competitive NMDA receptor antagonist CGP 40116 in a rat model of focal cerebral ischemia: an *in vivo* magnetic resonance imaging study. *J. Cereb. Blood Flow Metab.* **13**, 595–602. [*Chapter 1*]

Sauer D., Massieu L., Allegrini P.R., Amacker H., Schmutz M. & Fagg G.E. (1992). Excitotoxicity, cerebral ischemia and neuroprotection by competitive NMDA antagonists. In: 'Emerging Strategies in Neuroprotection' (eds, P.J. Marangos & H. Lal), pp. 93–105. Birkäuser: Boston. [*Chapter 1*]

Sauter, A. & Rudin, M. (1986). Calcium antagonists reduce the extent of infarction in rat middle cerebral artery occulusion model as determined by quantitative magnetic resonance imaging. *Stroke* **17**, 1228–1234. [*Chapter 2*]

Sauter, A. & Rudin, M. (1991). Visualization and quantification of quinolinic acid lesions in rat brain magnetic resonance imaging (MRI). *Experientia* **47**, A5. [*Chapter 2*]

Sawada, M. & Kawai, H. (1975). Clinico-pharmacological study of ifenprodil (FX-505). *Clin. Evaluation* **3**, 357–381. [*Chapter 6*]

Scatton, B., Giroux, C., Thenot, J.P., Frost, J., George, P., Carter, C. & Benavides, J. (1994b). Eliprodil hydrochloride. *Drugs Future* **19**, 905–909. [*Chapter 6*]

Scatton, B.S., Avenet, P., Benavides, J., Carter, C.J., Duverger, D., Oblin, A., Perrault, G., Sanger, D.J. & Schoemaker, H. (1994a). Neuroprotective potential of the polyamine site-directed NMDA receptor antagonists ifenprodil and eliprodil. In 'Direct and Allosteric Control of Glutamate Receptors' (eds, M.G. Palfreyman, I.J. Reynolds & P. Skolnick), pp. 139–154. CRC Press: London. [*Chapter 6*]

Scheyer, R.D., Cramer, J.A., Leppik, I.E., Pellock, J.M., Hochholzer, J.M., Spencer, M., O'Hara, K. & Muir, K.T. (1992). Remacemide elimination after initial and chronic dosing. *Clin. Pharmacol. Therap.* **51**, 189. [*Chapter 10*]

Schild, H.O. (1947). pA, a new scale for the measurement of drug antagonism. *Br. J. Pharmacol.* **2**, 189–206. [*Chapter 2*]

Schoemaker, H. & Pigasse, S. (1993). Developmental shift in the pharmacological properties of polyamine-sensitive [^{3}H]ifenprodil binding to the NMDA receptor. *Br. J. Pharmacol.* **108**, 84P. [*Chapter 6*]

Schoemaker, H., Allen, J. & Langer, S.Z. (1990). Binding of [^{3}H]ifenprodil, a novel NMDA antagonist, to a polyamine-sensitive site in the rat cerebral cortex. *Eur. J. Pharmacol.* **176**, 249–250. [*Chapter 6*]

Schoemaker, H., Oblin, A. & Langer, S.Z. (1991). Binding of [^{3}H]ifenprodil, an NMDA antagonist, to polyamine and sigma sites in the cerebral cortex. *FASEB J.* **5**, A703. [*Chapter 6*]

Schoemaker, H., Pigasse, S. & Zivkovic, B. (1992). The NMDA antagonist [^{3}H]ifenprodil binds to heterogeneous polyamine-sensitive sites in rat brain. *Soc. Neurosci. Abstr.* **18**, 1154. [*Chapter 6*]

Schoemaker, H., Pigasse, S., Caboi, F. & Oblin, A. (1994). Polyamines effects on radioligand binding to receptors and recognition sites. In: 'Neuropharmacology of Polyamines' (ed., C. Carter), pp. 107–154, Academic Press: San Diego. [*Chapter 6*]

Schoepp D.D., Salhoff C.R., Hillman C.C. & Ornstein P.L. (1989). CGS 19755 and MK-801 selectively prevent rat striatal cholinergic and GABAergic neuronal degeneration induced by *N*-methyl-D-aspartate and ibotenate in vivo. *J Neural Transm. Gen. Sect.* **78**, 183–193. [*Chapters 1, 7*]

Schoepp, D.D., Lodge, D., Bleakman, D., Leander, J.D., Tizzano, J.P., Wright, R.A., Palmer, A.J., Salhoff, C.R. & Ornstein, P.L. (1995). *In vitro* and *in vivo* antagonism of AMPA receptor activation by (3*S*,4a*R*,6*R*,8a*R*)-6-[2-(1(2)*H*-tetrazole-5-yl)ethyl]decahydroisoquinoline-3-carboxylic acid. *Neuropharmacology* **34**, 1159–1168. [*Chapter 7*]

Schoepp, D.D., Ornstein, P.L., Salhoff, C.R. & Leander, J.D. (1991). Neuroprotectant effects of LY274614, a structurally novel systemically active competitive NMDA receptor antagonist. *J. Neural Transm.* **85**, 131–143. [*Chapter 7*]

Schoepp, D.D., Salhoff, C.R., Fuson, K.S., Sacaan, A.I., Tizzano, J.P., Ornstein, P.L. & May, P.C. (1996). Selective protection against AMPA-and kainate-evoked neurotoxicity by (3*S*,4a*R*,6*R*,8a*R*)-6-[2-(1(2)*H*-tetrazole-5-yl)ethyl]decahydroisoquinoline-3-carboxylic acid (LY293558) and its racemate (LY215490). *J. Neural Transm.*, in press. [*Chapter 7*]

Schoepp, D.D. & Conn, P.J. (1993). Metabotropic glutamate receptors in brain function and pathology. *Trends Pharmacol. Sci.* **14**, 13–20. [*Chapter 3*]

Segerdahl, M., Ekblom, A. & Sollevi, A. (1994). The influence of adenosine, ketamine, and morphine on experimentally induced ischemic pain in healthy volunteers. *Anesth. Analg.* **79**, 787–791. [*Chapter 3*]

Shalaby, I.A., Chenard, B.L., Prochniak, M.A. & Butler, T.W. (1992). Neuroprotective effects of the *N*-methyl-D-aspartate receptor antagonists ifenprodil and SL-82,0715 on hippocampal cells in culture. *J. Pharmacol. Exp. Ther.* **260**, 925–932. [*Chapter 6*]

Sharp, F.R., Butman, M., Koistinaho, J., Aardalen, K., Nakki, R., Massa, S.M., Swanson, R.A. & Sagar, S.M. (1994). Phencyclidine induction of the hsp70 stress gene in injured pyramidal neurons is mediated via multiple receptors and voltage-gated calcium channels. *Neuroscience* **62**, 1079–1082. [*Chapter 6*]

Sharp, F.R., Butman, M., Wang, S., Koistinaho, J., Graham, S.H., Sagar, S.M., Noble, L., Berger, P. & Longo, F.M. (1992). Haloperidol prevents induction of the hsp70 heat shock gene in neurons injured by phencyclidine (PCP),MK801, and ketamine. *J. Neurosci. Res.* **33**, 605–616. [*Chapter 6*]

Sheardown, M.J. (1993). The triggering of spreading depression in the chicken retina: a pharmacological study. *Brain Res.* **607**, 189–194. [*Chapter 8*]

Sheardown, M.J., (1989). The quinoxalinediones, a new series of potent and selective non-NMDA receptor antagonists. *Drugs Future* **14**, 667–674. [*Chapter 8*]

Sheardown, M.J., Nielsen, E.Ø., Hansen, A.J., Jacobsen, P. & Honore, T. (1990). 2,3-dihydroxy-6-nitro-7-sulfamoyl-benzo(F)quinoxaline: a neuroprotectant for cerebral ischemia. *Science* **247**, 571–574. [*Chapter 8*]

Sheardown, M.J., Suzdak, P.D. & Nordholm, L. (1993). AMPA, but not NMDA, receptor antagonism is neuroprotective in gerbil global ischaemia, even when delayed 24 h. *Eur. J. Pharmacol.* **236**, 347–353. [*Chapter 8*]

Siddique, T., Juneja, T., Caliendo, J., Roos, R.P., Laing, N.G., Rinmler, J., Smith, R. & Pericak-Vance, M.A. (1994). Spectrum of the phenotype in familial ALS with SOD1 mutations. *Neurology* **44**(Suppl. 2), 157S, A165. [*Chapter 9*]

Sierocinska, J., Nikolaev, E., Danysz, W. & Kaczmarek, L. (1991). Dextrorphan blocks long but not short-term memory in a passive avoidance task in rats. *Eur. J. Pharmacol.* **205**, 109–111. [*Chapter 6*]

Sills M.A., Fagg G., Pozza M.F., Angst C., Brundish D.E., Hurt S.D., Wilusz E.J. & Williams M. (1991). [^{3}H]CGP 39653: a new *N*-methyl-D-aspartate receptor antagonist radioligand with low nanomolar affinity in rat brain. *Eur. J. Pharmacol.* **192**, 19–24. [*Chapters 1, 6*]

Simmonds J.T., Sailer T.L. & Moyer J.A. (1993). The effects of CGS 19755 in rat focal cerebral ischemia produced by tandem ipsilateral common carotid artery and middle cerebral artery occlusion. *Soc Neurosci. Abstr.* **19**, 1647–674.14. [*Chapter 1*]

Simon R. & Shiraishi K. (1990). *N*-Methyl-D-aspartate antagonist reduces stroke size and regional glucose metabolism. *Ann. Neurol.* **27**, 606–611. [*Chapter 1*]

Simon, R.P., Swann, J., Griffiths, T. & Meldrum, B.S. (1984). Blockade of *N*-methyl-D-aspartate receptors may protect against ischemic damage in the brain. *Science* **226**, 850–852. [*Chapter 6*]

Sinforiani, E., Bono, G., Nappi, G., Leotta, D., Bergamasco, B., Martucci, N., Agnoli, A., Fioravanti,

M., Manzoni, G.C. & Leto Di Prelo, S. (1988). Neuropsychological and clinical effects of ifenprodil in chronic cerebrovascular disorders (CCVD): a double blind study vs. placebo. *New Trends Clin. Neuropharmacol.* **2**, 315–320. [*Chapter 6*]

Singh, L., Oles, R.J., Vass, C.A. & Woodruff, G.N. (1991). A slow intravenous infusion of *N*-methyl-DL-aspartate as a seizure model in the mouse. *J. Neurosci. Methods* **37**, 227–232. [*Chapter 6*]

Sluka, K.A., Jordan, H.H. & Westlund, K.N. (1994). Reduction in joint swelling and hyperalgesia following post-treatment with a non-NMDA glutamate receptor antagonist. *Pain* **59**, 95–100. [*Chapter 8*]

Smith, S.E. & Chapman, A.G. (1993). Acute and chronic anticonvulsant effects of D(–)CPPene in genetically epilepsy-prone rats. *Epilepsy Res.* **15**, 193–199. [*Chapter 2*]

Smith, S.E. & Meldrum, B.S. (1993). Cerebroprotective effect of a non-*N*-methyl-D-aspartate antagonist, NBQX, after focal ischaemia in the rat. *Functional Neurol.* **8**, 43–48. [*Chapter 8*]

Smith, S.E. & Meldrum, B.S. personal communication. [*Chapter 5*]

Sonsalla, P.K. (1995). The role of *N*-methyl-D-aspartate receptors in dopaminergic neuropathology produced by the amphetamines. *Drug Alcohol Depend.* **37**, 101–105. [*Chapter 6*]

Sonsalla, P.K., Riordan, D.E. & Heikkila, R.E. (1991). Competitive and noncompetitive antagonists at *N*-methyl-D-aspartate receptors protect against methamphetamine-induced dopaminergic damage in mice. *J. Pharmacol. Exp. Ther.* **256**, 506–512. [*Chapter 6*]

Sosnowski, M. & Yaksh, T.L. (1989). Role of spinal adenosine receptors in modulating the hyperesthesia produced by spinal glycine receptor antagonism. *Anesth. Analg.* **69**, 587–592. [*Chapter 3*]

Spencer, P.S., Allen, C.N., Kisby, G.E. & Ludolph, A.C. (1994). On the etiology and pathogenesis of chemically induced neurodegenerative disorders. *Neurobiol. Aging* **15**(2), 265–267. [*Chapter 9*]

Spencer, P.S., Nunn, P.B., Hugon, J., Ludolph, A.C., Ross, S.M., Roy, D.N. & Robertson, R.C. (1987). Guam amyotrophic lateral sclerosis-parkinsonism-dementia linked to a plant excitant neurotoxin. *Science* **237**, 517–522. [*Chapter 9*]

Spencer, P.S., Roy, D.N., Ludoph, A., Hugon, J., Dwivedi, M.P. & Schaumburg, H.H. (1986). Lathyrism: evidence for role of the neuroexcitatory amino acid BOAA. *Lancet* **ii**, 1066–1067. [*Chapter 9*]

Sprengel, R. & Seeburg, P.H. (1993). The unique properties of glutamate receptor channels. *FEBS Lett.* **325**, 90–94. [*Chapter 3*]

Srinivasan, J., Richens, A. & Davies, J.A. (1995). The effect of the desglycinyl metabolite of remacemide hydrochloride (FPL 12495AA) and dizocilpine (MK-801) on endogenous amino acid release from mouse cortex. *Br. J. Pharmacol.* **116**, 3087–3092. [*Chapter 10*]

Stables, J.P. (1992). Letter from NIH, Epilepsy Branch, unpublished. [*Chapter 2*]

Stagnitto, M.L., Palmer, G.C., Ordy, J.M., Griffith, R.C., Napier, J.J., Becker, C.N., Gentile, R.J., Garske, G.E., Frankenheim, J.M., Woodhead, J.H., White, H.S. & Swinyard, E.A. (1990). Preclinical profile of remacemide: a novel anticonvulsant effective against maximal electroshock seizures in mice. *Epilepsy Res.* **7**, 11–28. [*Chapter 10*]

Stannard, C.F. & Porter, G.E. (1993). Ketamine hydrochloride in the treatment of phantom limb pain. *Pain* **54**, 227–230. [*Chapter 3*]

Steinberg G.K., Perez-Pinzon M.A., Maier C.M., Sun G.H., Yoon E., Kunis D.M., Bell T.E., Powell M., Kotake A. & Giffard R.G. (1994). CGS 19755 (selfotel): correlation of in vitro neuroprotection, protection against experimental ischemia and CSF levels in cerebrovascular surgery patients. In: 'Pharmacology of Cerebral Ischemia 1994' (eds, J. Krieglstein & H. Oberpichler-Schwenk), pp. 225–232. Wissenschaftlich: mbH, Stuttgart. [*Chapter 1*]

Stewart L., Bullock R., Jones M., Kotake A. & Teasdale G.M. (1993). The cerebral haemodynamic and metabolic effects of the competitive NMDA antagonist CGS 19755 in humans with severe head injury. 2nd International Neurotrauma Symposium, Glasgow, Abstract. [*Chapter 1*]

Stone, T.W. (1993). Neuropharmacology of quinolinic and kynurenic acids. *Pharmacol. Rev.* **45**, 309–379. [*Chapter 2*]

Subramaniam, S., Donevan S.D. & Rogawski, M.A. (1996). Block of the *N*-methyl-D-aspartate receptor by remacemide and its desglycine metabolite. *J. Pharmacol. Exp. Ther.* **276**, 161–168. [*Chapter 10*]

Sucher, N.J., Akbarian, S., Chi, C.L., Leclerc, C.L., Awobuluyi, M., Deichter, D.L., Wu, M.K., Yuan, J.P., Jones, E.G. & Lipton, S.A. (1995). Development and regional expression of a novel NMDA receptor-like subunit (NMDAR-L) in the rodent brain. *J. Neurosci* **15**, 6509–6520. [*Chapter 6*]

Sveinbjornsdottir, S, Sander, J.W.A.S., Upton, D. Thompson, P.J., Patsalos, P.N., Hirt, D., Emre, H., Lowe, D. & Duncan, J.S. (1993). The excitatory amino acid antagonist D-CPP-ene (SDZ EAA 494) in patients with epilepsy. *Epilepsy Res.* **16**, 165–174. [*Chapters 2, 6*]

Swedberg, M.D.B., Jacobsen, P. & Honore, T. (1990). NBQX: Anticonvulsant effects in mice. *Soc. Neurosci. Abstract.* **16**, 489.8. [*Chapter 8*]

Swedberg, M.D.B., Jacobsen, P. & Honore, T. (1991). Discriminative effects of NBQX (2,3-dihydroxy-6-nitro-7-sulfamoyl-benzo (f)quinoxaline). *Soc. Neurosci. Abstr.* **17**, 260. [*Chapter 8*]

Szatkowski, M. & Attwell, D. (1994). Triggering and execution of neuronal death in brain ischemia: two phases of glutamate release by different mechanisms. *Trends Neurosci.* **17**, 359–365. [*Chapter 5*]

Takizawa S., Hogan M. & Hakim A.M. (1991). The effects of a competitive NMDA receptor antagonist (CGS 19755) on cerebral blood flow and pH in focal ischemia. *J. Cereb. Blood Flow Metab.* **11**, 786–793. [*Chapter 1*]

Tamura, Y., Sato, Y., Yokota, T., Akaike, A., Sasa, M. & Takaori, S. (1993). Ifenprodil prevents glutamate cytotoxicity via polyamine modulatory sites of *N*-methyl-D-aspartate receptors in cultured cortical neurons. *J. Pharmacol. Exp. Ther.* **265**, 1017–1025. [*Chapter 6*]

Tanabe, Y., Masu, M., Ishii, T., Shigemoto, R. & Nakanishi, S. (1992). A family of metabotropic glutamate receptors. *Neuron* **8**, 169–179. [*Preface*]

Tecoma, E.S., Monyer, H., Goldberg, M.P. & Choi, D.W. (1989). Traumatic neuronal injury *in vitro* is attenuated by NMDA antagonists. *Neuron* **2**, 1541–1545. [*Chapter 6*]

Teitelbaum, J.S., Zatorre, R.J., Carpenter, S., Gendron, D., Evans, A.C., Gjedde, A. & Cashman, N.R. (1990). Neurologic sequelae of domoic acid intoxication due to the ingestion of contaminated mussels. *New Engl. J. Med.* **322**, 1781–1787. [*Chapter 9*]

Thomson, A.M., West, D.C. & Lodge, D. (1985). An N-methylaspartate receptor-mediated synapse in rat cerebral cortex: a site of action of ketamine. *Nature* **313**, 479–481. [*Preface*]

Toulmond, S., Duval, D., Serrano, A., Scatton, B. & Benavides, J. (1993a). Biochemical and histological alterations induced by fluid percussion brain injury in the rat. *Brain Res.* **620**, 24–31. [*Chapter 6*]

Toulmond, S., Serrano, A., Benavides, J. & Scatton, B. (1993b). Prevention by eliprodil (SL 82.0715) of traumatic brain damage in the rat. Existence of a large (18 h) therapeutic window. *Brain Res.* **620**, 32–41. [*Chapter 6*]

Troupin, A.S., Mendius, J.R., Cheng, F. & Risinger, M.W. (1986). MK 801. In 'New Anticonvulsant Drugs', pp. 191–201 (eds, B.S. Meldrum & R.J. Porter). John Libbey: London. [*Chapter 2*]

Trujillo, K.A. & Akil, H. (1991). Inhibition of morphine tolerance and dependence by the NMDA receptor antagonist MK-801. *Science,* **251**, 85–87. [*Chapter 5*]

Trullas, R. & Skolnick, P. (1990). Functional antagonists at the NMDA receptor complex exhibit antidepressant actions. *Eur. J. Pharmacol.* **185**, 1–10. [*Chapters 5, 6*]

Trullas, R., Folio, T., Young, A., Miller, R., Boje, K. & Skolnick, P. (1991). 1-aminocyclopropanecarboxylates exhibit antidepressant and anxiolytic actions in animal models. *Eur. J. Pharmacol.* **203**, 379–385. [*Chapter 5*]

Turski, L., Jacobsen, P., Honore, T. & Stephens, D.N. (1992). Relief of experimental spasticity and anxiolytic/anticonvulsant actions of the alpha-amino-3-hydroxy-5-methyl-4-isoxazolepropionate antagonist 2,3-dihydroxy-6-nitro-7-sulfamoyl-benzo(F)quinoxaline. *J. Pharmacol. Exp. Ther.* **260**, 742–747. [*Chapter 8*]

Tutka, P., Turski, W.A., Czuczwar, S.J. & Kleinrok, Z. (1992). Anticonvulsant activity of the competitive NMDA antagonist D-CPPene and the magnesium valproate in mice-a comparison. *Pol. J. Pharmacol. Pharmacy* **44**, 234. [*Chapter 2*]

Urwyler, S. & Puente, M. (1989). Modulation of the release of endogenous glutamate from rat brain synaptosomes by putative presynaptic receptors. *Soci. Neurosci. Abstr.* **15**, 1161. [*Chapter 2*]

Vajda, F.J.E. (1992). New anticonvulsants. *Curr. Opin. Neurol. Neurosurg.* **5**, 519–525. [*Chapter 3*]

Van Luijtelaar, E.L.J.M. & Coenen, A.M.L. (1995). Effects of remacemide and its metabolite FPL 12495 on spike-wave discharges, electroencephalogram and behaviour in rats with absence epilepsy. *Neuropharmacology* **34**, 419–425. [*Chapter 10*]

Vender, J.R., Nair, S.N. & Lehmann, J.C. (1995). Pharmacotherapeutic potential for compounds acting at NMDA receptors: update 1995. *Exp. Opin. Invest. Drugs* **4**, 475–485. [*Chapter 10*]

Vezzani, A., Wu H-Q. & Samanin, R. (1987). [^{3}H]Norepinephrine release from hippocampal slices is an in vitro biochemical tool for investigating the pharmacological properties of excitatory amino acid receptors. *J Neurochem.* **49**, 1438–1442. [*Chapter 7*]

Voltz, C., Fage, D. & Carter, C. (1994). Synergism between the NMDA receptor antagonistic effects of ifenprodil and the glycine antagonist, 7-chlorokynurenate, *in vivo. Eur. J. Pharmacol.* **255**, 197–202. [*Chapter 6*]

Von Lubitz, D.K.J.E., Lin, R.C.S., McKenzie, R.J., Devlin, T.M., McCabe, T.R. & Skolnick, P. (1992). A novel treatment of global cerebral ischemia with a glycine partial agonist. *Eur. J. Pharmacol.* **219**, 153–158. [*Chapter 5*]

Wafford, K.A., Bain, C.J., Le Bourdelles, B., Whiting, P.J. & Kemp, J.A. (1993). Preferential co-assembly of recombinant NMDA receptors composed of three different subunits. *Neuroreport* **4**, 1347. [*Chapter 6*]

Wahl, F., Allix, M., Plotkine, M. & Boulu, R.G. (1993). Effect of riluzole on focal cerebral ischemia in rats. *Eur. J. Pharmacol.* **230**(2), 209–214. [*Chapter 9*]

Walker, J.M., Bowen, W.D., Walker, F.O., Matsumoto, F.R., De Costa, B. & Rice, K.C. (1990). Sigma receptors: biology and function. *Pharmacol. Rev.* **42**, 355–402. [*Chapter 6*]

Walker, J.M., Matsumoto, R.R., Bowen, W.D., Gans, D.L., Jones, K.D. & Walker, F.O. (1988). Evidence for a role of haloperidol-sensitive σ receptors in the motor effects of antipsychotic drugs. *Neurology* **38**, 961–965. [*Chapter 6*]

Wamil, A.W., Cheung, H., Harris, E.W. & McLean, M.J. (1996). Remacemide HCl and its metabolite, FPL 12495AA, limit action potential firing frequency and block NMDA responses of mouse spinal cord neurons in cell culture. *Epilepsy Res.* **23**, 1–14. [*Chapter 10*]

Wang, S., Yang, Q., Moller, C.J., Sharp, F.R. & Haglid, K.G. (1995a). Eliprodil prevents expression of the 70kDa heat shock protein in MK-801 injured neurons. *Pharmacol. Toxicol.*, in press. [*Chapter 6*]

Wang, S., Zhou, D., Fischer, J.B., Knapp, A.G. & Holt, W.F. (1995b). CNS 1102 protects against brain damage due to hypoxic ischemia (HI) in the neonatal rat. *Soc. Neurosci. Abstr.* **21**, 995. [*Chapter 4*]

Warach, S., Dashe, J.F. & Edelman, R.R. (1996). Clinical outcome in ischemic stroke predicted by early diffusion-weighted and perfusion magnetic resonance imaging: a preliminary analysis. *J. Cerebr. Blood Flow Metab.* **16**, 53–59. [*Chapter 4*]

Watkins, J.C. (1978). Excitatory amino acids. In: 'Kainic Acid as a Tool in Neurobiology' (eds, E.G. McGeer, J.W. Olney & P.L. McGeer) pp 37–69, Raven Press: New York. [*Chapter 2*]

Watkins, J.C. (1980). NMDA receptors: new light on amino acid-mediated synaptic excitation. *Trends Neurosci.* **3** 61–66. [*Chapter 2*]

Watkins, J.C. (1991). Structure/activity relations of competitive NMDA receptor antagonists. In: 'Excitatory Amino Acid Antagonists' (ed. B.S. Meldrum), pp. 84–100, Blackwell: Oxford. [*Chapter 6*]

Watkins J.C. & Collingridge, G.L. (1994). Phenylglycine derivatives as antagonists of metabotropic glutamate receptors. *Trends Pharmacol. Sci.* **15** 333–342. [*Chapter 2*]

Watkins, J.C. & Evans, R.H. (1981). Excitatory amino acid transmitters. *Annu. Rev. Pharmacol. Toxicol.* **21** 165–204. [*Preface, Chapter 2*]

Watkins, J.C. Krogsgaard-Larsen, P. & Honoré, T. (1991). Structure–activity relationships in the development of excitatory amino acid receptor agonists and competitive antagonists. The pharmacology of excitatory amino acids. A special report. *Trends Pharmacol. Sci.* 4–12. [*Chapter 2*]

Watkins, J.C., Krogsgaard-Larsen, P. & Honoré, T. (1990). Structure–activity relationships in the development of excitatory amino acid receptor agonists and competitive antagonists. *Trends Pharmacol. Sci.* **11**, 25–33. [*Preface, Chapter 5*]

Watson, G.B. & Lanthorn, T.H. (1990). Pharmacological characteristics of cyclic homologues of glycine at the N-Methyl-D-Aspartate receptor-associated glycine site. *Neuropharmacol.* **29**, 727–730. [*Chapter 5*]

Weiss, J., Goldberg, M.P. & Choi, D.W. (1986). Ketamine protects cultured neocortical neurons from hypoxic injury. *Brain Res.* **380**, 186–190. [*Chapter 6*]

Westbrook, G.L. & Mayer, M.L. (1987). Micromolar concentrations of Zn^{2+} antagonise NMDA and GABA responses of hippocampal neurones. *Nature* **328**, 640–643. [*Chapter 6*]

Westergren, I. & Johansson, B.B. (1993). Blockade of AMPA receptors reduces brain edema following opening of the blood-brain barrier. *J. Cereb. Blood Flow Metab.* **13**, 603–608. [*Chapter 8*]

Wieloch, T., Lindvall, O., Blomqvist, P. & Gage, F. (1985). Evidence for amelioration of ischemic neuronal damage in the hippocampal formation by lesions of the perforant path. *Neurol. Res.* **7**, 24–26. [*Chapter 6*]

Willets, J., Balster R.L. & Leander, D.J. (1990). The behavioral pharmacology of NMDA receptor antagonists. *Trends Pharmacol. Sci.* **11**, 423–428. [*Chapter 5*]

Willets, J., Balster, R.L. & Leander, J.D. (1990). The behavioral pharmacology of NMDA receptor antagonists. *TIPS* **11**, 423–428. [*Chapter 10*]

Williams, K. (1993a). Effects of Agelenopsis aperta toxins on the *N*-methyl-D-aspartate receptor: polyamine-like and high-affinity antagonist actions. *J. Pharmacol. Exp. Ther.* **266**, 231–236. [*Chapter 6*]

Williams, K. (1993b). Ifenprodil discriminates subtypes of the *N*-methyl-D-aspartate receptor: selectivity and mechanisms at recombinant heteromeric receptors. *Mol. Pharmacol.* **44**, 851–859. [*Chapter 6*]

Williams, K. (1994). Modulation of the N-methyl-D-aspartate receptor by polyamines: molecular pharmacology and mechanisms of action. *Biochem. Soc. Trans.* **22**, 884–887. [*Chapter 6*]

Williams, K. (1995). Pharmacological properties of recombinant *N*-methyl-D-aspartate (NMDA) receptors containing the ε4 (NR2D) subunit. *Neurosci. Lett.* **184**, 181–184. [*Chapter 6*]

Williams, K., Kashiwagi, K., Fukuchi, J. & Igarashi, K. (1996). An acidic amino acid in the *N*-methyl-D-aspartate receptor that is important for spermine stimulation. *Mol. Pharmacol.* **48**, 1087–1098. [*Chapter 6*]

Williams, K., Russell, S.L., Shen, Y.M. & Molinoff, P.B. (1993). Developmental switch in the expression of NMDA receptors occurs *in vivo* and *in vitro*. Neuron **10**, 267–278. [*Chapter 6*]

Williams, K., Zappia, A.M., Pritchett, D.B., Shen, Y.M. & Molinoff, P.B. (1994). Sensitivity of the *N*-methyl-D-aspartate receptor to polyamines is controlled by NR2 subunits. *Mol. Pharmacol.* **45**, 803–809. [*Chapter 6*]

Willner, P., Towell, A., Sampson, D. & Muscat, R. (1987). Reduction of sucrose preference by chronic unpredictable mild stress, and its restoration by a tricyclic antidepressant. *Psychopharmacol.* **93**, 358–364. [*Chapter 5*]

Winslow, J.T., Insel, T.R., Trullas, R. & Skolnick, P. (1990). Rat pup isolation calls are reduced by functional antagonists of the NMDA receptor complex. *Eur. J. Pharmacol.* **190**, 11–21. [*Chapter 5*]

Witkin, J.M. & Steele, T.D. (1992). Effects of strychnine-insensitive glycine receptor ligands on discriminative stimulus effects of N-Methyl-D-Aspartate (NMDA) channel antagonists. *Soc. for Neurosci. Abstr.* **18**, 447. [*Chapter 5*]

Witkin, J.M. & Tortella, F.C. (1991). Modulators of N-Methyl-D-Aspartate protect against diazepam or phenobarbital-resistant cocaine convulsions. *Life Sci.* **48**, PL51-PL56. [*Chapter 5*]

Witkin, J.M., Brave, S., French, D. & Geter-Douglass, B. (1995). Discriminative stimulus effects of R-(+)-3-amino-l-hydroxypyrrolid-2-one, [(+)-HA-966], a partial agonist of the strychnine-insensitive modulatory site of the N-Methyl-D-Aspartate receptor. *J. Pharmacol. Exper. Ther.* **275**, 1267–73. [*Chapter 5*]

Wong, E.H., Kemp, J.A., Priestley, T., Knight, A.R., Woodruff, G.N. & Iversen, L.L. (1986). The anticonvulsant MK-801 is a potent *N*-methyl-D-asparate antagonist. *Proc. Natl Acad. Sci. USA* **83**, 7104–7108. [*Preface*]

Wong, E.H., Knight, A.R. & Woodruff, G.N. (1988). [3H]MK-801 labels a site on the *N*-methyl-D-aspartate receptor channel complex in rat brain membranes. *J. Neurochem.* **50**, 274–281. [*Chapter 6*]

Wong, E.H.R. & Kemp, J.A. (1991). Sites for antagonism on the *N*-methyl-D-aspartate receptor channel complex. *Annu. Rev. Pharmacol. Toxicol.* **31**, 401–425. [*Chapter 4*]

Woolf, C.J. & Thompson, S.W. (1991). The induction and maintenance of central sensitization is dependent on N-methyl-D-aspartic acid receptor activation; implications for the treatment of post-injury pain hypersensitivity states. *Pain* **44**, 293–299. [*Chapter 3*]

Woolf, C.J. (1989). Recent advances in the pathophysiology of acute pain. *Br. J. Anaesth.* **63**, 139–146. [*Chapter 3*]

Woolf, C.J. (1991). Central mechanisms of acute pain. In 'Pain Research and Clinical Management, Vol. 4, Proc. VI World Congress on Pain' (eds, M.R. Bond, J.E. Charlton & C.J. Woolf), pp. 25–34. Elsevier: Amsterdam. [*Chapter 3*]

Wright, R.A., McDonald, J.W. & Schoepp, D.D. (1994). Distribution and ontogeny of 1*S*,3*R*-ACPD-sensitive and quisqualate-insensitive [^{3}H]glutamate binding sites in the rat brain. *J. Neurochem.* **63**, 938–945. [*Chapter 7*]

Xue, D., Huang, Z.-G., Barnes, K., Lesiuk, H.J., Smith, K.E. & Buchan, A.M. (1994). Delayed treatment with AMPA, but not NMDA, antagonists reduces neocortical infarction. *J. Cereb. Blood Flow Metab.* **14**, 251–261. [*Chapter 8*]

Xue, D., Huang, Z.G. & Buchan, A.M. (1993). The effect of the anticonvulsant remacemide on transient focal cerebral ischemia. *Can. J. Neurol. Sci.* **20**(Suppl. 2), S75. [*Chapter 10*]

Yaksh, T.L. (1989). Behavioral and autonomic correlates of the tactile evoked allodynia produced by spinal glycine inhibition: effects of modulatory receptor systems and excitatory amino acid antagonists. *Pain* **37**, 111–123. [*Chapter 3*]

Yamaguchi, S. & Rogawski, M.A. (1992). Effects of anticonvulsant drugs on 4-aminopyridine-induced seizures in mice. *Epilepsy Res.* **11**, 9–16. [*Chapter 10*]

Yamaguchi, S., Donevan, S.D. & Rogawski, M.A. (1993). Anticonvulsant activity of AMPA/kainate antagonists: comparison of GYKI 52466 and NBQX in maximal electroshock and chemoconvulsant seizure models. *Epilepsy Res.* **15**, 179–184. [*Chapter 8*]

Yamamoto, T. & Yaksh, T.L. (1992). Comparison of the antinociceptive effects of pre- and posttreatment with intrathecal morphine and MK801, an NMDA antagonist, on the formalin test in the rat. *Anesthesiology* **77**, 757–763. [*Chapter 3*]

Yamazaki, M., Mori, H., Araki, K., Mori, K.J. & Mishina, M. (1992). Cloning, expression and modulation of a mouse NMDA receptor subunit. *FEBS Lett.* **300**, 39–45. [*Chapter 6*]

Yoneda, Y., Suzuki, T. & Ogita, K. (1994). Differential profiles of binding of a radiolabeled agonist and antagonist at a glycine recognition domain on the *N*-methyl-D-aspartate receptor ionophore complex in rat brain. *J. Neurochem.* **62**, 102–112. [*Chapter 6*]

Young, A.R., Bouloy, M., Boussard, J.F., Edvinsson, L. & Mackenzie, E.T. (1981). Direct vascular effects of agents used in the pharmacotherapy of cerebrovascular disease on isolated cerebral vessels. *J. Cereb. Blood Flow Metab* **1**, 117–128. [*Chapter 6*]

Zarnowski, T., Kleinrok, Z., Turski, W.A. & Czuczwar, S.J. (1994). The competitive NMDA antagonist, D-CPPene, potentiates the anticonvulsant activity of conventional antiepileptics against maximal electroshock-induced seizures in mice. *Neuropharmacology* **33**, 619–624. [*Chapter 2*]

Zeevalk, G.D. & Nicklas, W.J. (1990). Action of the anti-ischemic agent ifenprodil on *N*-methyl-D-aspartate and kainate-mediated excitotoxicity. *Brain Res.* **522**, 135–139. [*Chapter 6*]

Zeevalk, G.D. & Nicklas, W.J. (1992). Developmental differences in antagonism of NMDA toxicity by the polyamine site antagonist ifenprodil. *Brain Res. Dev. Brain Res.* **65**, 147–155. [*Chapter 6*]

Zhou, L.M., Szendrei, G.I., Fossom, L.H., Maccecchini, M.L., Skolnick, P. & Otvos, L. (1996). Synthetic analogues of conantokin-G: NMDA antagonists acting through a novel polyamine-coupled site. *J. Neurochem.*, **66**, 620–628. [*Chapter 5*]

Zuccarello, M., Lewis, A.I., Upputuri, S., Farmer, J.B. & Anderson, D.K. (1994). Effect of remacemide hydrochloride on subarachnoid hemorrhage-induced vasospasm in rabbits. *J. Neurotrauma* **11**, 691–698. [*Chapter 10*]

Zukin, R.S. & Bennett, M.V.L. (1995). Alternatively spliced isoforms of the NMDAR1 receptor subunit. *Trends Neurosci.* **18**, 306–313. [*Chapter 6*]

Index

Edwards Brothers Malloy
Ann Arbor MI. USA
May 29, 2013